T0249784

A Manual for
Wildlife Radio Tagging

A Manual for Wildlife Radio Tagging

Robert E. Kenward
Centre for Ecology and Hydrology
Winfrith Technology Centre
Dorchester, Dorset DT2 8ZD

ACADEMIC PRESS

A Harcourt Science and Technology Company

San Diego San Francisco New York Boston
London Sydney Tokyo

Copyright © 2001 by ACADEMIC PRESS

Academic Press
A Harcourt Science and Technology Company
Harcourt Place, 32 Jamestown Road, London NW1 7BY, UK
http://www.academicpress.com

Academic Press
A Harcourt Science and Technology Company
525 B Street, Suite 1900, San Diego, California 92101-4495, USA
http://www.academicpress.com

ISBN 0-12-404242-2

Library of Congress Catalog Card Number: 00-103147

A catalogue for this book is available from the British Library

Typeset by Mackreth Media Services, Hemel Hempstead, UK

Transferred to Digital Printing 2007

Contents

Preface

Biologists use animal radio tags for two main purposes: to locate study animals in the field, and to transmit information about the physiology or behaviour of wild or captive animals. These uses can be described, respectively, as 'radio tracking' and 'radio telemetry', the latter term being derived from the Greek words for distance and measurement.

Wildlife radio tags were first used for telemetry. One of the earliest projects, inspired by the use of physiological telemetry on US Navy test pilots, resulted in an implanted transmitter to monitor chipmunk heart-rates (Le Munyan *et al.*, 1959). This study closely coincided with a publication from Norway describing an externally-mounted transmitter for telemetering heart and wing beats from mallard (Eliassen, 1960). The construction of these first tags, in the late 1950s, was crucially dependent on the development of the transistor.

The first radio tags transmitted a continuous signal, with physiological changes being indicated by slight changes in the signal frequency. Similar frequency modulation (FM) of a continuous carrier signal is still widely used for medical telemetry and animals in laboratories, because it is a very accurate way of conveying subtle changes in muscle potentials, neural activity, joint pressures and other physiological parameters.

For wildlife radio tags, however, the cell life is greatly extended by transmitting the signal as brief pulses of the carrier frequency. In theory, at least, one 20 ms pulse could be repeated every second for 50 times as long as a continuous signal from the same cell. Moreover, a faint pulsed signal is easier for the human ear to detect, against the continuous background noise from wideband FM broadcasts or cosmic radiation, than a continuous whine. Wildlife radio tracking has therefore, since its start in the early 1960s (e.g. Cochran and Lord, 1963; Marshall and Kupa, 1963), been based almost entirely on pulsed signals.

During the next two decades, tags with constant pulse rates were tracked to sample animal locations or for radio surveillance. Radio surveillance involved using the tag to find an animal so that it could be watched, captured or monitored in other ways. Some tags also had signal pulses modulated by a variety of simple sensor subcircuits, to telemeter temperature, posture, movement, compass orientation and other aspects of animal activity. A great deal of information was published on the subject, in proceedings from conferences in Europe (e.g. Amlaner and MacDonald, 1980; Cheeseman and Mitson, 1982), from the International Conferences on Wildlife Biotelemetry held in North America, in an introductory guide (Mech, 1983), in technical notes and in hundreds of other scientific papers. This book's forerunner was published in 1987, to provide a single source of simple advice on choosing equipment, collecting observations and analysing the data from pulsed-signal radio tags.

Wildlife Radio Tagging: Equipment, Techniques and Analysis (Kenward, 1987) described the construction and use of VHF (very high frequency) tracking and telemetry systems, gave some BASIC software routines for elementary data analyses and noted the potential for automation in data-storage tags and UHF (ultra high frequency) tags for tracking by satellites. During the 12 years since that book, the hardware and software have improved substantially in performance and reliability. Remote-sensed maps have come into use. Global positioning system (GPS) equipment has become widely available, and can be used in wildlife tags to estimate locations. DNA techniques and auto-recording of microtransponders have become practical alternatives to radio-tracking in some situations.

During the last decade, there have also been some useful review papers (e.g. S. Harris *et al.*, 1990), a comprehensive chapter by Samuel and Fuller (1994) in the Wildlife Society's 5th edition manual on Research and Management Techniques for Wildlife and Habitats, and several more conference proceedings (e.g. Amlaner, 1989; Uchiyama and Amlaner, 1991; Priede and Swift, 1992; Mancini *et al.*, 1993; Cristalli *et al.*, 1996; Penzel *et al.*, 1998). White and Garrott (1990) provided a thorough consideration of analysis techniques, and a recent symposium updates their work (Millspaugh and Marzluff, in press). Nevertheless, *Wildlife Radio Tagging* has remained the only book to attempt a comprehensive review and guide for wildlife biologists at all levels of experience.

Andy Richford, the scientific editor of Academic Press in London, suggested in 1996 that a revision was due. However, the abundance of recent publications, together with many notes collected on changing practices, would have made a simple revision inadequate. Apart from the start of this preface, much of Chapters 5, 7 and a few other sections of text, this book is new. Many original plates and figures have been retained, but there is 70% more text, more than twice as many references and the new name: *A Manual for Wildlife Radio Tagging*.

As before, Chapter 1 is set at the most basic level, for those who are wondering whether radio tags might help in their research. It compares radio tagging with other marking techniques, describes the performance to be expected of radio tags, and discusses the planning that is necessary for successful projects. For those committed to a project, there is then advice in Chapters 2, 3 and 4 on choosing and obtaining equipment and software. There are approximate prices for 1999 in United States dollars, converted at rates of UK£1 = US$1.6 and DM1 = US$0.5.

The production of reliable modern transmitters requires surface-mount facilities and continual revision to replace components that become obsolete. It is therefore no longer worthwhile for biologists to built their own transmitters. However, projects can still prepare tags, by adding cells, antennas and attachment materials to basic transmitters, as described in Chapter 5. This approach permits rapid redesign or refurbishment of tags, as well as saving money.

Chapter 6 continues the preparation theme by discussing a wide variety of tag mounting techniques. There is a brief review of implantation, which is the best way to tag some species and is now the rule for physiological telemetry. Chapter 7 describes the principles and techniques used for radio tracking on foot, or in cars, boats and aircraft. Last, but far from least because their subjects are so crucial to the production of satisfactory results, Chapter 8 contains greatly extended sections on systematic collection of data, followed by reviews of the analyses for these behavioural and demographic data in Chapter 9 and Chapter 10. For convenience, the computing advice in these and other chapters is based mainly on using Ranges V software (Kenward and Hodder, 1996), but is equally relevant to other packages if they provide comparable functions.

On the whole, the sequence of information is aimed at the first-time user, who needs to know about the controls on receivers before using them in the field, and who might not understand a discussion of data analysis before learning about tracking techniques. The reader who has already done some radio tagging, however, may wish to start well into the book, to learn about the types of tag which suit a new study species (Chapters 2, 3 and 6), or to review the fast-changing field of data analysis (Chapters 9 and 10). However, Chapter 1 contains thoughts that remain relevant to many projects. Moreover, sections on software and data handling (Chapters 4 and 6–8) should be read before starting new fieldwork. Otherwise, their suggestions and recommendations may come too late.

At least a hundred papers are now published each year on radio-tagged wildlife. This book concentrates on those that describe methods. I apologise to those whom I have omitted to cite, and for falling back on my own studies when there may be better examples elsewhere. Please send me your reprints so I know about your successes. There are no apologies for including many references from the first edition; originators are now too often forgotten because their early papers are omitted from digital search systems.

Although it is 40 years since the first attachment of radios to wildlife, consensus is still developing on the most appropriate signal formats, attachment methods, data collection and analysis techniques. There will surely be exciting hardware developments in the next couple of decades, especially for automated collection of data. Automated analyses, based on expert-system processing of the data to determine the most appropriate techniques, will be needed to handle the volume of data generated. These changes may eventually lead to radio tagging fulfilling a crucial role in the development of modelling, for species, food webs and eventually complete ecosystems. The resulting spatially explicit predictive models will be needed during the next century if conservation is to move away from merely reacting to problems, and predict effectively how developing human populations and changing climate will affect living systems.

It is a real pleasure to thank all the people who have contributed in many different ways to the knowledge in this book and its forerunner. I have learned much from advisors and supervisors, including Mohammed al Bowardi, Jack Dempster, Nick Fox, Alan Gray, John Goss-Custard, David Jenkins, Hans Kruuk, Vidar Marcström, Ian Newton, Chris Perrins and Jeremy Thomas. I'm also grateful to the colleagues from many countries who have been a pleasure to work with over the years, especially Evgeny Bragin, Rolf Brittas, John Cooper, Frank Doyle, Kjell Einvik, Steve Freeman, Robin Fuller, Ilppo Hanski, Jonathon Hardcastle, Ben

Kenward, Anatoli Levin, Steve Knick, Ian Newton, David Macdonald, Rupert McShane, Santi Mañosa, Mick Marquiss, Pat Morris, Alan Morriss, Byron Morgan, Mike Nichols, Maarit Pahkala, Ralf Pfeffer, Kent Riddle, Rob Rose, Vic Simpson, Tony Tyack, Tomas Willebrand and Nick Williams.

My knowledge of transmitter design owes much to Mike Dolan, who first showed me how to build them, to John French, and especially to Brian Cresswell who has continually improved my understanding of tags and receivers. Others who have given technical advice include Charlie Amlaner, Fred Anderka, Bill Cochran, Tom Dunstan, Jim Lotimer, Shane Nelson, Kevin Nicholas, Ray Rafarel, Trevor Storeton-West, Colin Sunderland, and engineers from many companies. As a biologist, I would have been lost without all their help.

Comments leading to hardware and software improvements have been provided especially by Frank Doyle, Ian Hill, Jessica Holm, Mats Karlbom, Tim Parish and Craig Weatherby, together with many Biotrack customers, while Nicholas Aebischer, Christine Bunck, Bruce Don, Steve Knick, Vicky Meretsky, Steve Tapper, Gary White and Bruce Worton have aided my understanding of mapping and analysis techniques. Ben Kenward, Rob Rose and Brian Cresswell kindly allowed me to use unpublished data from tracking blackbirds and squirrels. Brian, together with Mike Fenn, David Macdonald, Ron Mitson, Andrew Roy and Sean Walls, also substantially improved drafts of the two books and Jessica kindly let me use her cartoons. The best figures were prepared by Jessica Holm and Stephen Hayward.

I am deeply grateful to you all. I have particularly valued discussion and support since I started radio tracking from Mark Fuller, and for many years from Nicholas Aebischer, Ralph Clarke, Kathy Hodder, Pete Robertson, Torgeir Nygård, Sean Walls and, above all, from my very patient wife Bridget.

1 First Questions

In an age of electronic gadgets, radio tagging is a seductive technique. For shy or elusive animals, and those too small to carry conspicuous colour tags or other visible markers, radio tracking is often the only way to collect data systematically on behaviour and some aspects of demography. Moreover, automatic radio telemetric recording can be used to record data from large numbers of unseen animals 24 hours a day. Radio tracking can be used to find domestic animals that are lost. It can even save human lives, and allow memory-impaired patients freedom to roam without constant supervision (McShane *et al.*, 1998).

Nevertheless, much effort and money has been wasted on poorly planned radio tagging, and many research projects still fail to take elementary steps that decide between success and failure. Perhaps this is why the leading international scientific journals, *Nature* and *Science*, have published only about five papers based on wildlife radio tagging among their thousands of letters and reports in the last 10 years. It seems that those with the brightest ideas mainly seek other ways to collect their data, typically choosing species that rapidly yield results because they are easy to observe visually. On the other hand, radio tagging has become a conservation mainstay, for collecting year-round data on relationships between habitats, survival and dispersal for all those species that are rare or elusive.

The challenge now is to make better use of radio tags in theory-driven studies, and thus in basic as well as applied research. Such work no longer risks lack of data due to equipment failures, because 95% of VHF tags can now be expected to reach their quoted cell life. The converse problem in some early projects, of drowning in data, is soluble with recent software. Questions about optimal analysis techniques (White and Garrott, 1990) are still receiving attention, and there is still no small and accurate automatic location system, but progress continues in these areas. The greatest need for improvement is in the *biological* questions being asked, in the planning to address them and in the training needed to implement those plans.

1.1 BIOLOGICAL QUESTIONS

Although the development of new techniques is itself a field of study for some people, most projects are interested in radio tagging primarily as a tool for answering biological questions. As well as being the only effective way to record some environmental and physiological data from wild animals, radio tagging has two main advantages over other techniques for recording behaviour and demography. It makes animals accessible to systematic sampling, thus reducing many sources of bias, and it does this for specific individuals, for which other attributes can be recorded during capture for tagging and through subsequent monitoring.

These advantages have been exploited to answer questions about descriptive, relational and causative issues. Early studies were mainly descriptive. They measured sizes of home ranges, what habitats animal prefer, how often they interact with others, how often they kill or are killed, or simply what they do all the time. Such studies still have an important role to play, especially for discovering the basic biology of elusive species. Descriptive studies can also include calibration of other techniques that are more suitable for widespread use by volunteers, for example to determine what proportion of animals are missed by visual census techniques,

However, many ecologists now investigate how demography and behaviour respond to changing environments. They seek relationships between animal performance, in terms of survival, dispersal or productivity, and factors like habitat, weather, or pressure from other animals of the same or different species, including humans. Relationships of albatross foraging with weather patterns on a global scale (Jouventin and Weimerskirch, 1990) or between predation and abundance of sea-mammals (Estes *et al.*, 1998) gave rise to recent papers in the leading international science journals. An ultimate aim of relational studies is to understand mechanisms that link demography to environment, and thus to predict effects of human activities and climate change. Relationships between demography and environment were previously based on multi-area studies, but an alternative approach is to build models based on differences between individuals in a single very variable area. Individual-based demographic modelling has been pioneered from observations of highly visible birds (Goss-Custard, 1996; Sutherland, 1996), but radio tagging makes this modelling practical for the majority of species that are not easily observed year-round.

Finally, the ability to collect data systematically from individuals also makes radio tagging very suitable for experimental studies. The most widespread have been release projects, to test the hypothesis that animals can survive again after factors causing an original extinction have been removed. Tests with radios of hypotheses about foraging of penguins (Ancel *et al.*, 1992), social behaviour of rodents (Ims, 1988) and effects of predator increase (Thirgood and Redpath, 1997), have also produced recent publications in the top science journals. Radio tagging has much unexploited potential for testing hypotheses, and thereby overcoming the criticism that lack of such tests makes ecology an immature science (Peters, 1991).

1.2 PLANNING

All too often, radio tagging suffers from what might be called the 'James Bond' syndrome, from the film character's attitude that sophisticated equipment is 'pick-up-and-go' (with no similar inference for the ornithologist whose name was used by Ian Fleming!). Those who occasionally use radio tracking, for example to find lost people or domestic animals, are especially vulnerable. When faced with the loss, they are naturally anxious, and anxiety does not favour effective use of unfamiliar equipment. In safety-critical applications, training to use the equipment is indispensable.

The syndrome can also affect research supervisors, who neglect training needs and overestimate ultimate capabilities. Beware! There are now plenty of experienced radio trackers to provide advice during planning; otherwise they are as likely to provide criticism in review. Radio tracking can look easy on television, or in the hands of a skilled practitioner, and is sometimes dismissed as 'only a tool'. A laser-scalpel is a tool too, but no one would put it in the hands of a novice surgeon and expect immediate brain surgery.

It is a particular irony that those with some of the most interesting biological questions to ask, through career experience using techniques other than radio tagging, are especially at risk from the James Bond syndrome. It is so easy to underestimate the skill required in other people's work. A request that you pay careful attention to the planning and training issues raised in the next six sections, and throughout this book, is an attempt to immunise you against this syndrome.

1.2.1 Is radio tagging the best approach?

There is much satisfaction to be gained by radio tagging a relatively unstudied species. Tracking a few animals and thus gaining an intimate picture of their habits can be great fun, but is only a preliminary to more serious science. You will need data from statistically meaningful samples of animals in order to make valid comparisons between sexes, age classes, areas and experimental treatments. Pilot studies are an important part of any radio tagging (see section 1.2). However, after gaining preliminary insights with radios, you may consider that it would be more efficient to put visual markers on many animals, instead of radio tagging a few, or to analyse hundreds of faeces instead of watching a few radio-tagged individuals feed. In these days of limited resources for wildlife science, judgement of your projects will probably depend more on how efficiently those resources were used than on how much fun you had. It is therefore worth considering the relative advantages and disadvantages of alternatives to radio tagging.

Attached visual markers, including coloured or numbered collars, bands and streamers (reviews by Stonehouse 1978; Bub and Oelke 1980; Parker *et al.*, 1990; Calvo and Furness, 1992; Nietfeld *et al.*, 1994; Murray and Fuller, in press), or light-emitters at night (Buchler, 1976; Wolcott, 1977; Batchelor and McMillan 1980), have been available for many years. Together with subjective or photographic recognition of individuals by their natural markings (Clutton-Brock and Guiness, 1975; Scott, 1978; Pennycuick, 1978), the attached markers are often adequate for making systematic observations of animals which live in the open, and can be better than radio tags for identifying individual animals among others in a group. For species that are recorded frequently by volunteers, visual records or band reports can give useful survival and movement data from wide-ranging species without the expense of radio tracking. Nevertheless, VHF tags on raptors have indicated substantially higher survival than band records (Kenward, 1993) and satellite-tracked UHF tags can reveal much more than band recoveries about long-distance movements (Fuller *et al.*, 1984; Jouventin and Weimerskirch, 1990; Meyburg *et al.*, 1995; Brodeur *et al.*, 1996; Meyburg and Meyburg, 1998).

New technologies continue to provide alternatives to radio tagging, but with restrictions. For example, DNA analysis of material extracted from hairs, faeces and other traces (Taberlet and Bouvet, 1992; Morin and Woodruff 1996; Kohn and Wayne, 1997) can give data on movements and territory boundaries of many individuals, but is expensive. The feeding of indigestible beads (Randolf, 1977; Kruuk, 1978) for detection in faeces is inexpensive but best for groups of animals. Excretion of a particular low-activity radioactive isotope (Kruuk *et al.*, 1980; Jenkins, 1980; Crabtree *et al.*, 1989) can only be used to map movements of one animal in an area. The tracking of small animals with radioactive tags (Bailey *et al.*, 1973; Ricci and Vogel, 1984) is discouraged by short detection distances and by welfare concerns (including human health and safety restrictions). This technique also identifies only one or two individuals (with different strength radiation sources) in the same area at a time. The same lack of ability to distinguish many individuals applies to the use of tiny harmonic radar transponders for recording location and speeds of flying insects (Riley *et al.*, 1996), although these 2 mg tags on tiny animals give good (line of sight) range with powerful transceivers. Larger transponders for tracking by radar can be coded for individual identification (French and Priede, 1992).

An idea that seems to have been little used since an original publication was the attachment of spools of very long, very fine thread to record a complete movement trajectory for small animals (Miles *et al.*, 1981). Fastening on vegetation along the animal's path, this thread leaves a three-dimensional movement record that is much more accurate than could be obtained by radio tracking. A similar approach uses a container of four different fluorescent pigments to leave a dust trail (Mullican, 1988), which can be combined with radio tracking of slow-moving animals to recapture them and refill the container (Goodyear, 1989). However, neither of these ingenious approaches have been widely used.

It is two other, relatively new technologies, that are the most widely-used alternatives to UHF/VHF radio tagging. Long life at low range is given by passive integrated transponders (PITs), which have been used in wildlife for more than a decade (Fagerstone and Johns, 1987). The smallest PITs are now about 2×7 mm (see Appendix I for addresses) and can therefore be implanted in very small animals. These tags emit a weak resonant signal if they are activated by a powerful nearby transmitter on the correct frequency, and are programmed to return an individual code. Their range to portable receivers is up to about 15 cm in air and 30 cm in water, but slightly greater ranges can be obtained with specialised antennas. Although PITs cannot compete with radio tags for giving detailed data on movements and fates, they can be used to record visits automatically to any site with restricted access, such as a nest entrance, feeder, travel tunnel for mammals or water passage for migratory fish. They can be used to record predation by tagging large numbers of prey, and then scanning remains at nests or even scanning predators that are radio tagged, such as snakes (Reading and Davies, 1996). They can also be used to trigger events, such as cameras to record behaviour, or to permit access of particular individuals to feeders. PITs therefore have considerable potential for experimental studies. A possible future development is the extension of the antenna in such tags to increase their range. Ranges of about 1 m per 10 mm of antenna have been available from diode transponders developed without individual codes to find avalanche victims.

The second type of tag that offers some competition for radio tags, but also complements them, stores data from sensors in battery-backed memory. Advantages of data-storage tags are their ability to acquire information in the absence of an observer and in places unsuitable for transmission of radio signals. Thus, they have given unique data from diving animals (Hill *et al.*, 1983; Cairns *et al.*, 1987). They can also record locations estimated by miniature Global Positioning System (GPS) receivers (Rempel *et al.*, 1995; Rempel and Rodgers, 1997; Moen *et al.*, 1996). Less accurate location records for migrants might be obtained by recording sunrise and sunset times for comparison with those expected at different longitudes and latitudes (Wilson *et al.*, 1992), but the many factors that influence light intensity records have hindered implementation of this approach. Data storage devices are now often combined with radio tags, either to aid recovery of the tags or to avoid the need for recovery by re-transmitting the data at convenient times. In this context, they are given more consideration in Chapter 3.

1.2.2 Do I have time for radio tagging?

There are four stages to a radio-tagging project. These are needed to: (i) obtain equipment; (ii) tag animals successfully; (iii) collect satisfactory data and (iv) analyse the results. Each of these stages needs careful planning at the start of the study. In particular, analysis is not something to leave until the end of the study, as is done in so many student projects. Preliminary analysis will be needed at the least to fine-tune your data collection, and may possibly modify your need for equipment (e.g. to get tags with longer life) or the timing of your tagging.

It is desperately easy to underestimate the delays that can beset each stage of the study. For a start, time is needed to select and obtain appropriate equipment. Advice on how to select equipment that gives adequate detection range, life, reliability and attachment requirements can be obtained from other users, from equipment manufacturers and in Chapter 2. However, radio tags are generally built to order, and often require specific components for permitted radio frequencies that differ from country to country. It takes a manufacturer time to obtain the necessary parts and build the tags, so that delivery is typically 6–8 weeks after placing orders. If you have a particular marking 'window', for example because birds leave the nest at a particular time, equipment orders must be planned well in advance or you risk losing a whole field season (Fig. 1.1).

You may also need time to develop a tagging technique that has no impact on the study species. Even if the species has been tagged before, previous workers may not have taken the trouble to record or report drawbacks to their tagging. Even if the tagging did not affect their results, or cause unacceptable harm to the animals, it may prove to be a consideration that requires modification of equipment or techniques in your case, especially if the data you need are different.

Many projects have failed to appreciate how long it can take to develop field techniques. For instance, you may need to learn how best to approach tagged animals without disturbing them, or how to interpret their behaviour from radio signal characteristics. It may take several months to discover that it is impossible to collect data as you intended, and more time to develop another way of gathering the information. By that time, your sample of tagged animals may have become inadequate, because of deaths, dispersal and radio failures; there could then be a serious delay before you can trap and tag some more.

As well as needing to analyse data from an early stage to fine-tune your fieldwork, early analysis may raise further biological questions. There may be the chance of really making a breakthrough if you could only collect some slightly

Figure 1.1. Equipment orders should be planned well in advance.

different data, or tag another ten animals. Again, early analysis and a little time in hand become a big advantage.

It is important to realise that experienced researchers often obtain few useful data in the first year of radio tagging a new species. For projects at the boundaries of the technology, development delays may take even longer. Radio tagging is therefore definitely not a technique to grasp at in the last year of a tricky doctoral thesis. In fact, it is a very good idea to set aside an initial period for a pilot study, in which to develop and refine your techniques (S. Harris *et al.*, 1990). If your biological questions can be answered in only one season of the year, you may well need a year for pilot work and two more to collect adequate data, making a three-year project. If you need a question answered in one to two years, because of a conservation crisis or commercial issue, you would be better advised to employ experienced personnel than to start a student project, especially if the species has not been tagged before.

1.2.3 Can I catch enough animals?

There is no point in obtaining large quantities of expensive radio equipment before you know that you can catch your animals. Even the well-proven techniques often seem easier when described on paper than when you are trying them out in the field. It is not just a matter of how the trap works, but of where best to place it, how best to bait it, how often to visit it, and the little bit of stick that needs to be wedged in just the right place to stop the wind setting it off! If at all possible, get someone who has used the trap to demonstrate it for you. You may need to go on a proper instruction course, for example if you are going to climb cliffs or trees to reach nests, or to use mist nets, or anaesthetic darts which contain dangerous drugs (you can radio track the darts too, to find the sedated animal or shots that miss). You may even need to pass an exam to be licensed for an invasive tagging technique. Consider buying only the minimum radio equipment, or borrowing at least the receiving system, until you are sure that you can tag enough animals for your study. This is another part of your pilot study.

Make sure, too, that your capture technique is not unacceptably biased. If you wish to tag adult animals, you could waste a lot of time finding that only the inexperienced juveniles readily enter the traps. Bear in mind that trapping may select the hungrier or more active segments of the population. This is an important consideration if you need representative data on mortality or emigration rates. Moreover, take care that neither the capture technique nor the tags themselves unduly shock the animals, or cause other adverse effects (see Chapter 4).

In a predation study, where there is the possibility of tagging either a predator or its prey, ease of capture may affect which species is tagged. Tagging the predator can provide a variety of data on its ecology and diet, with opportunities to recover fresh kills to study selection effects, and needs relatively few tags, but requires additional data on predator and prey densities for estimating the predation impact (Kenward, 1980). If you want to tag the prey, you must be able to catch and tag large numbers of them, and there may be difficulty attributing deaths to a particular predator, but you can gather data on general ecology of the prey and total mortality.

1.2.4 Can they be radio tagged?

Two important constraints to radio tagging are the signal propagation conditions and the animal's size. For instance, the propagation of radio waves through water depends strongly on its conductivity (Fig. 1.2). With water of low conductivity, such as in rivers and freshwater lakes with a conductivity less than 0.01 S m⁻¹ (mho m⁻¹), the surface range of a radio tag may be reduced by only 50% at a depth of 10 m. However, the range drops sharply with depth as conductivity increases. Acoustic tagging (sonar) is therefore used when conductivity is greater than about 0.5 S m⁻¹ (Keuchle, 1982), in all brackish and salt water (reviews in Stasko and Pincock, 1977; Ireland and Kanwisher, 1978; Priede, 1980; Harden Jones and Arnold, 1982; Priede 1992). On the other hand, radio tracking is preferable in rivers, unless they are very large and dirty. It avoids the inconvenience of having to immerse hydrophones in the water; moreover, the detection range of acoustic tags is low in turbulent and aerated water, especially if it is shallow. Migrating fish can be radio located with rapid searches along riverside roads or from aircraft.

Most invertebrates and a few terrestrial vertebrates are too small to be radio tagged. The smallest crystal-controlled VHF tags weigh 350–500 mg. Recommendations have been made to keep loading below about 5% of animal weight (Brander and Cochran, 1971; Cochran, 1980). However, a given percentage loading

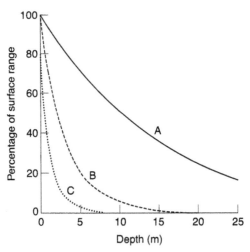

Figure 1.2.   The surface range of radio tags in fresh water reduces as their depth increases. The curves show theoretical range loss as conductivity increases from 7×10^{-3} S m⁻¹ (A), through 30×10^{-3} S m⁻¹ (B) to 80×10^{-3} S m⁻¹ (C). Redrawn, with permission, from Keuchle (1982).

has a greater impact on large animals (Caccamise and Hedin, 1985; Pennycuick 1989). Small birds can add fat loadings of more than 50% for migration, and with the smallest bats seem able to fly well with packages at 10% of their weight (Graber and Wunderle, 1966; Stebbings, 1982; 1986). It may therefore be reasonable to use the smallest tags on animals of 5–10 g. Nevertheless, an absolute limit of around 5% is probably appropriate for animals with mass above about 10 g (Cochran 1980). Moreover, although flying species can manage a 5% loading at 70 g without loss of agility (Aldridge and Brigham, 1988), it is worth noting that a 3% loading gives an equivalent effect on flight power for a 200 g bird (Caccamise and Hedin, 1985). A limit of 2–3% is therefore probably wise for larger bats and for birds that migrate or otherwise depend greatly on flight, and for any tag attached to a bird away from its centre of lift.

Tags may become slightly smaller, but their size is ultimately limited by the size of the cells, crystals, inductors and capacitors that they contain. Cells and crystals are the heaviest parts. The smallest silver-oxide cells weigh 140–190 mg, with VHF crystals now available at about 90 mg. One must allow 100–200 mg for other components, antenna and potting materials. Tags without crystal control can be slightly smaller, but are inefficient and generally illegal. The small components are not made especially for radio tags, but for other electronic devices with much larger markets. Small size is advantageous in watches and small communications devices, such as pagers and hearing aids. Future military needs, such as for smart munitions, may drive further miniaturisation.

The sort of biological questions that can be asked for small animals is limited by the restriction that tag size places on transmission life and ease of detection. Tag life can be increased by reducing the rate and width of signal pulses. However, signal rates of less than about 50/minute hinder direction finding, and pulse widths less than 10–20 ms may reduce detection distances. Tags weighing 350 mg can transmit short signals for up to a week. However, 500 mg tags with a more efficient cell and circuitry manage 2–3 weeks, which is enough to record a seasonal home-range estimate and find a nest or den, so that range variables can be linked to productivity. A larger cell in a 1 g package gives a more useful 5–8–1 week life, which is enough to see how foraging changes while rearing young. The year of transmission possible from 4 g tags (Fig. 1.3) enables a study of seasonal changes in foraging and survival.

The discharge characteristics of 1.5 V silver-oxide cells are suitable for small tags. They have a lower self-discharge rate than 1.35 V mercury-oxide cells, which were used widely in tags until the 1990s, but are now becoming unavailable for environmental reasons and are already illegal in some countries. More powerful tags use lithium cells, primarily lithium-thionyl-chloride, which readily give low currents for long periods and are available to military specifications for high reliability. These 3 V cells power extra circuitry to give detection distances that are routinely double those of 1.5 V tags. In theory, to transmit twice as far requires four times the power (by the inverse square law), such that tags would require four times as much cell capacity to double their range. In reality, the 3 V tags tend to have

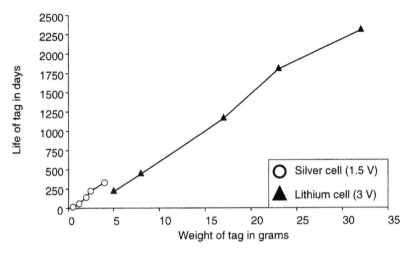

Figure 1.3.   The life in days for continuous operation of some typical radio tags for wildlife, as a function of their weight.

higher conversion efficiencies (of electrical energy to radio signal) than 1.5 V designs, so that doubling the range may reduce life by only 30–50% at the same tag weight. Although the energy density of small lithium cells is appreciably less than for larger cells, the life increase of 70 days per gram (over a size range typical for 3 V tags) compares well with the 90 days per gram of equivalent tags with silver-oxide cells (Fig. 1.3). Tags of 15–17 g can now last reliably for 2–3 years, for studies of age-specific survival and dispersal.

Maximum detection ranges are also strongly dependent on antenna format and dimensions. Thus, the 1.5 V tags have maximum line-of-sight ranges at ground level of only 100 m with integral tuned-loop antennas, but up to 3 km with 25 cm whip antennas of fine wire. Using the most sensitive hand-held receiving equipment, line-of-sight ranges of more than 20 km are possible to birds flying with these tags, and more than 80 km from aircraft to birds in flight with 3 V tags. The importance of using long whip antennas cannot be stressed too strongly, and it is also worth noting that a thick antenna radiates slightly better than a thin one (Fig. 1.4).

All these considerations mean that some projects are impractical because although the animals can carry VHF tags, they cannot carry ones with enough power or life to address the biological questions. Power can be increased on small tags, at the expense of transmission life, by using two silver-oxide cells in series. However, the main problem for small animals is often the inability to support a long antenna without welfare concerns. This is especially relevant when small animals have very large foraging ranges (e.g. some bats and small birds), but can also affect larger species whose tags cannot be equipped with efficient antennas (e.g. implanted or ingested tags for amphibia and snakes).

The smallest UHF tags for tracking by satellite are currently about 20 g. These transmit continuously for only about 400 hours, but sophisticated timing of the

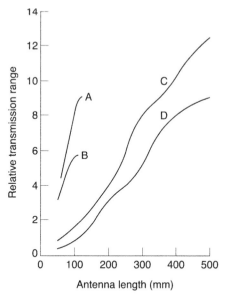

Figure 1.4. The range of radio tags increases with antenna dimensions. Moreover, at a given circumference, tuned loops made from 8 mm brass strip (A) gave longer ranges than loops of 1 mm multistrand copper wire (B), and whips of 2 mm steel cable (C) were more efficient radiators than whips of 0.2 mm nickel guitar wire (D). The range curves were estimated assuming no propagation losses, from the power output of a 150 MHz transmitter which was tuned to each antenna. Antennas were orientated for peak signal transmission, and were in air; ranges are 30–50% less from loops on animals.

transmissions can provide data at weekly intervals for a year (see Chapter 3). The tags are relatively bulky, and need little overhead cover if their signals are to be received adequately. They are therefore most suitable for use on birds with mass that exceeds 500 g. The locations from types used in the present ARGOS tracking system are much less accurate than from most VHF tracking, so they are most suited to studies of migration or of very wide-ranging foragers, like albatrosses (Jouventin and Weimerskirch, 1990; Prince *et al.*, 1992).

1.2.5 Can I collect appropriate data?

If you have reasonable numbers of animals marked with reliable modern tags, you may be able to collect plenty of data, but then be balked by difficulties with the analysis. A major problem is that a large number of records collected from one tagged individual cannot be considered to be statistically independent data. This is clearly the case if observations are made in rapid succession, because a resting animal may remain in the same place, or a walking animal may have had no time to move out of a large habitat patch. Autocorrelation analysis is available to test for a sampling interval that gives low spatio-temporal dependence (Swihart and Slade, 1985a), but this does not overcome the dependence that arises through repeated use of favourite feeding or resting sites, or other time-tabling behaviour.

It therefore becomes necessary to summarise all the data from each animal as one record, of travel speed, or range area estimated with a given method, or proportion of locations in a particular habitat (Kenward, 1992; Aebischer *et al.*, 1993b; Otis and White, 1999). Analyses become based on numbers of tagged animals. This is one of the most crucial considerations in radio-tagging studies, and all too often neglected. Many students are very pleased to have collected several hundred locations from two or three animals. It can then be devastating to discover that, although home ranges and other behaviours can be described, they can do no valid quantitative analyses.

It is therefore important to think very carefully at the pilot stage of the study about what sample sizes you need to answer particular questions. Note that three in each category is a minimum for significance at the 5% level in a rank-based (non-parametric) test, assuming no overlap of measures. So, even if you only want to compare range characteristics of males and females, planning to monitor less than five of each makes you a hostage to the occurrence of death or dispersal. If you want to look at proportions of animals that experience particular events, such as mortality, you will need samples of at least 20–30 individuals (Pollock *et al.*, 1989). If collection of data is automated, gathering information from many individuals may be no problem. However, if data collection requires fieldwork, you will probably want to maximise efficiency by minimising the number of observations from each tagged individual. That minimum number of observations will depend on the summary statistics needed to answer your biological questions, on the species and on the situation (see Chapter 6). This means that you will only be able to plan the collection of sufficient observations if you have appropriate analysis software at the beginning of the project.

If you are planning work that involves analysis of interactions between animals and maps of habitats or other resources, you will need to consider how to obtain the map data. For large areas, you may need data imaged by overflights of aircraft or satellites. However, you should not imagine that purchase of a Landsat scene, for example, will fulfil your requirements. The data are provided as spectral reflectances, and require much additional work to turn them into maps of land cover that are suitable for analysis with animal location data. Even if you only need a map of vegetation in a small area, the time required to record and digitise it should not be underestimated (see Chapter 4).

1.2.6 Can I afford it?

Neither the tags nor the receivers made especially for VHF radio tracking have increased greatly in price during the last decade. A lightweight receiver adequate for 10 tag frequencies now costs about US$430, or upwards of US$800 for receivers to use with up to 100 tags. For large numbers of tags, a workable alternative to the more expensive receivers is a commercial frequency synthesis scanner, such as the Yupiteru MVT7100 or AOR AR8000 (available from Maplin, see Appendix I for addresses) which cost US$500–700. However, beginners would be wise to stick with equipment designed for wildlife tracking, because it will come with advice on how best to use it and will have features that make it more weatherproof or otherwise more appropriate for the purpose. You will need a tracking antenna as well as the receiver, and headphones if they are not also part of a package, so allow another US$160 for these extras.

VHF radio tags can be brought from a number of firms, as ready-to-mount packages or as basic transmitters to which batteries and antennas must be added. Ready-to-mount tags cost US$115–300. Basic transmitters have been built successfully in a handful of studies during the last two decades, but this is not really practical any longer. Designs have to be modified continually to replace obsolete components, which adds to the difficulty of learning to assemble electronic components reliably (Fig 1.5).

However, financial savings of 20–50% are possible by adding cells, antennas and mounting materials to basic commercial transmitters, which start at about US$80. This approach, covered in Chapter 4, also gives some flexibility in package design. The greatest saving can be had by building mammal collars, which is straightforward but especially time-consuming for manufacturers, and therefore expensive. However, it is important to realise that cells and the transmitters themselves can be damaged by careless soldering, while clumsy potting can allow moisture to penetrate or add unnecessary weight (Fig. 1.6). Unless you have some experience, are using a simple tag design and need to use and re-use large number of tags, you may be wise to rely on the manufacturer's packaging skills.

Figure 1.5. An ability to create practical designs is one of the skills needed to build transmitters for wildlife.

Figure 1.6. Clumsy potting can add unnecessary weight.

UHF tags for tracking by satellite cost US$2000–4000). To that must be added the cost of having them tracked. The ARGOS system currently charges about US$3 for each day of tracking, depending on the type of data required from the system, which totals to US$1000 per year.

Analysis software can be obtained free or at minimal cost, at least for making basic estimates of home range area or survival (see Appendix II). However, if you plan detailed analyses of how animals interact with each other or with habitat features, or need to display data from many animals at the same time, or want to save time by running analyses on many animals as one operation, then you may be wise to buy specialist software. Software packages dedicated to analysis of radio-tracking data cost US$500–1000. More general Geographic Information System (GIS) software can cost even more, and may be necessary for handling map data, although the US Army's GRASS is free. You may also benefit from software to help record the data in the field, or need to buy remote-sensed map data.

A final cost that can be a serious embarrassment unless planned adequately is that of travel. Systematic tracking of species that range over several square kilometres can require 200–300 km of vehicle travel to obtain 2–3 locations a day on 5–10 individuals. Thus, recording a home range with 30 locations each may require 3000 km of travel, or more if the individuals are marked over a large area. Hire of aircraft can make serious holes in a project budget too; helicopters are only for those with really generous funding.

1.3 TRAINING

With so many considerations involved in the use of radio tags, one might expect there to be lots of training opportunities. Unfortunately, this is not the case. Good advice on equipment choice is available from manufacturers, and one firm (Biotrack) even provides an expert-system that can be downloaded from its Internet site. Courses in particular software packages are sometimes run at meetings. However, training in field techniques tends to remain up to the student. If you are that student, this book is intended to help as much as possible, but you will do even better if you also talk to your nearest experienced radio-tracker. Moreover, be certain to get help with any delicate tagging operations, to practise with both hardware and software before you start, and to seek advice on study design from a statistician now. Please don't wait for that analysis advice until you are halfway through preparing the final report or thesis.

If you feel that all the above considerations are a bit daunting, that is the intention. If you begin to understand the care required for efficient research based on radio tagging, you are much more likely to plan your budget and data collection adequately, buy the right equipment for collecting that data, get any available training and allow plenty of time for teething problems. You will have given yourself the best chance of success.

1.4 CONCLUSIONS

1. Radio tagging is a very sophisticated tool for research and management of wildlife, especially for studies that require data without bias and from individual animals. It is ideal not only for defining the life-styles of rare or elusive species, but also for collecting relational data that can define mechanisms to be used in predictive models, and for field experiments to test those mechanisms and models.
2. However, successful use of radio tagging requires: (i) sound biological questions; (ii) careful planning to address those questions and (iii) training to use the planned techniques. The failure to provide adequate questions, planning and training has often made radio tagging an ineffective tool, which can deter funding.
3. Project planners must ask themselves whether radio tagging is their best marking technique, among alternatives that include short-range transponders, visual markers, radioactive and DNA tracers, harmonic and coded radar, thread and fluorescent trails, sonar and storage tags. Are animals above the absolute thresholds of 5–10 g for marking? Is there adequate time to obtain equipment, tag large enough samples of animals without bias that may invalidate results, collect data and analyse results? Can adequate training be provided?

2 Basic Equipment

This chapter provides a review of the basic VHF transmitting and receiving equipment used in wildlife radio tracking and radio telemetry. UHF transmitters and the ARGOS system for tracking by satellite are described in Chapter 3, along with other systems for automated tracking and telemetry. The many designs for mounting all types of transmitters, packaged in tags with the necessary cells and antennas, are considered in Chapter 5, which also gives tips on building those that can readily be assembled by biologists.

Until the 1980s, crystal-controlled transmitters in wildlife radio tags were of two main types. In the simplest 'single-stage' circuits, an elementary oscillator circuit

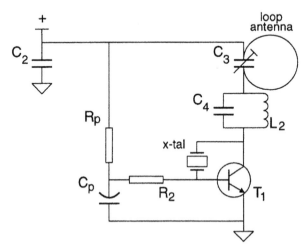

Figure 2.1. A circuit for a 'single-stage' transmitter with a tuned-loop antenna. The single-transistor (T_1) design generates signal pulses by charging capacitor C_p at a rate determined by resistor R_p, other components being the crystal and inductor L_2. This simple oscillator is relatively unstable but has been used very widely in wildlife radio tags.

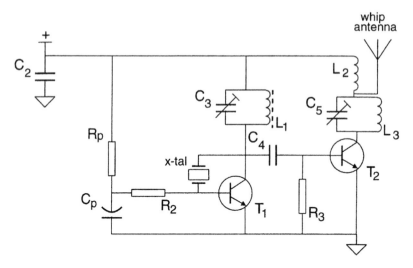

Figure 2.2. A circuit for a 'two-stage' transmitter with a whip antenna. A simple oscillator generates signal pulses, but the separate output stage with transistor T_2 gives enhanced stability.

that resonated with the crystal to give the desired radio frequency (RF) also charged a single large capacitor to generate the signal pulses (Fig. 2.1). Pulse width and interval were governed by the 'pulse capacitor'. With all else held constant, increasing the value of this capacitor increased the pulse duration and the interval between pulses. The same form of pulse control was used in 'two-stage' transmitters, but was to a greater extent separated from RF tuning in a second stage, which improved tuning and power output by better impedance matching with the antenna (Fig. 2.2).

A disadvantage of these early designs is that pulse width and interval (hence rate) are not set independently. Moreover, these characteristics are affected by the RF tuning and *vice versa*. Pulse rates can vary appreciably if animals earth the antenna, for example by touching wet ground, which can even stop a poorly-tuned tag. Variation in voltage during discharge of the pulse capacitor can cause frequency variation (squelching) that reduces tag efficiency.

Nevertheless, the simple single-stage circuits can be built without circuit boards, by fixing components to the crystal can and soldering fine wires between them to give a very compact transmitter. A tag weight of 350 mg is possible with the smallest crystals, cells and other components (Fig. 2.3).

More precise control of frequency and pulses, for example permitting separate sensors to modulate pulse width and interval, became possible in transmitters that contain multivibrator pulse generators (Fig. 2.4) separate from the RF tuning circuitry (Taylor and Lloyd, 1978; Anderka, 1980; Lotimer, 1980; Thomas, 1980). In the late 1980s, it looked as if hybrid technology might be the best way to miniaturise the improved transmitter designs, by adding a few large components

Figure 2.3.   Large insects, such as this rhinocerus beetle, can easily carry the smallest tags.

(e.g. crystals and coil inductors) to boards that contained integrated multivibrator circuits. However, improvements in industrial CMOS (complementary metal oxide semiconductor) technology began to provide complex integrated circuits for low voltages in surface-mount chips. Manufacturers therefore turned mainly to surface-mounting, which gives greater design and construction flexibility than with hybrids.

The result has been transmitters that can have frequencies and pulse parameters set more precisely than in the original one-stage and two-stage designs. The greater precision can be used not only to define pulse widths and intervals for telemetry (e.g. temperature, activity, mortality), but also to identify tags and extend their life. Pulse intervals can be set so that three to four different individuals on the same frequency can be identified (Fig. 2.5A), by timing ten signal pulses with a stopwatch. Very short pulses can be generated, in combinations (Fig. 2.5B) that can code for many different tags on the same frequency (Lotimer 1980, Howey *et al.*, 1988). The latter approach can in its simplest form be identifiable by ear, but usually requires pulse detection circuitry in the receiving system. Such circuitry can also detect individual pulses that are too short to be audible when they are faint: tags with very short pulses have very long life.

The adoption of surface mount construction techniques has also affected the design of receivers, directly by enabling a reduction in the size and power consumption and indirectly by providing scope for decoding and pulse-stretching facilities. The result has been improved reliability and efficiency, from tags with longer transmission lives and more sophisticated receivers. Designs of both tags and receivers are also dependent on their radio frequencies, which are considered in the next part of this chapter.

Figure 2.4.   An example of a modern transmitter circuit in which generation of signal pulses in an integrated astable multivibrator permits their characteristics to be set with high precision and circuit stability.

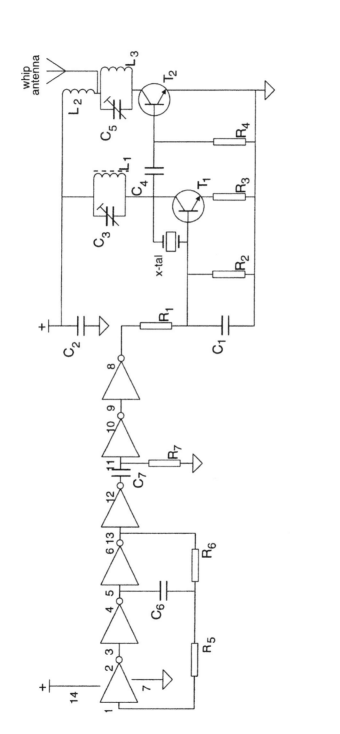

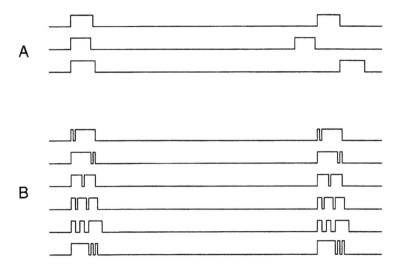

Figure 2.5. A schematic view of how transmitters with similar power requirements, either using slightly shorter pulses with shorter intervals (A) or simple coding of pulse trains with similar total length (B), can be used to distinguish signals from tags on similar radio frequencies.

2.1 FREQUENCY

Many countries allocate a frequency band of 150 kHz–2 MHz for wildlife radio tags, mostly in the Land Mobile Band at 138–174 MHz. Most European countries have bands at 150–151 MHz, with 142 MHz also used in Austria and 148–149 MHz allocated in France, Holland, Spain and Switzerland. Tags are restricted to 142 MHz in Norway, 230 MHz in Finland and 173 MHz in the United Kingdom, the latter frequency also being used in the Czech Republic, Greece, Poland, Portugal and Spain. Bands at 150–151 MHz are used in North America, where the USA also permits tags on 40 MHz, 165 MHz and 216–220 MHz. Further use of 150–151 MHz is in Australasia, with 165 MHz available too in New Zealand. The 216 MHz band has been used without problems in Asia and the Middle East.

By 2008, a new harmonised VHF band may be provided throughout the European Union for wildlife radio tagging. If allocated at all, possible benefits for standardising equipment or work on migratory species will depend on its width. At least 100 separate tag frequencies are needed for moderately ambitious projects. Separation of 10 kHz is used routinely for tags on different animals, although a 5 kHz separation can be practical with the most modern tags and receivers. Whether the allocation is national or international, an allocation of at least 500 kHz is desirable, although a much narrower band may have value for tags sending coded signals on the same frequency.

As there is continual growth in the number of radio applications competing for frequencies, governments often cannot allocate even 500 kHz for protected use in radio tagging. The allocation may have to be shared with other low-power transmitters. This is the case in the United Kingdom, which has a band at 173.20–173.35 MHz shared with narrow-band industrial telemetry and telecontrol equipment, and a further band at 173.70–174.00 MHz shared with users of radio-microphones. The lower band has been satisfactory for wildlife biologists who work only in rural areas, but is often quite crowded in built-up areas. Where other narrow-band transmissions are strong, biologists must avoid having tags on the same 3–5 kHz. In the upper band, there is little evidence of interference from radio microphones. However, a few powerful transmitters licensed to broadcast on other frequencies produce appreciable signals on harmonics within the band. Intermittent voice and data signals from these powerful transmitters are a nuisance when searching for wildlife tags from aircraft.

In a world with continual increase in radio traffic, some governments are already charging commercial groups for allocation of frequencies. It therefore behoves biologists to be vigilant in defence of their frequency requirements. The price of not paying for frequency is the risk of having it allocated for another use; as a minority application, wildlife radio tagging is at particular risk. Government biologists are best placed to take the lead in negotiating for frequency, and can improve their negotiating position by partnership with other groups that can share bands. VHF animal tags are of such low power that they are unlikely to interfere with most other users of the same frequency band. Any problems that arise are liable to be suffered by the biologists, when their tracking of weak signals is prejudiced by other users. Good partners are those who use low-power equipment mainly at other times or places than biologists. This is the case with the mainly urban and evening application of radio microphones. Sharing with military communications has been practical in some countries.

Where there is a choice of frequencies, several factors must be taken into account, including antenna efficiency and directional accuracy, habitat effects on signal propagation, and ease of transmitter construction. Antenna dimensions are often expressed as fractions of a wavelength. These may be readily calculated from frequency, since radio waves travel at about 300×10^6 m s^{-1}, by the formula:

$$\text{wavelength } (\lambda, \text{ in m}) = \frac{300}{\text{frequency } (f \text{ in MHz})}$$

For instance, at 150 MHz the wavelength is 2 m. A simple receiving antenna for 150 MHz is a $\lambda/2$ dipole, which is about 1 m long. However, the most common hand-held receiving antenna at this frequency is actually a three-element Yagi antenna (Fig. 2.6), which is more bulky than a dipole although its longest element is little more than 1 m. This antenna, unjustly called after the translator of the original Japanese design, by Uda and Mushiake (1954), has about twice the gain of the dipole and is far better for taking bearings. Various twin-dipole H-shaped antenna designs are slightly more compact than a three-element Yagi, but give ambiguous (bi-directional) bearings and have slightly less gain.

Just as efficient receiving antennas on 142–230 MHz are convenient to carry, so too are the transmitting antennas on these frequencies efficient at small size.

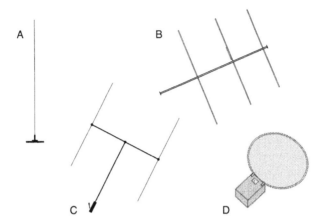

Figure 2.6. The antennas most commonly used for mobile radio tracking: a dipole for mounting on a car roof (A), a three-element Yagi (B), an H-Adcock (C) and a loop antenna mounted on a receiver (D).

Nevertheless, it is lower frequencies that propagate best through water. Tags have also been used on 27 MHz, 40 MHz and 102–104 MHz for tracking fish. However, the longest element of a λ/2 Yagi at 104 MHz is nearly 1.5 m, which is cumbersome, and would be about 5.6 m at 27 MHz. The most common directional receiving antenna at the lowest frequencies is a λ/6 loop, which has about half the range of a λ/2 dipole and is 0.5 m in diameter at 27 MHz.

For non-aquatic species, the disadvantage of increased signal absorption by vegetation at high frequencies is normally considered a minor inconvenience, compared with the benefits of improved antenna efficiency, at least for frequencies up to 230 MHz. Frequencies closer to 142–150 MHz may have a slight advantage in habitats where the vegetation is very dense and damp, such as rain-forests. In compensation, tags at 216–230 MHz need only 60–70% the length of antenna for the same transmission efficiency as the lower frequencies, which can be an important consideration for small species, especially in dry, open terrain. On the whole, however, frequencies between 142 MHz and 230 MHz are a good compromise for animal radio tagging, and are also convenient frequencies for the electronic components that are used in lightweight tags with low power consumption. Although tags have also been built on the UHF frequency at 400 MHz for use on wildlife in open country, they have not yet reached the efficiency and reliability of VHF tags.

2.2 RECEIVERS

2.2.1 Receivers for small projects

Radio-tag receivers can be divided into three main categories, which are sometimes available as different models from the same supplier. The most basic category is sold primarily for locating trained raptors or dogs. These receivers, which start at about US$300, are not usually as robust as the receivers built for regular research use; they also cover a frequency band of only 30–200 kHz. None the less, these receivers can be just as sensitive as the more sophisticated models, and some are small enough to fit into a large pocket. They can be used for pilot studies or low-budget projects in which no more than 3 (with 30 kHz bandwidth) to 20 (with 200 kHz) tag frequencies will be required, especially if the emphasis is on radio surveillance. It is possible to put two or three tags with different signal pulse rates on each frequency, but this creates problems when trying to detect weak signals or to take bearings near another tag on the same frequency. A tag frequency separation of about 10 kHz is therefore preferred in most projects. This enables tags to be identified individually despite frequency drift: some tag frequencies shift from their intended value with changes in temperature and cell voltage, typically by 1–3 kHz, because of variation in crystals or other components. It is also possible to use a tag separation of 5 kHz, provided: (i) that the receivers have high selectivity and drift no more than 1 kHz from –20°C to +40°C; (ii) that the tag crystals have high calibration accuracy with low temperature drift (expensive) and (iii) that tags are made with different pulse rates on alternate frequencies. Nevertheless, this puts a lot of pressure on the tag maker. Note that the repeated re-tagging of 10 small animals typically requires 15 to 20 tag frequencies, if exhausted tags are to be replaced rapidly without using the same frequency twice.

2.2.2 Receivers for large projects

Receivers for up to about 50 tags can now be bought for US$560, or from US$800 for a 1 MHz band that is adequate for 100 tag frequencies. Four reliable and widely used models are shown in Fig. 2.7. They typically differ by no more than about 3 dBm in sensitivity, at around –147 dBm to –150 dBm (0.015–0.010 μV) for the lowest signal strength that can be heard, although some individual receivers perform worse than others of the same type. Signal strength of –148 to –150 dBm is about the faintest that can be detected against background radiation and noise from components. These receivers weigh 1–2.5 kg with internal batteries, which is convenient for carrying round the neck; much more than about 2 kg can become uncomfortable.

The Mariner Radar M-57 receiver (Fig. 2.7, top left) has integral dial illumination for nocturnal researchers, is almost completely watertight and has survived being run over by a Land Rover. However, it is also the heaviest (2.5 kg including batteries), and covers only 150 kHz or 300 kHz bands. The Custom Electronics CE-12 (top right) is based on the well-proven LA-12 design but with improved electronics. It covers a basic 300 kHz band with 12 channels of 25 kHz but can have other bands added to cover 1.5 MHz with 60 channels. The Wildlife Materials TRX-1000 (bottom left) is based on more modern frequency-synthesis electronics, which provide digital tuning over an initial 1 MHz, with a second 1 MHz band also available. Lids on these receivers protect them during storage, but

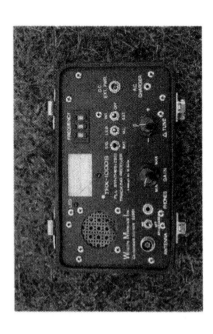

are inconvenient during use, such that fieldworkers resort to operating them through transparent plastic bags in rain. Another frequency-synthesis receiver with basic controls and a 2 MHz band is the Regal 2000 from Titley Electronics in Australia. One of the most widely used is the Telonics TR-4 (bottom right). These receivers cost US$650–850.

These receivers are all straightforward to operate, although it is easiest to find tags quickly with digital tuning; otherwise one needs to know the channel number and the fine-tune position which is equivalent to each frequency, or to set a large sweep dial with considerable accuracy. This makes the receivers in the bottom row most convenient for projects with more than 20–30 tagged animals. All the models have panel meters to indicate signal strength and the voltage in their 9–12 V supplies, which can be primary cells, rechargeable NiCads, or an external source such a vehicle battery. For intensive work, it is best to buy receivers with rechargeable cells, to avoid the cost and inconvenience of frequently renewing primary cells, but to carry a small separate battery pack as a back-up power supply. All have integral loudspeakers, as well as jackplug sockets for headphones or earplugs. Since the receivers are similar in performance and reliability, many users base their choice as much on ease of servicing, price (including possible shipping and import charges) and familiarity, as on real differences between the models.

2.2.3 Programmable receivers

The third category of receiver can be programmed, from an integral or external memory, to scan through a number of tag frequencies. This facility is indispensable during mobile searches for numbers of lost animals, or for automatic recording of data from many tags at one site, and is always convenient for rapid tuning to tag frequencies.

In projects where several animals at a time may go out of range from normal tracking points, for instance through dispersal movements or being carried away by predators, it becomes very tiresome to set tuning controls repeatedly for each tag frequency. When driving a car or moving fast across country in an aircraft, such a task creates unacceptable pressure on both the operator and the equipment. A programmable receiver can be set to tune each frequency automatically, usually for 3–10 s, in a repeating sequence that can be stopped when a signal is detected. This saves wear on the tuning controls and leaves the searcher's hands and mind free to concentrate on getting from place to place. Since non-programmable receivers must be tuned by hand to each tag frequency, they are normally used for automatic recording of signals from only one animal at a time. The presence or absence of several tags on the same frequency can be determined in low budget projects by listening through tape-recorded audio output, provided that each tag has a different

---

Figure 2.7. Four receivers that have been used widely for radio tracking. The M-57 (top left) is European whereas the CE-12 (top right, photo from Custom Electronics), TRX-1000 (lower left) and TR-4 (lower right) are made in the United States.

pulse rate. However, this is a tedious process that requires much time. A programmable receiver can be used to cycle repeatedly through a number of tag frequencies, recording for the same preset sample period on each. Analogue audio signals may be output to simple recording devices, including chart or tape recorders, or converted to digital format for more sophisticated data-logging equipment (see Chapter 3).

With three operating modes of: (i) receiving; (ii) scanning and (iii) logging, three main types of programmable receiver have evolved. The simplest approach is to have a separate memory/scanning module, which may also contain some simple signal logging capability. An example is the Telonics TR-2 receiver (Fig. 2.8, top left), costing US$1,800, which can be separated from the TS-1 memory and control unit (US$1,100) and is then suitable for normal tracking work. The Wildlife Materials Falcon-5 receiver with DL-2000 data logger adopts a similar approach, and can also be connected to a separate computer for sophisticated data logging; this is a combination of receiving, scanning and logging functions in separate units.

A second approach, originating in Cedar Creek Bioelectronics Laboratory in the University of Minnesota but refined as the R2000 receiver by Advanced Telemetry Systems, keeps the receiving and scanning electronics within one unit, with simple controls for entering frequencies to memory and scanning them manually or automatically (top right). This receiver covers a 2 MHz band for US$2150, with versions covering 4 MHz and 8 MHz bands also available. The R2100 and R4000 models, at $2400–2800, can be connected to a separate D5041 data logger costing US$3100. A similar combination of receiver, memory and scanning is found in the much smaller Australis 26K receiver from Titley Electronics at only US$1300, but scan options are restricted and the membrane switches used for all controls except volume and gain are not quite so convenient for frequency manipulations. Membrane switches are also used on the Lotek STR-1000, which is similar in size to the ATS range but also contains circuitry for estimating strength and rate of the incoming signal pulses.

The third approach is to combine receiving, scanning and logging functions in a single unit. The Lotek SRX 400 adopts this approach in a relatively large format (Fig. 2.8, bottom left), with a basic cost of US$5750 (W9 option). A number of other hardware and software options for this receiver provide automated switching between numbers of antennas, an elegant ultrasonic-VHF converter that enables sonar tags to be detected instead of radios, logging of pulse-modulated sensor data and use of pulse coding to detect multiple tags on the same frequency.

The second receiving system in this final category is the most innovative receiver of the last decade, the Televilt RX900 (bottom right). This $3500 hand-held system weighs only 700 g, with the battery in a detachable section for convenient renewal. It has a four-row display on a back-lit LCD screen that is angled for viewing from two

---

Figure 2.8. Four programmable receivers and associated logging systems. The TR-2 (top left) and ATS R200 (top right) have separable logging units, whereas the Lotek SRX 400 (lower left) and Televilt RX900 (lower right) can be hand-held.

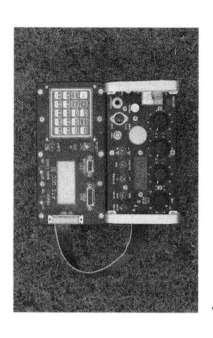 

 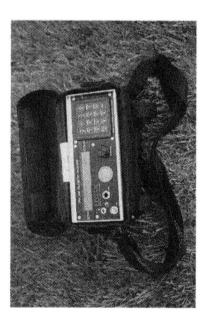

directions, with a 24-button membrane, and is controlled by a thumb-wheel. As well as storing data for download through an RS232 interface, the system can control either the two antennas for which BNC connectors are provided, or up to nine antennas that can give limited locating ability in a small area or record passage times along a short length of river. Another strength of this system is its ability to stretch very short signal pulses from long-life tags. However, the use of an electronic signal detector for pulse-stretching and logging is also an Achilles heel, because it is not yet as sensitive as the human ear and results in reduced detection distances.

Storage of frequencies and ability to step through them, with sensitivity that is quite adequate for radio tracking, is also provided by some commercial 'scanning' receivers, such as the $500 Yupiteru MVT7100 or AOR AR8000. The scanning function of these receivers is for rapid search and locking on to the carrier wave of broadcast signals, rather than the ability to dwell for several seconds through a cycle of pre-set of frequencies. The Yupiteru gain control lacks an adequate adjustable range, so that the receiver is swamped by strong signals from nearby tags and loses ability to indicate the changes in signal strength that indicate direction. A solution is to attach a signal attenuator between the receiver and the antenna cable, to reduce the signal strength by 20 dB or 40 dB when such swamping occurs. It also tunes tags on both side of their central frequency, which can cause confusion between tags spaced 10 kHz apart. The more expensive AOR 8000 has a better gain control, single side-band (SSB) operation and can also be set to step through frequencies stored in memory. Although these receivers are not as easy to use as the purpose-built tracking receivers, the AOR 8000 at least is practical for experienced trackers on low budgets, or as an emergency standby.

The problem of inadequate manual gain control for close-range tracking also affects some radio-tracking receivers. In some cases the manual gain control is supplemented by a signal attenuator switch. In this case, it is extremely important that the switch cannot be accidentally flicked into the 'attenuator on' position. The user of such equipment should be aware of the need to check the position of the attenuator switch before searching for distant tags.

2.3 RECEIVING ANTENNAS

2.3.1 Portable designs

At frequencies above 140 MHz, the three-element Yagi is the most commonly used antenna for tracking on foot. When well tuned, this antenna has a gain of 6–7 dB over the $\lambda/2$ dipole. The decibel is a logarithmic measure, and a gain of 6 dB represents a theoretical doubling of the reception range. The three-element Yagi therefore has about twice the range of the dipole.

The Yagi's other big advantage is its signal reception pattern (Fig. 2.9B). There are peaks in signal reception along the line of the boom, with the strongest peak to the front of the antenna. Such an antenna is said to have a good 'front-to-back' ratio, which makes it relatively easy to distinguish between true and reverse bearings. In contrast, the dipole has symmetrical signal peaks at right angles to its main axis, with nulls along the line of the antenna (Fig. 2.9A). It is therefore difficult to obtain an unambiguous bearing with a horizontal dipole, but a vertical dipole has the advantage of being omnidirectional in the horizontal plane. Another aspect of antenna performance is the 'sharpness' of the peaks and nulls in the reception pattern, which contribute to the taking of accurate bearings. With a three-element Yagi, for instance, bearing accuracy in moderately open country averages about 5°, compared with 5–10° for a dipole (Amlaner, 1980; Cederlund *et al.*, 1979).

Another portable antenna is the H-Adcock design (Taylor and Lloyd, 1978), which has nearly 6 dB greater gain than a dipole (Amlaner, 1980). The signals from the two elements are fed to the receiver with a 180° phase difference (using a balun match), so the reception pattern has a sharp null when the elements are equidistant from the source (Fig. 2.9C). This gives the Adcock antenna an angular accuracy of about 3° (Macdonald and Amlaner, 1980), somewhat better than a three-element Yagi. However, its reception pattern is like the dipole's in producing an ambiguity between true and reverse bearings. This is usually resolved: (i) by taking a cross bearing from another position; (ii) by reference to the animal's position shortly beforehand or (iii) by reference to landscape features. For instance, the ambiguity from dipole or Adcock antennas is not very important when tracking fish from riverbanks.

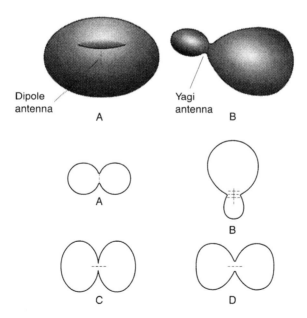

Figure 2.9. Patterns that use distance from the centre of the antenna to show strength of the signal radiated (or received) in different directions from dipole (A), Yagi (B), H-Adcock (C) and loop (D) antennas. The patterns are three-dimensional (e.g. doughnut-shaped for a dipole), but are normally shown in horizontal cross-section.

Another type of H antenna is available from some equipment suppliers. It differs from the H-Adcock in having one element shorter than the other, with the two elements phased to give maximum signal reinforcement when the shorter element is towards the transmitter. The antenna therefore functions like a two-element Yagi, without the Adcock's ambiguity between true and reverse bearings, but with slightly less gain and accuracy than the three-element Yagi. The design's advantage over the Yagi is its smaller size. However, the Yagi's greater accuracy, especially in the vertical plane, is essential for precision tracking of arboreal species (i.e. when both vertical and horizontal bearings are required).

Loop antennas are preferred for tracking on foot at frequencies below 100 MHz. The tuned $\lambda/6$ loop has 3 dB more gain than a $\lambda/6$ dipole, with a reception pattern which gives a null at each side along the loop's axis (Fig. 2.9D). The $\lambda/6$ loop antenna thus resembles dipole and Adcock antennas in its tendency to give reverse bearings, with a directional accuracy which is at best about 5° (Cederlund *et al.*, 1979). Loop antennas can also be useful for short-range work at higher frequencies: at 150 MHz the $\lambda/6$ loop is small enough (20 cm diameter) to be fixed to the outside of a board like a table-tennis bat, where it is much less likely than a Yagi, H-Adcock or dipole to get snagged in dense cover.

Since a very strong signal is obtained from a transmitter within the loop circumference, small open loops are handy for finding tags which have dropped off animals, for instance under muddy water (Solomon and Storeton-West, 1983). With the antenna on a long coaxial (shielded) cable, a fishing rod or extendable pole can also be used to locate live tagged animals precisely when they are hiding in ground cover, without disturbing the animal or vegetation by a close approach (Ostfeld, 1986). At high frequencies, a small dipole antenna may be more useful than a loop.

For convenience when travelling, the ungainly shape of Yagi or H-antennas is often available as a folding design. To obtain optimal tuning, hence maximum gain and accuracy, it is important that the antennas are fully extended before use. They are typically not quite as well tuned (impedance matched) as fixed antennas. A recent innovation, using internal tuning to overcome this problem, is an antenna with elements of thin metal strip that is curved across its long axis, to stiffen it in the same way as an extendable metal tape-measure but also to allow bending (Biotrack). Such antennas are very convenient when tracking in cover or frequently getting in and out of vehicles. Contrasting but less convenient approaches are to protect antennas from breakage by casing them in plastic piping (Livezey, 1988) or making them as 'cats cradles' that can be unravelled for use (Bosak, 1992).

All these antennas can be obtained at reasonable prices from manufacturers listed in Appendix I. The antenna should be obtained with a coaxial cable of suitable length for connection to the receiver. As there are two connection systems commonly used for this purpose, either with screw-on UHF connectors or with smaller, bayonet-type BNC connectors, be sure to order a cable that matches the receiver socket.

For those running projects on a shoestring, there are instructions for building a variety of suitable types in Amlaner (1980) and in the manual of the American Radio Relay League (ARRL, 1984). Note that the gains of antennas in some texts are given with reference to a theoretical isotropic point source (dB$_i$), rather than with reference to a $\lambda/2$ dipole (dB$_d$). The former units are 2.14 dB greater than the latter (dB$_d$ = dB$_i$ − 2.14 dB).

2.3.2 Antennas on vehicles

Cars, minibuses and lorries can be equipped with bulkier but more powerful direction-finding systems than can be carried on foot (see e.g. Verts, 1963; Proud, 1969; Hallberg *et al.*, 1974; Kolz and Johnson, 1975; Cederlund and Lemnell, 1980). For instance, adding elements to a Yagi antenna increases both its gain (Fig. 2.10) and its accuracy. Thus, a twelve-element Yagi has a gain of about 14 dB over a dipole, and some 7 dB over a three-element Yagi. This gain is not achieved without cost, however, because the higher the gain the narrower the angle (beamwidth) in which it can be obtained. The peak gain of a Yagi is obtained with the antenna pointing directly at the signal source, and this drops by 3 dB when a three-element Yagi points about 30° to either side of this line, giving a total half-power beamwidth of 60° (Fig. 2.11). The half-power beamwidth of a twelve-element Yagi is much narrower, only 30°. The narrow beam-width can be an advantage, in that it increases bearing accuracy, but can also be a nuisance: the narrower the beam, the easier it is to miss a pulsed signal if the antenna is swung round too fast. A six-element Yagi is therefore a popular compromise for vehicle-mounting (Fig. 2.12), giving a 3 dB better gain than a three-element design, and an accuracy of about 3° (Cederlund *et al.*, 1979).

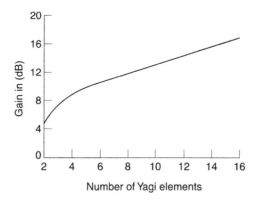

Figure 2.10. The gain of a Yagi antenna increases with the number of elements. The gain is given in decibels, and shows the increase with reference of a 1 dB dipole. Each gain increase of 6 dB represents a range doubling: a ten-element Yagi has about twice the range of a three-element Yagi. The gain increase from three to six elements is about 3 dB, which represents a power increase of 10^{-3}, and thus a range increase of $\sqrt{10^{-3}}$, i.e. 40%. Redrawn with permission from the American Radio Relay League.

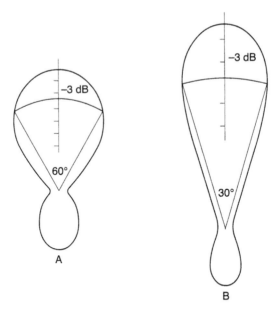

Figure 2.11. The ideal 3 dB 'beamwidth', through which the signal falls by 3 dB on each side of the peak, is 60° for a three-element Yagi antenna (A), compared with 30° for a 12-element antenna (B). To obtain ideal beamwidths, the antennas must be well tuned and at least 2λ from any surface, with their elements vertical. Figure 2.9B shows a more typical beamwidth, for a hand-held Yagi with its three elements horizontal, of about 100°.

Even greater accuracy, without loss of beamwidth, can be obtained by combining a pair of vertical 4–6 element Yagis in a 'null-peak' system (Banks *et al.*, 1975). A switch box is used to feed the signals from the two antennas exactly in phase, thus producing the optimum gain for finding the approximate direction of a tag, or out of phase to produce a very sharp null when the array points straight at the tag (Fig. 2.13). Such systems can give a bearing accuracy of close to 1° (Hallberg *et al.*, 1974; Cederlund *et al.*, 1979), but tend to be too damage-prone for vehicle-mounting in rough or wooded terrain.

When there is a risk from low branches, cables and other obstacles, the most convenient mast-mounted antenna for a vehicle is a Yagi with horizontal elements that are close to the roof of the vehicle when the mast is lowered. It is good to be able to lock such an antenna in a forward-pointing position from inside the vehicle for general searches, but also to be able to rotate it by hand for taking rough bearings as one passes a nearby tag. In this way, you can gain an impression of a tag's position without stopping a vehicle to take accurate bearings. Alternatively, for searching from vehicles that do not normally carry antennas, whips with magnetic base-plates can be very convenient, albeit with less gain than a Yagi. When this omnidirectional dipole picks up a signal, a car can be stopped to take bearings with a hand-held antenna. The choice of antenna

Figure 2.12. A diesel-engined Landrover, equipped with a horizontal Yagi and repeater compass on a six-meter telescopic mast. The mast is raised by an air-compressor powered by the starter battery.

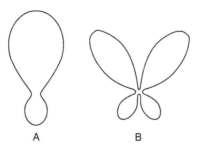

Figure 2.13. The radiation/reception patterns, in the horizontal plane, for twin six-element Yagi antennas with vertical elements. When the antennas are connected to are receiver in phase (A), the single lobe could have a 3 dB beamwidth of about 50°, compared with the 20° beamwidth for the null when the antennas are switched out of phase (B).

for tracking in aircraft depends on the wing position: a Yagi on a strut under each wing is ideal, but aircraft design, pilot preference or insurance regulations may mean that a whip is the only option. Mobile antenna systems are discussed more fully in Chapter 7.

2.3.3 Fixed stations

For non-directional recording of activity data or other telemetered information, many projects use a vertical whip with ground plane elements (Amlaner, 1980; ARRL, 1984) placed near the centre of the study area. Sometimes, however, the recording site is to one side of the study area, because of local topography or because of a convenient building in which to keep the receiving system. In this case it is often wise to use a Yagi antenna. The distance and width of the study area will determine the optimum number of Yagi elements, bearing in mind that gain increases at the expense of beamwidth. In other words, if the study area subtends an angle of about 60° at the receiving antenna, you should not use a Yagi with more than three or perhaps four elements.

On the other hand, to record the presence of a bird at a tree nest from ground level, it might be best to use a twelve-element Yagi pointed straight at the nest, with λ/4 or λ/8 dimensions if small size is more important than high gain. Another alternative would be to put a small loop antenna just under the nest. However, this could present reception problems if it required a long cable to the receiving equipment: without very effective shielding, the connecting cable would pick up signals from birds off the nest. Long cables from fixed antennas to receiving systems can also result in an appreciable signal loss. Each cable type has a figure for its loss per unit length at particular frequencies, and a preamplifier should be sited next to the antenna if the loss will exceed about 3 dB. This is the loss in about 10 m of most coaxial antenna cables at 138–174 MHz.

Direction-finding stations suitable for simple tags have usually been based on rotating Yagis at two or more mast sites. Automatic systems can be built, using microprocessors to assess and record the signal peaks from each site (Cochran *et al.*, 1965; Deat *et al.*, 1980; French *et al.*, 1992). However, many more projects have recorded the bearings manually, either using two people who communicate between fixed sites, or a manned mast site in contact with a mobile station, or a master site which can control the antenna rotation at a slave site and monitor the signals received there (Smith and Trevor-Deutsch, 1980). Manned stations are best for detecting faint signals, for high accuracy using null-peak direction finding and for rapid rechecking of 'improbable' fixes.

Other types of radio-location system can be built without rotating antennas. For instance, systems tested in the United States (Patric and Serenbetz, 1971), Australia (Yerbury, 1980, Spencer and Savaglio, 1996) and Sweden (Lemnell *et al.*, 1983) have measured nanosecond time differences of arrival (TDOA) of signals at separate receiving sites. An alternative approach is to estimate directions from phase differences as a signal arrives at an array of dipole antennas. To date this has only been implemented with relatively long pulses (Angerbjorn and Becker, 1992; Schober *et al.*, 1992) and with powerful transmissions because the active-array dipoles have less gain than a Yagi and the sensitivity of signal detection is poor. Greater gain is obtained by direction-finding from relative signal strengths across a fixed radial array of 6–8 outward-pointing Yagis (Larkin *et al.*, 1996), but this does not give very accurate bearings. Fixed station recording systems are covered more extensively in the next chapter.

2.4 TRANSMITTERS

Although the terms 'transmitter' and 'radio tag' are often treated as synonyms, in this book I maintain the convention of using the term 'transmitter' for the electronic circuitry alone, which is combined with a power supply, antenna and mounting materials to make a 'radio tag'.

Most transmitters are now surface-mount designs, with stages for frequency generation and antenna matching separated from a pulse-forming circuit. These can be divided into low-voltage models that work on 1.35–1.5 V and types that typically operate with one or more 3–3.5 V cells. A new type of transmitter includes surface-mount microcontroller circuitry to provide an array of functions that is still expanding. Perhaps the most common applications are to enhance tag life, by reducing or stopping transmissions at times when animals are not being tracked, and to produce complex identity coding within a pulse (Fig. 2.5B) that can identify many tags on one frequency. However, microcontrollers can control much more complex tasks, such as the averaging or accumulation of inputs from different sensors for transmission in modulated pulses at different times of day, or continually in coded pulse trains.

2.4.1 Construction and performance

Surface-mount components are pasted with 'solder cream' on pre-soldered circuit boards in batches under a microscope, and then baked in special ovens that melt the solder paste without harming the components. Crystals that can tolerate only low oven temperatures are usually added later by hand, and all joints inspected before electronic testing. On a board that is only 0.2 mm thick, a transmitter with an astable multivibrator circuit can now weigh as little as 200 mg before potting (Fig. 2.14).

The crystal is a thin slice of quartz, which oscillates on a very precise frequency. This fundamental frequency is inversely proportional to the thickness of the quartz slice. Since the quartz cannot be cut thin enough for a fundamental much above 20 MHz, higher frequencies are obtained by cutting so that the crystal oscillates more on an overtone than on the fundamental frequency. Thus, a crystal for a 150 MHz transmitter would probably have a fundamental frequency of 16.7 MHz, but the quartz would be cut in such a way that it tended to oscillate on the third overtone: it would be supplied as a 50 MHz crystal. The transmitter circuitry would then

Figure 2.14. Surface-mount transmitters, showing upper, lower and coated views of a simple 3 V design (top), views of a 200 mg design for 1.5 V before and after coating (right) and a programmable transmitter (left).

'multiply' this frequency by a whole number, in this case 3, to give 150 MHz. However, there would also be a tendency for signal production at other harmonics of 50 MHz: at 50 MHz (first harmonic), 100 MHz (second harmonic), 200 MHz (fourth harmonic) and even higher values.

These harmonics can be examined on a spectrum analyser, as shown in Fig. 2.15. The spectrum on the left of this figure is from a transmitter that was radiating strongly on several harmonics, and produced more signal on the fourth harmonic than on the third, for which it was intended. The transmitter was grossly inefficient, wasting power on unwanted harmonics. In contrast, the transmitter on the right met the requirements for Type Approval to the 1980s MPT 1309 and MPT 1312 specifications in the UK, by radiating at least 25 dB more on the intended third harmonic than on any other. Transmitters in the European Union should now comply with new, tighter EMC regulations, which result in even greater efficiency.

The standardised construction techniques give more consistent transmitter performance than the free-standing construction described for single-stage transmitters in the first edition, or even the circuit board approach used for two-stage designs (Kenward, 1987). There have also been improvements in the consistency and performance of lithium cells. The result is a more than eightfold improvement in the performance of tags that were considered adequate in the

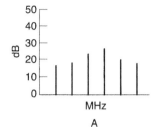
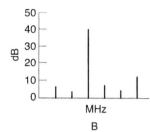

Figure 2.15. Spectrum analyser output from two radio tags. One tag (A) is poorly tuned: it radiates strongly on all six harmonics that are visible on the display, and more strongly on the fourth harmonic than on the intended third harmonic. The other tag (B) is well tuned, radiating much more strongly on the third harmonic than at any other frequency.

1980s. Thus, the 14 g single-stage tags with 1/2AA cells for goshawks in the early 1980s had a detection range of up to 20 km with λ/4 whip antennas and lasted 9–12 months. Tags of the same weight in the 1990s can be detected in excess of 40 km with λ/5 whips, and last more than two years (Fig. 1.2). Not only has tag life doubled, but the doubled detection range is equivalent (by the inverse power law) to a fourfold increase in power output. Greater reliability is reflected in the ability to track 90% of such tags on buzzards for at least a year, compared with only 32–56% of the goshawks (Kenward and Walls, 1994).

2.4.2 Long range or long life?

Getting the optimum balance of tag range and life is an important question in the planning of many tagging projects. Low-voltage tags are often as complex as more powerful types, and at least as difficult to make, but tend to cost less because their small cells and other materials are often less expensive. However, it is mainly the weight requirement that dictates their use.

Low-voltage and other tags can be built to maximise life by reducing the width and interval of their signal pulses. By changing the length of audio-signals in the laboratory, Kolz and Johnson (1981) produced a calibration graph indicating that detectability of signals declined rapidly at lengths below 30 ms (Fig. 2.16). However, field tests show that the detection ranges of pulses from new tag types with tracking receivers in the field do not decline as rapidly as expected. The maximum detection distance for 10 ms pulse widths is about 75% of that at 30 ms (B.H. Cresswell and S.S. Walls, unpublished) compared with a prediction of about 60% from Fig. 2.16. With 10 ms pulses at 1200 ms intervals (50/minute), a 4 g low-voltage tag can give an acceptable range for a year. Tags of only 2 g can give lives of about 9 months with 5 ms pulses, but their signals are clicks. Such short signals do have appreciably reduced audibility, which not only provides a detection range about half that of 30 ms signals, but also makes it difficult to discern changes in volume and thus to track the tags with a directional antenna. Nevertheless, such signals are adequate for short-range telemetry.

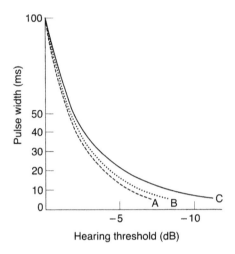

Figure 2.16. The audibility of sound pulses declines as the pulse duration is reduced, but less severely for high pitched tones of 500 Hz (A) to 200 Hz (B) than at 1000 Hz (C). Tracking is normally done with a high pitched reception tone. Pulses of 10 ms should then be about 3 dB less audible than those at 30 ms, giving a range reduction of 40%. However, field tests with tags indicate a range reduction closer to 25%.

The self-discharge rate of silver-oxide cells is relatively low, typically about 7% *pa* at 20°C (although 7% in 20 days at 60°C), which is another reason for preferring them to mercury cells. However, they are not built in large sizes, so it is usual to switch to lithium cells for tags weighing much more than about 4 g. Storage lives of the better lithium cells are in the order of 10 years.

Small 3V tags can be built, powered by two silver-oxide cells in series, if the emphasis is on detection range rather than weight. The smallest practical 3 V transmitters weigh about 1 g, and a 3 V tag of about 1.5 g can have a good range with a life of 10 days, or about 60 days at 2.5 g. Typical uses for small 3 V tags are for diminutive, wide-ranging species, such as birds or bats which are migrating or foraging very far from their roosts, or if antennas must be short. And in some cases, where low-voltage tags are adequate for mobile tracking, their signals may not be strong enough for automatic recording equipment, which detects weak signals less well than the human ear.

If you cannot decide in advance whether low-voltage transmitters will suffice, it is worth trying these as well as the 3 V types, provided that time is available and the animals are reasonably easy to catch. It is sometimes handy to have the extra range of a more powerful tag at the start of a project. Later on, when you are familiar with the technique and the animals' habits, you can gain tag life by changing to low-voltage designs.

2.4.3 Life enhancement with microcontrollers

Another option for extending tag life is the use of a microcontroller to vary the tag's on-off duty cycle and its power output. This option is based on the development of surface-mount 8-pin microprocessors with small programmable read-only memories (PROMs) and calendar circuits, known as PICs (programmable integrated circuits). The simplest approach has been to stop transmissions when tracking is not required. The microcontroller circuit continues to draw a current of 5–10 μA when the transmitter is off, but this is much less than the 80–200 μA typically drawn when transmitting 10–25 ms signals at 1200 ms intervals.

Consider the case of a tag based on a '10–28' lithium cell (10 mm diameter, 28 mm long), which can be obtained weighing 5 g with a highly reliable specification. A standard tag with 10 ms signals at 1200 ms intervals would last about 14 months with this cell and weigh about 7 g with minimal potting and lightweight antenna. With a microcontroller, such a tag would weigh about 8 g, but could be set to transmit for only 4 months in the year, extending its life to 3 years.

However, this approach is not quite as simple as it seems. When the tag is switched on, it starts running the programme in its PROM, for example to power the transmitter for 8 months, then stop for 8, and repeat until the cell is exhausted. When the tag stops transmitting for 8 months, but continues to power its microcontroller, after a while the cell adjusts internally to give the low current required but to resist supplying a higher current, a process known as passivation. When the microcontroller switches the transmitter on again, the much higher current required during a signal pulse can result in the cell voltage dropping to a level that adversely affects the microcontroller. Depending on the severity of the voltage drop, the processor may then either re-set completely and start running the program again, or lose its place in the programme with unpredictable results ('brown out'). Since the program started with the transmitter 'on', each reset may result in it trying to run until the cell readjusts to give a higher current, or brown-out may lock-up the controller completely. This problem is likely to be worst at low temperatures, which reduce battery performance, and when tags are run for very short periods before being shut down for several months.

There are several ways round the passivation problem. One is to shut the transmitter down for periods that are too short for appreciable passivation, such as overnight; this requires a highly accurate calendar chip if the tag is to transmit in synchrony with daylight for several years. Longer shut-down periods, for example of 10 days off and 5 days on, are tolerated in some tags if low temperatures can be avoided, for example in back-pack tags under the feathers of birds. However, it is important for biologists to start a set of tags in synchrony if they wish to search for them all in the same five-day period. A third approach is not to shut the transmitter down completely during a period of winter hibernation, but to reduce its current consumption by, for example, increasing the pulse intervals to several seconds.

Design improvements will probably result in PIC tags that run reliably to give extended lives without brown-out problems. The risk of adverse effects from brown-outs can also be reduced if the program starts with tags 'on'. Such tags should still produce signals if the microcontroller re-sets, though there is still the risk that a brown-out may cause a crash. If the routine starts 'off', then not only is passivation at its worst but repeated re-setting of the controller to 'off' may prevent any transmissions.

An approach that avoids passivation is to reduce the strength of signals permanently after the tag has run for an initial period. For example, buzzards make long dispersal movements that favour use of strong transmissions, but complete these movements in their first 2 years and then move little from relatively small home ranges (Walls and Kenward, 1998). Their tags can therefore be switched after 2 years to give low power transmissions, which are not ideal for tracking but are adequate for monitoring survival and attempts to breed. The power reduction should increase the life of a 30 g tag from 4 years to about 8 years, without problems from passivation because there will never be an increased power requirement to cause a voltage drop.

Life enhancement from microcontrollers is most useful below the size at which tags can give the required life without extra circuitry. Projects that are not funded for more than 3 years will benefit most in species that cannot carry tags with 12 g cells, because 12 g is the weight of 2/3AA cells that can power tags continuously for three years. Microcontrollers can extend life usefully in very small tags only if very short ranges or periods of transmission are acceptable. The smallest microcontroller tag is 0.8 g (Lotek), with an active transmission life of 14 days for coded signal transmitted every 5 seconds.

It is worth remembering, too, that other techniques can be used to shut tags down when they are not being tracked. Tags with CMOS latching circuitry can be switched on and off to conserve cell life by photosensors (at night), by temperature sensors (during torpor), by fixed-interval timers (Smith, 1980), or by seawater between exposed metal contacts (Broekhuizen *et al.*, 1980). The disadvantage with these options is the risk of signal loss if, for example, animals with photosensors enter holes or die with the tag covered.

Tags with microcontrollers are in their infancy, offering great promise but still requiring work to overcome teething problems that reduce reliability. Moreover, although tags without microcontrollers seem reliable with lives up to about 5 years, and show much promise for long-term studies of behaviour and demography, new problems may reduce reliability beyond 5 years, due to slow effects that would not affect tags with shorter lives. These effects could include water penetration through photodegraded potting, fatigue of antennas or collar material and unforeseen problems with cell chemistry. Tags that transmit for more than 5 years take a long time to test.

2.5 TAG DESIGNS

The performance of the transmitter is crucial to the success of a tag. Nevertheless, failures of tags are now least often caused by the transmitter electronics, more often due to power supplies, and most often a result of inadequate design or materials used for the antenna and packaging that complete the tag. The next section of this chapter outlines these other aspects of tag design. There is guidance on the principles that dictate choice of tag type, life and other options, but decisions on these issues must also involve tag manufacturers, who have specific information on their products and on projects that have used them. Advice on buying tags, and on building tags from purchased transmitters, is given in Chapters 4 and 5. However, the need for timely consultation with manufacturers, and of placing orders well before the transmitters or completed tags will be needed, cannot be stressed too highly.

2.5.1 Power sources: primary cells

Most wildlife tags are driven by primary cells, typically silver oxide cells and organic or inorganic lithium cells (reviewed in Ko, 1980). Table 2.1 shows the voltages of these cells, and compares the energy densities of some which are used for radio tags in the United Kingdom. Suppliers are given in Appendix I.

Cells with a lithium anode and thionyl-chloride cathode material have the highest voltage (3.6 V) and energy densities, deliver high currents and work efficiently in a temperature range of –20°C to +50°C. They can be made with layers of electrode and electrolyte paste in a cylindrical spiral, to provide cells with a high ratio of generating volume to protective outer casing. In this design, the casing does not start to affect weight disproportionately until capacity is reduced below 500 mAh (the 10–28 size, with 10 mm diameter and 28 mm length). Those with a capacity of at least 500 mAh have energy densities of 0.6–0.9 W h ml^{-1} (>0.3 W h g^{-1}), which makes them preferable for powering tags of more than 7 g. The smaller lithium/thionyl chloride cells not only have reduced energy density due to the relatively greater weight of casing, but are also difficult to manufacture reliably and therefore prone to early failure.

Other cell types are made with generating materials in flat layers, typically in a button shape. Silver cells, with zinc anodes and silver oxide cathodes, have energy densities of 0.3–0.4 W h ml^{-1} (<0.12 W h g^{-1}) in the 0.2 g to 2.3 g sizes. These cells are preferable for powering the smallest tags. Like the lithium/thionyl-chloride cells, the casing accounts for a disproportionate fraction of the mass below 0.2 g. They deliver reasonable currents, but the smaller sizes in particular have sufficient reduction in performance to stop the more powerful tags below 0°C. Cells with mercury oxide anodes are mostly obsolete, but sensitivity to low temperature, high self-discharge rate and environmental problems make them undesirable anyway.

There are a number of lithium/manganese-dioxide cells with similar capacity to the silver cells. These 3 V cells are most easily built relatively flat, which keeps energy densities to little better than the silver cells at 0.3–0.5 W h ml^{-1} (<2 W h g^{-1}) for cells below 2 g. Although they are more tolerant than silver cells of low temperatures, most types are not designed to deliver the currents required by radio tags and their relatively large diameters are inconvenient for small packages. The 3 g 1/3N cell is more compact, but relatively unreliable in tags. These cells are best used only where low temperatures are an important consideration, or to power tags

Table 2.1. The dimensions, voltages, capacities and lives at typical working currents of some cells commonly used in wildlife radio tags

Type	Weight(g)	Cap (mAd)	Diam(mm)	Height(mm)	Cap (mAh)	Voltage	Cap (Wh)	Vol (ml)	W h g⁻¹	W h ml⁻¹	life at 0.03 mA	life at 0.07 mA
Silver button												
Ag335	0.14	0.21	5.8	1.25	5.04	1.5	0.007	0.035	0.049	0.199	7 days	3 days
Ag317	0.19	0.48	5.8	1.65	11.52	1.5	0.017	0.047	0.089	0.361	16 days	7 days
Ag364	0.3	0.79	6.8	2.2	18.96	1.5	0.028	0.086	0.093	0.325	26 days	11 days
Ag397	0.51	1.54	7.9	2.7	36.96	1.5	0.055	0.143	0.107	0.384	51 days	22 days
Ag392	0.57	1.79	7.9	3.6	42.96	1.5	0.064	0.191	0.112	0.335	60 days	26 days
Ag393	1.13	2.9	7.9	5.35	69.6	1.5	0.104	0.284	0.092	0.366	3.0 mos.	41 days
Ag386	1.7	5	11.6	4.3	120	1.5	0.18	0.493	0.105	0.365	5.4 mos.	2.3 mos.
Ag357	2.27	7.3	11.6	5.35	175.2	1.5	0.262	0.614	0.115	0.426	7.8 mos.	3.4 mos.
Lithium button												
BR2032	2.5	7.9	20	3.2	189.6	3	0.568	1.092	0.227	0.52	6.6mo.	25days
Lithium spool												
10-25	5.5	20.8	10	25	499.2	3.6	1.797	2.133	0.326	0.842	1.4yr.	3.2mo.
1/2AA	8.6	37.5	15	26	900	3.6	3.24	4.992	0.376	0.649	2.5yr.	5.5mo.
AA	21	87.5	15	52	2100	3.6	7.56	9.985	0.36	0.757	6yr.	1.1yr.
C	56	219	26	53	5256	3.6	18.921	30.579	0.337	0.618	>10yr.?	2.7yr.
D	115	583	35	62	13992	3.6	50.371	64.823	0.438	0.777	>10yr.?	7.3yr.

that could accommodate a cell between the 2 g of the Ag375 and the 5 g of the 10–28. The 2.5 g BR2032 is a suitable cell in this size range. However, when 3 V are required to increase transmitter power in the smallest tags or for microcontrollers, tag makers generally prefer to use two silver cells in series. The use of two silver cells in parallel to provide increased capacity is discouraged, because their voltage is too low for effective diode isolation (see Chapter 5).

On the whole, dividing the specified cell capacity by the average transmitter current tends to overestimate tag life, but not too seriously for cells in Table 2.1 apart from those mentioned as having questionable reliability. If you are putting your own packages together and wish to use other types, you should check with the manufacturers of cells or transmitters that the cells are reliable for providing 2–3 mA in 20 ms pulses at one-second intervals.

2.5.2 Solar power

As an alternative to primary cells, photoelectric panels can power wildlife tags for several years (Patton *et al.*, 1973; Church, 1980; Andersen, 1994). Tags weighing only 6 g (low-voltage) to 15 g (3 V) can be powered by photoelectric panels alone, but stop transmitting in low light intensities. Such tags are not very satisfactory for systematic studies, because loss of signal may mean that the transmitter is heavily shaded rather than out of range, and the signal also fails at night, in nest-holes, or if the animal lies dead on its back. Slightly heavier packages have been be built with 20 mA h rechargeable cells, which are charged by the solar panels to provide power in poor light. However, rechargeable cells that are subject to frequent unregulated charging tend to malfunction within a couple of years as a result of passivation and other effects, after which the tag depends solely on the solar panels (Ko, 1980; Keuchle, 1982). Moreover, solar cells must remain exposed to radiation, in some cases requiring plastic shields to prevent a covering of feathers (Snyder *et al.*, 1989) that would reduce drag and visibility of other tag designs. Poor survival of game birds with solar-powered tags (Marks and Marks, 1987; Burger *et al.*, 1991) may stem from reflected light flashes that attract avian predators.

Improved transmitter efficiency means that reliable lives in excess of 2 years can be obtained from tags with primary cells at weights that previously required solar cells. Even longer lives can be obtained by using microcontrollers. Since solar-powered tags are also not satisfactory unless they can be attached dorsally, to diurnal species that live in fairly open habitats, their use is becoming restricted to animals that require very powerful transmitters. With microcontrollers to regulate the charge-discharge regime, solar power still has advantages for some sizes of UHF tag (see Chapter 3).

2.5.3 Antenna format

The main types of antenna used for animal tags are loaded loops, tuned loops and whips. Whips are the simplest antennas to construct. Nickel guitar wire, nylon-coated multistrand stainless steel fishing trace or new super-flexible alloys are commonly used for small tags, with multistrand steel cabling covered by tough heat-shrink tubing for larger animals. Systems with a main whip and a ground-plane antenna are the most efficient antennas commonly used on animal tags, provided that whip length can be at least λ/8. The ground-plane antenna is typically a second whip, ⅔ the length of the main whip antenna, and perpendicular to it or in the opposite direction (Fig. 2.17).

Simple tuned-loop antennas are the most practical compact alternative to whips for the 142–230 MHz band. A loop of wire or brass strip is used to form a collar on small to medium-sized mammals, the upper size limit being determined by the maximum size that is easily tuned to resonance at the frequency concerned. This limit is a diameter of about 5.5 cm for a 10 mm wide strip of 1 mm thick brass at 173 MHz, being greater at lower frequencies and vice versa. Using thinner collar materials raises the limit somewhat, but also reduces the antenna efficiency at a given diameter. Tuned loops tend to be most advantageous for mammals of 300 g to 3 kg that destroy whip antennas, squirrels being a good example. Smaller animals tend increasingly to be equipped with whip antennas, even if these have to be totally enclosed within the collar and therefore relatively short. A major reason is that tuned loops are harder to attach with good tuning than whips on cable-tie collars; further detuning of loops, through capacitance effects due to proximity and movements of the animal, causes considerable performance loss in modern tags with efficient antenna matching. At best, the loop signals are radiated with horizontal peaks and nulls (see Fig. 2.9), which can hinder tracking, especially if you are relying on changes in signal strength registered by a probe antenna. Tuned-loop tags also stop transmitting if cut by a predator, which is most likely on small animals. Some studies of small mammals have found that whips extending from a collar along the back are not chewed as much as was feared.

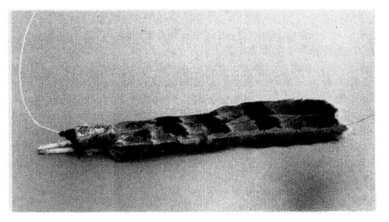

Figure 2.17. A tag on two goshawk tail feathers. Note the free-standing ground-plane antenna at the front of the tag; the main antenna is tied and glued along the shaft of one feather.

For implants, or other tags with embedded antennas, loops are less prone to detuning than they would be as exposed antennas. The most common antenna format for implants remains a single- or double-loop antenna, around the whole tag or at one end. The larger tags can also be built with a spiral whip antenna just below the surface of the potting. Loaded loop antennas, made of iron-dust or ferrite cores wound with several turns of enamelled copper wire, have been used for the smallest tags. Implants have also sometimes used loop or whip antennas extending from the tag. Even whip antennas through the skin to maximise power output have been used. However, the less compact forms raise increased concern about animal welfare.

2.5.4 Mounting techniques

In choosing tags designed for a particular attachment technique, one needs to be aware of the associated advantages and limitations. The mounting technique affects the size of tag that can be attached, its radiation efficiency, its influence on the animal and the skill needed to attach it. These aspects of project planning are discussed now, leaving the details of tag construction and attachment until Chapters 5 and 6.

Tags are attached to birds mainly with harnesses, tail-mounts, leg-mounts and ponchos or necklaces, all of which have whip antennas. Harness-mounted backpacks are the only feasible tags for many projects, primarily where ponchos or necklaces have adverse effects and longer tag lives are required than practical with leg-mounts and tail-mounts. Backpacks are placed close to the bird's centre of lift, and may therefore be heavier than the alternative types.

Unfortunately, harness-mounted tags have been shown to have an adverse effect on plumage, behaviour and even survival in a wide range of species (Nicholls and Warner, 1968; Boag, 1972; Ramakka, 1972; Greenwood and Sargeant, 1973; Lance and Watson, 1977; Johnson and Berner, 1980; Perry, 1981; Small and Rusch, 1985; Massey *et al.*, 1988; Marcström *et al.*, 1989; Hooge, 1991; Burger *et al.*, 1991; Patton *et al.*, 1991; Foster *et al.*, 1992; Pietz *et al.*, 1993; Ward and Flint, 1995). Nevertheless, extensive tests on some raptors have been unable to show serious adverse effects of harnesses (Vekasy *et al.*, 1996; Walls and Kenward, 1998), so the technique may be safe on some species if used carefully (see Chapter 6). Adhesive mounting of backpacks is an alternative to harnesses (Graber and Wunderle, 1966; Raim, 1978; Perry *et al.*, 1981; Heath, 1987; O'Connor *et al.*, 1987; Wilson and Wilson, 1989; Sykes *et al.*, 1990; Johnson *et al.*, 1991). Glues and resins have been used to attach tags usefully for several days to game bird chicks (Kenward *et al.*, 1993a), and for several months to species whose feathers do not pull out easily (Wanless *et al.*, 1988; Kalås *et al.*, 1989). Although other birds (especially passerines) have shed such tags too quickly (Johnson *et al.*, 1991), rapid tag detachment can be an advantage if many nests or roosts are to be found by re-using a small number of tags. A sacral mount using a leg harness (Rappole and Tipton, 1991) may also prove satisfactory, but is removed by some species (C. Winchell, pers. comm.). Tags can be sutured to bird backs (Martin and Bider, 1978; Mauser and Jarvis, 1991; Wheeler, 1991), but tend to be shed prematurely (Houston and

Greenwood, 1993; Rotella *et al.*, 1993); countries concerned about animal welfare are likely to license such an invasive technique only as a last resort.

Tail-mounts are preferable to backpacks, provided that the tail feathers are large and firm enough, and the tags need not outlast the moult (Bray and Corner, 1972; Dunstan, 1973; Fuller and Tester, 1973; Fitzner and Fitzner, 1977; Kenward, 1978; Giroux *et al.*, 1990; Sodhi *et al.*, 1991; Taylor 1991). These tags should not weigh more than about 2% of bodyweight. Compared with backpacks, weight is saved through lack of a harness, and by using lightweight antenna wire if this is secured by binding it along the shaft of a rectrix (Fig 2.17). Like glue-on backpacks, tail-mounts also detach safely (through moulting), so that tags which are still transmitting can be recovered and re-used.

Leg-mounts have been used on large, long-legged birds, such as storks (A. Burnhauser, personal communication) and cranes (Melvin *et al.*, 1983). They are also useful for tagging nestlings (Fig. 2.18), whose growing feathers might be damaged by harnesses or tail-mounts (Kenward, 1985). However, leg-mounts often have less range than the other types, because they are closer to the ground and have shorter antennas that are also damage-prone.

Figure 2.18. A goshawk nestling tagged on the leg. If backpacks are put on hawks at this age, the young birds habituate to them before fledging. (Photo by M. Karlbom.)

Ponchos, which evolved from a visual marker worn round the neck, are a good long-term tag for some birds (Amstrup, 1980). The movements and survival of ruffed grouse with ponchos were better than for grouse with harness-mounted tags (Small and Rusch, 1985). Similarly, a necklace design has proved much simpler than harnesses or tail-mounts to attach to pheasants (Fig. 2.19), having less adverse affect on the birds than the former, and less tendency to detach prematurely than the latter (Marcström *et al.*, 1989). The necklaces also give the best detection ranges, partly because their main antennas are vertical and partly because of a ground-plane antenna in the necklace cord. Ponchos or necklaces seem ideal for game-birds. However, the main antenna can get tucked under the wing and thus irritate birds which spend much time in flight, and these designs would be unsafe for birds which swallow food items nearly as large as their heads. Like backpacks, necklaces may cause problems for diving species (Sorenson, 1989) and must not be too heavy or bulky (Marks and Marks, 1987; Cotter and Gratto, 1995). They should also probably be avoided for water-birds at high latitudes, where neckband visual markers can accumulate lethal masses of ice (Zicus *et al.*, 1983) and decrease survival (Samuel *et al.*, 1990; Castell and Trost, 1996).

Other tag-mounting techniques for birds include the marking of diving species with nasal saddles (Swanson and Keuchle, 1976) or implants (Boyd and Sladen, 1971; Woakes and Butler 1975; Korschgen *et al.*, 1984; Olsen *et al.*, 1992; Korschgen *et al.*, 1996a), to avoid the drag and possible insulation loss from harnesses. Subcutaneous implants have recently become popular for marking the

Figure 2.19. A tag attached round the neck of a pheasant (*Phasianus colchicus*). The necklace cord contains a ground-plane antenna to increase radiation efficiency, and must be loose enough for the bird to swallow the largest likely food item. This type of tag has frequently been used without adverse effects on game birds, but is unsuitable for birds that fly frequently.

precocial chicks of ducks and other game birds (Korschgen *et al.*, 1996b, c; Riley *et al.*, 1998). However, high rates of unexplained loss of ducklings with these tags coincide with high premature failure rates, and necrosis associated with cell leakage, from similar implants in captive doves (Schulz *et al.*, 1998). Radios mounted as patagial tags have been satisfactory on condors (Ogden, 1985; Wallace and Temple, 1987; Wallace *et al.*, 1994), but are probably only suitable for species that flap their wings slowly and relatively little. Finally, transmitters have been built into dummy eggs, either to record incubation parameters (Howey *et al.*, 1977; Schwartz *et al.*, 1977; Boone and Mesecar, 1989) or to help identify nest predators (Willebrand and Marcström 1988).

The typical mammal tag is a collar (Fig. 2.20). For large mammals, the tag itself is usually bound, bolted or riveted to a collar of leather, braided nylon strapping, or other flexible plastic. If grooming is likely to damage an exposed whip antenna, a whip may be sandwiched between two strips of collar material, or an antenna embedded within the tag, for example by running a zigzag wire whip between two strips of leather. For large mammals that must be collared young, or whose necks may expand during courtship, collars with a variety of expandable pleats or sliding mechanisms have been developed (Beale and Smith, 1973; Garcelon, 1977; Follman and Buitt, 1978; Jackson *et al.*, 1985; Jullien *et al.*, 1990; Hölzenbein, 1992). However, young animals sometimes become entangled in such collars (Weber and Meia, 1992). An alternative approach for growing young is to attach

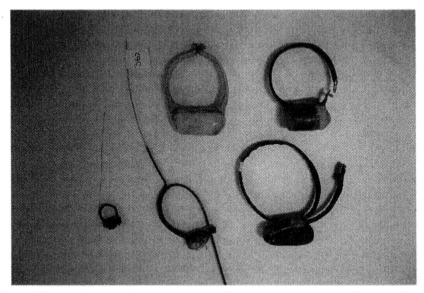

Figure 2.20.　A selection of tags for mammals. The top row are 15 g tuned-loop tags for red squirrels (*Sciurus vulgaris*), with side-opening (left) and top-opening (right) designs. The lower row are 1 g (left) and 3 g (centre) cable-tie collars for small mammals, and a Teflon-sleeved sliding-brass design that is closed round the neck of a mink (*Mustela vison*) by crimping, all with whip antennas.

tags initially to their ears (Serveen *et al.*, 1981; Garrott *et al.*, 1985), and if necessary recapture the animals to attach collars when they are large enough. Smaller mammals, provided they would not damage exposed whip antennas or be impeded by them, are often equipped with necklace tags based on cable-ties or other self-locking plastic materials, from which an antenna protrudes at the top of the neck. Tubing is used to hold the antenna and protect the animal from abrasion or over-tightening, and may have a soft coating to minimise stress for easily-shocked species like hares.

For small mammals that cannot have exposed whips of any length, it is often most efficient to use a tuned-loop collar. The smallest can be made of PTFE-coated wire, which is tightened round the neck with a metal crimp or heat-shrink tubing (Fig. 2.21). Animals above about 150 g can carry collars of fine brass strip or multi-strand brass wire (picture wire), which give stronger signals than a fine wire collar with a similar circumference. The most efficient tuned-loop collars have a fixed circumference, and are bolted or crimped together round the neck. Brass-strip collars that contain several bolt holes, so that their circumference can be varied to suit a range of head and neck sizes, have less precise tuning than fixed-circumference types. Some have variable capacitor that can be 'tweaked' once they are on the animal, but this is no easy task unless the creature is effectively immobilised.

The use of collar tags is difficult or impossible on mammals with very short or tapering necks, especially mustelids and species streamlined for swimming or burrowing. On mink, for instance, supple leather collars with sandwiched whip antennas seem less likely to be shed than brass tuned loops, which are more efficient but too rigid to give a really close fit (N. Dunstone, personal communication). Moreover, it is best not to use collars with inflexible brass strips on mammals that live in rocky holes. Rapid loss of signals from such tags on wild martens (fishers), despite satisfactory operation on animals in enclosures (Skirnisson and Feddersen, 1985), suggests that the tags were either damaged or fatally wedged in the dens.

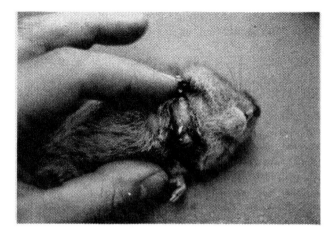

Figure 2.21. A bank-vole (*Arvicola terrestris*) with a 'sliding-8' collar.

Since collars will not stay on otters or many seals at all, tags for these animals have been attached to harnesses or glued to the fur using epoxy resins (Broekhuizen *et al.*, 1980; Fedak *et al.*, 1983; Mitchell-Jones *et al.*, 1984). It is difficult to build harnesses which will not eventually abrade these animals, and adhesive mounting cannot be relied on to last more than 2 or 3 months, at best. In such cases, intraperitoneal implants appear to be the best solution for long-term tagging (Melquist and Hornocker, 1979; Eagle *et al.*, 1984; Davis *et al.*, 1984; Guynn *et al.*, 1987; Ralls *et al.*, 1989). Marine mammals have also had tags attached to their fins or flippers (Butler and Jennings, 1980; Garshelis and Siniff, 1983), or to hooks or miniature harpoons in the blubber of whales (Watkins *et al.*, 1981; Mate *et al.*, 1983, 1997; Goodyear, 1993). The tags are attached where they will be most exposed when the animal surfaces.

Adhesive mounting is favoured for tagging small bats (Williams and Williams, 1967; Stebbings, 1982), typically using epoxy resin to glue tags with whip antennas to the back, close to the centre of lift. A necklace provides a durable attachment for the larger bats, and can be equipped with a ground-plane antenna to increase the signal strength.

Some mammals have horns or other protuberances to which tags can be attached. Pasty silicone sealants give flexibility for attaching transmitters to hedgehog spines (Morris, 1980), while Anderson and Hitchins (1971) embedded tags in rhinoceros horns and Pages (1975) wired them to pangolin scutes. Tags have also been fixed to mammal tails (Gorman *et al.*, 1992).

Many large reptiles have horny scutes which can be drilled to anchor attachment wires. Tags have been bound and glued in this way to turtle and tortoise carapaces (Schubauer, 1981; Priede and French, 1991), and to the heads or backs of crocodiles (Smith, 1974; Yerbury, 1980). Small reptiles are less easy to tag, although harnesses have been used on lizards (Fullagar, 1967). Adhesive mounting, with surgical tape and cyanoacrylate glue, has been used for the temporary attachment of tags to snakes and small lizards (A. Cooke and A. Gent, personal communication). Although slim external transmitters have been sutured to the tails of snakes without obvious adverse affects (Ciofi *et al.*, 1992), long-term radio monitoring has most often used surgical implantation (Barbour *et al.*, 1969; Madsen, 1984; Weatherhead and Anderka, 1984). The same applies to amphibians (Stouffer *et al.*, 1983; Smits, 1984) whose skin is not only too damp for adhesive mounting but is also rather easily damaged by harnesses. However, frogs and toads and snakes have been persuaded to swallow transmitters in some projects (Osgood, 1970, Fitch and Shirer, 1971; Brown and Parker, 1976). Ingestion of a large tag might influence subsequent behaviour (Lutterschmidt and Reinert, 1990), but anurans have swallowed tags and continued feeding normally in at least one study (Oldham and Swann, 1992).

Four main techniques have been used for radio tagging fish (Winter *et al.*, 1978). Tags have been attached externally to the dorsal surface, either beside the dorsal fin or mid-dorsally on a saddle mount behind this fin, by sewing absorbable sutures (Solomon and Storeton-West, 1983; Beaumont *et al.*, 1996) or 'ground-plane' wires through the dorsal musculature. Such tags are streamlined and have short whip antennas, which are relatively more efficient in water than in air (Weeks *et al.*, 1977). Tags can also be inserted into the stomach, where they have been retained for several weeks by salmon in fresh water (Solomon and Storeton-West, 1983). Careful insertion is required to avoid damage to the stomach, and the risk of rupture increases as the stomach lining atrophies while salmon move upstream to spawn (Haynes, 1978). Since external attachments can abrade the skin and increase the risk of fungal infection, the most satisfactory technique for long-term tagging of most species appears to be implanting in the abdominal cavity. Implants can have whip antennas threaded with a cannula under the skin (Winter *et al.*, 1978), in which case the reception ranges may be only 10% less than for external tags, provided that they are properly tuned.

Implanting thus appears to be the most humane and effective technique for tagging fish and amphibia, both of which have easily-damaged skin. However, it is also used increasingly for streamlined aquatic animals, and some other species that risk entanglement, lethal heat loss or starvation if equipped with harnesses.

2.5.5 Sensors

Tags for movement or temperature telemetry are straightforward to construct (Chapter 4), and are therefore widely available commercially. They use a mercury tilt-switch or a thermistor, respectively, to alter the signal rate.

Mercury tilt-switches, containing a drop of mercury that rolls to bridge internal contacts, are used to indicate movement, posture and mortality. In the simplest tags, a low-value pulse capacitor gives a short and frequent pulse when the switch is open. Closure of the switch puts a second capacitor in parallel with the first, thereby increasing the pulse duration and decreasing the pulse rate. Such switches have long been used to indicate general activity and diurnal patterns (Swanson and Keuchle, 1976; Gillingham and Bunnel, 1985; Palomares and Delibes, 1991; Kunkel *et al.*, 1992). In these applications, a switch in which mercury rolls longitudinally can be mounted at a shallow angle, say 5° to the horizontal. On a slow moving animals such as a tortoise or hedgehog, a transverse switch position gives a steady slow signal when the animal rests, but an irregular pulse as its rolling gait slops the mercury, thus opening and closing the switch. On fast-moving mammals the switch may be set fore-and-aft in a collar, at 5° to the horizontal and closed when the collar hangs level. The slight accelerations and decelerations when the animal moves are enough to make the signal irregular. Some firms make tags which are sensitive to movement in any direction: they contain tipover switches (jitter switches), where the mercury can roll in any direction from central electrodes.

More sophisticated information on posture can be obtained with longitudinally positioned tilt-switches. For example, posture switches have been set at about 45° to the horizontal in transmitters on goshawk tails (Fig. 2.22). When a hawk is perched, with its tail close to the vertical, the mercury switch is closed and a signal

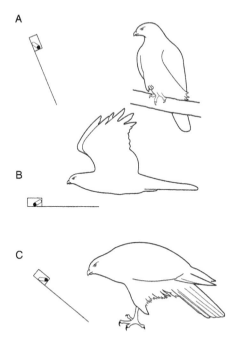

Figure 2.22. Mercury switches are used as posture sensors in tags tail-mounted on raptors, The tail points downward on resting birds, and the switch is set for a slow signal pulse (A). In flight the mercury moves to its other position, and the signal pulses are fast (B). When the raptor pulls at food, the tail is raised and lowered, thus slopping the mercury from one position to the other and giving an irregular signal (C).

is transmitted about once per second, a convenient rate for radio location while the hawk is in its most common posture. When a hawk flies, the tail is horizontal and the switch open, about 1.5 signals are transmitted each second, a noticeably faster rate. The flight signal also varies in amplitude as the hawk's direction and hence the antenna orientation changes, or as obstacles come between hawk and receiver. This signal can then be distinguished by ear from the steady fast pulse given by an incubating hawk or the weak, fast signals from one lying dead. Feeding produces a characteristic signal too, alternating between a slow pulse as the hawk arches its back to pull at food, and a fast pulse when the tail becomes more horizontal as the bird relaxes to swallow (Kenward *et al.*, 1982). Similar tags can be mounted as backpacks. Douglas and Pickard (1992) used two switches at different angles to record different types of duck feeding behaviour.

   In motion-sensing tags, a steady signal when the animal should be active may mean that it is dead, and this feature is exploited in special mortality-sensing tags. These change the pulse rate after a pre-set delay in which the mercury switch has not moved (Kolz, 1975; Keuchle, 1982). A delay of hours rather than minutes should be chosen, or there will be many false alarms as the animal takes short rests. Combined activity/mortality tags can now be built, inserting extra pulses in a slow resting rhythm to indicate activity, and changing to a continuous fast rate after prolonged lack of movement.

A problem with the use of these tags in some countries (e.g. Sweden) is that mercury switches have been made illegal on environmental grounds. It remains to be seen whether replacements with solid elements are as durable as mercury switches, the best of which now operate for years in posture-sensing tags. Moreover, the clicking made by the new designs has already been accused of fatally disrupting behaviour for animals with mortality tags on their ears (Swenson *et al.*, 1999). On the other hand, mercury switches operate poorly if at all in temperatures below −20°C (Maier *et al.*, 1996).

Temperature-sensing tags are based on thermistors. These are semiconductors that act as heat-sensitive resistors, typically with a negative temperature coefficient: the resistance decreases as the temperature rises. Since the thermistor replaces one resistor in the pulse-forming circuit, such a transmitter need weigh no more than in an unmodulated tag. In the simplest tags, the drop in resistance allowed a greater charging current to flow into the pulse capacitor, thus increasing the signal rate as the temperature increased. However, such tags stopped altogether if circuit current decreased too much at low temperatures. This is much less a problem with signal pulses formed in multivibrator circuits (Amlaner, 1978; Anderka, 1980).

An obvious use for thermistor tags is to measure body temperatures, using implanted tags, ingested tags or external tags with implanted sensors. Since the tag current changes with the signal rate, it is important to optimise the trade-off between the temperature sensitivity, the range of temperatures measurable, and the cell-life. Assume that the thermistor tag has a 2.5% change in pulse-rate per °C, enough for 1–2°C accuracy in field temperature measurements, and operates over a 40°C range. If the thermistor has a more or less linear temperature response in that range, the circuit current and pulse rate at 40°C will be double that at 0°C (since 40 × 2.5% is a 100% increase). If the tag operates mainly at the higher temperature, for instance in a mammal which only occasionally enters torpor, then the cell life will be about half that of the unmodulated tag. Doubling the sensitivity of such a tag would halve the cell-life again.

Another use of thermistors is to measure ambient temperature, perhaps to see what microclimate an animal prefers, but also to monitor behaviour. For instance, thermistors attached to harness straps under a bird's wing are cooled when the bird takes flight, from close to body temperature to ambient air temperature (Kenward *et al.*, 1982; Cresswell and Alexander, 1992). This is useful for monitoring the activity of species whose tails and bodies have the same tilt in flight and at rest. Similarly, thermistor collars on small mammals can indicate whether they are curled up in warm nests or out foraging (Osgood, 1980; Kenward, 1982a), and thermistor tags on all warm-blooded animals also indicate mortality (Stoddart, 1970). In these applications, in which the temperature is generally high, cell-life can be prolonged by engineering pulse width to decline as signal rate increases (Cresswell, 1992); alternatively, signal rate can rise when temperature falls.

There are a number of other sensors that have been used less widely in wildlife radio tags. Changes in resistance or capacitance caused by urine or salt water can reveal marking or aquatic behaviour (Charles-Dominique, 1977; Garner *et al.*, 1989). Ambient light intensity can be monitored with photovoltaic panels (Althoff *et al.*, 1989), and proximity to other objects or animals can be detected if they are equipped with magnets to operate a reed-switch on the radio tag. Pressure transducers can be used to indicate altitude of flying birds by change in air pressure from ground level (Bögel and Burchard, 1992). Sophisticated circuitry is needed in tags whose pulse rate is to be changed by pressure sensors (Keuchle, 1982), strain gauges (Keuchle *et al.*, 1987) or compasses (in acoustic tags: Mitson *et al.*, 1982), and this is also true for tags triggered by heart beats (Gessamen, 1974; Wolcott, 1980; Woakes and Butler, 1989; Klugman and Fuller, 1990). It is difficult to ensure that the latter tags are triggered only by ventricular contractions, and not by the muscle potentials resulting from auricular contractions and trunk musculature as well. With long leads from implanted electrodes to external transmitters, it is hard to maintain accurate pulse triggering for more than a few days.

Complex circuits are also required for tags which can telemeter information from more than one sensor, for instance by modulating pulse duration and pulse interval, or by sending a train of pulses in which each is modulated by a different sensor (Smith, 1974; Standora, 1977; Lotimer, 1980). The commercial availability of receiving equipment capable of interpreting such signals is likely to increase the use of multisensor data from microcontrolled tags. Similarly, there is scope for growth in the use of frequency-modulated tags, to transmit calls or feeding noises from microphones (Greager *et al.*, 1979; Gautier, 1980; Alkon and Cohen 1986; Alkon *et al.*, 1989) or electrocardiograms (Smith and Moore, 1989) and even electroencephalograms (Schmidt *et al.*, 1989) from wildlife.

2.5.6 On-off options

Many home-built long-life tags are transmitting when they are potted up, shortly before use. Others are simple to start in the field, for instance by joining a tuned collar loop, or by exposing solar-powered tags to the light. However, there are at least three more options for switching on tags.

The most elementary way of completing the transmitter circuit is to solder two exposed wires. The join should then be coated with a small amount of potting compound, or other suitable sealant, to prevent moisture penetrating the package along the wires. A portable soldering iron is needed to start short-life tags in the field: if starter wires are merely twisted together, their junction resistance may eventually increase enough to stop the transmitter. The Oryx butane-powered soldering iron (from RS or Biotrack; see Appendices I for addresses) is lighter and smaller than battery-powered types, also running for longer and being instantly 'rechargeable'.

Tags can also be built with a little protruding wire loop, which makes a connection to shut down the oscillator circuit transistor (Macdonald and Amlaner,

1980). Such a tag transmits when the wire loop is severed. It is simple both to test the tag, by cutting and then rejoining the loop, and to start the tag in the field, by removing the loop entirely. The drawback is that a transistor leakage current continues to flow through the circuit when it is shut down, at 2–4% of the operating current, so this option is only suitable for short-life tags if they are to be used fairly promptly. It is worth noting that a similar leakage current flows through tuned loop tags, which start by completing the loop.

For a small weight penalty, tags can be equipped with a magnet-operated reed switch. The normally-closed (NC) switch remains open while a magnet is in a marked position on the tag's outer surface, and no current flows in the transmitter circuit until the magnet is removed. Although starting tags in this way saves labour, it is not an entirely safe luxury. The contact surfaces of reed switches occasionally degrade with time. The resulting increase in resistance then reduces tag voltage, either reducing signal output or even stopping tags prematurely. Tags stored with magnets in place can also switch on if magnets are accidentally pushed together with opposing poles. Some tags, such as those in waterproof casings, must be equipped with reed-switches, in which case using two switches in parallel can reduce the risk of tag failure. However, for projects that must maximise tag reliability, reed switches are inadvisable.

2.6 TAGS AS CAPTURE AIDS

Radio tags can save a great deal of time spent in checking isolated traps (Hayes, 1982; Nolan *et al.*, 1984). A simple approach is to use a magnet-operated transmitter: a cord from the trap's door or other moving part pulls the magnet off the transmitter when the trap is activated. For fail-safe operation the radio beacon should have a normally-open (NO) reed switch, to turn the signal off when the magnet is pulled away.

Tags that fit within anaesthetic darts (Lovett and Hill, 1977), or around the needle of darts without a suitable compartment, can be obtained commercially from some manufacturers listed in Appendix I. Such tags, with ranges of 150–500 m, are useful not only for finding animals which travel some distance before succumbing to the drug, but also for finding darts, containing dangerous drugs like Immobilon, which miss the target.

Radio tags can also be used to locate dens and nests, for instance by laying out tagged carcasses which were scavenged to fox dens (Nicholls *et al.*, 1981; Voight and Broadfoot, 1983), or by dropping beacon tags to help ground-crews during aircraft searches for raptor nests. Once radio tagged, animals can sometimes be re-caught at roost sites, or with drugged food (Huempfner *et al.*, 1975; Kenward, 1976). The ultimate development in this line is a large-mammal collar containing anaesthetic syringes, which are activated by small charges on receipt of a coded radio signal. These collars have been used to recapture wolves, deer and bears (Delgiudice *et al.*, 1990; Mech *et al.*, 1990).

2.7 CONCLUSIONS

1. Frequencies between 142 MHz and 230 MHz are allocated nationally for wildlife radio tags. These frequencies are convenient for antenna design, tag construction and propagation in wildlife habitats, but vigilance may be required to maintain them in the face of commercial pressures on governments for radio frequency allocations.

2. A wide variety of reliable receivers are now available, from basic types suitable for tracking a handful of tags up to designs capable of sophisticated pulse interpretation and data storage.

3. Tracking on foot is done mostly with three-element Yagi antennas, but also H-antennas, loops and various probes. Mast-mounted Yagis with more elements, sometimes in null-peak systems, are popular for vehicles and direction finding stations.

4. Transmitters that generated signal pulses with simple single-stage and two-stage oscillator circuits have mostly been replaced by CMOS astable multivibrator designs on surface-mount boards. The new designs give improved consistency and reliability, with tag lives enhanced by greater efficiency and the discovery that detection distances are little reduced by shortening signal pulses to 10 ms.

5. The smallest tags operate at 1.5 V with silver oxide cells; more powerful tags use 3–3.5 V lithium cells, which give more reliable long life than solar power. Tag antennas are whips, tuned loops and tuned coils.

6. Sensors are available for modulating tag signals to indicate posture, activity, temperature, moisture, pressure, direction and physiological parameters. The advent of microcontrollers in commercial tags provides scope: (i) for combining sensor data in coded pulses; (ii) for complex coding to identify tags; (iii) for extending tag lives by shutting them down when tracking is not required and (iv) for much other ingenuity.

7. Tags on traps, darts and recapture collars are also used to help obtain animals for marking.

3 Automated Systems

The idea of data being collected automatically in all weathers and hours of day and night is often appealing. Three types of system have been developed for automated radio tracking of wildlife. Locations can be estimated: (i) by satellites from tag signals; (ii) by tags that receive transmissions from satellites or other sources or (iii) by ground-based systems. The following sections describe these methods. There is then a discussion of factors that have constrained the use of automated location systems. The final section considers the collection of telemetry data that do not include locations.

3.1 SATELLITE TRACKING

Tags for tracking by satellite have the ungainly title of platform transmitter terminals, but are generally known as PTTs. They were first used on wildlife, for location by the Nimbus 3 and Nimbus 6 weather satellites, in the early 1970s. The early PTTs weighed 5–11 kg and were therefore fitted only to large animals, such as wapiti (Buechner *et al.*, 1971), polar bears (Kolz *et al.*, 1980; Schweinsburg and Lee, 1982), turtles (Timko and Kolz, 1982) and basking sharks (Priede, 1980). Many other large mammals were tagged during the 1980s (R. Harris *et al.*, 1990).

During the last two decades, the size of PTTs has been steadily reduced. Tags for marine operation, with pressure protection and transmission only when surfaced, have been developed by Telonics, more recently also by Sirtrack. A programme to build PTTs as small as possible was encouraged by sponsorship from the US Army through Colonel Bill Seegar. Work at John Hopkins University initially produced tags of 170 g that were used on eagles, swans and giant petrels (Fuller *et al.*, 1984; Strikwerda *et al.*, 1986). Subsequent reduction by Microwave Telemetry, to 28 g in the early 1990s and now 20 g, permitted use on falcons (Fig. 3.1) and other mid-sized birds of prey (Fuller *et al.*, 1995; Seegar *et al.*, 1996). There were also some innovative early uses in Europe by Mariner Radar (Priede and French, 1991; French and Goriup, 1992). Basic transmitters weighing 15 g without power supplies

Figure 3.1. A tag from Microwave Telemetry for tracking by the Argos system, harness-mounted on a saker falcon (*Falco cherrug*).

are available from NTT/Toyocom of Japan for US$1500, but biologists usually buy PTTs as complete tags. These cost US$2000–4000, depending on complexity.

Tracking by satellite works on the Doppler principle. A frequency shift in each received signal indicates the satellite's speed relative to the tag, and the tag's bearing from the satellite's track is computed from the ratio of this speed to the satellite's ground speed (Fig. 3.2). A location can be estimated if there are at least two uplinks (i.e. two bearings) during each pass, which may take as little as 10 min (horizon to horizon). At least four uplinks over a period exceeding 420 s are needed for the most precise locations. Since the tag's position could be on either side of the satellite's track, it can only be located unambiguously (i) if it can be recorded again from the different track on another orbit, or (ii) with reference to a recent previous fix, or (iii) if one of the two computed positions is impossible (e.g. for a whale on dry land).

Since 1978, animal tracking has been primarily through the ARGOS system, with equipment designed and operated by Centre Nationale d'Études Spatiales (CNES) in Toulouse. For many years, the receptor units were carried on two Tiros satellites of the US National Oceanic and Atmospheric Administration (NOAA). A second-generation satellite was added in 1994. As a result of co-operation between France, the USA and Japan, a further satellite was launched during 1999 to provide third-generation capabilities, with provision for sending Global Messaging System instructions to PTTs as well as receiving their signals. The ability to instruct PTTs when to transmit increases their efficiency for sending streams of stored data. The number of reception units on new satellites has been increased from four to eight, with other improvements to give a 3.5-fold capacity increase and a 7 dB increase in sensitivity (Taillade 1992; ARGOS newsletters).

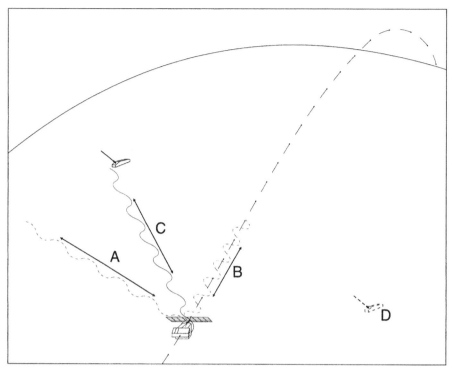

Figure 3.2. A satellite registers the nominal frequency of a tag only if the tag is perpendicular to the path of the satellite (A). The registered frequency is raised maximally for a tag far along the track in front of the satellite (B), with the true vector to the tag (C) estimated from the relative change in frequency. The ambiguous location (D) is eliminated by reference to a previous location.

ARGOS charges about US$1000 for daily data from each tag through a year. Signal output of 1–2 W was required initially, but tags with effective radiated power (ERP) of 100–150 mW can now be detected, although they provide relatively few location estimates and low accuracy (see section 3.4), especially since UHF signals are appreciably attenuated by vegetation and dense cloud cover. The low power prolongs transmission life, which may be only 400–500 hours for a 30 g tag, but can produce locations for 9–10 months if the duty cycle is on for only 8 hours every 5 days. Signals must include an initial 160 ms of constant carrier (at 401.65 MHz) for the Doppler vectoring, followed by a pulse train of at least 360 ms. The location process requires a very high frequency stability. Signals are rejected if they shift more than 2 Hz during a satellite pass, or 24 Hz between orbits (Priede, 1986). Compare this with the drift of perhaps 50 Hz per °C in standard VHF tags. Early PTTs used heating circuits to maintain thermal stability, but crystal-controlled oscillators are now adequate, thanks to advances in small transmitter technology for mobile phones.

Tracking by satellite is the most economic technique for wide-ranging marine animals, although the uplink requirements mean that locations have been obtained quite infrequently for species that are rarely on the surface. This is because the number of satellite passes has been as low as seven per day at the equator, with up to 15 at high latitudes. The doubling in number of satellites and receptors per satellite should increase the number of locations for species, like whales, that may spend only 5% of their time at the surface (Mate *et al.*, 1997). Tags on marine species must be immensely robust, tolerating pressures of up to the 101 atmospheres found at 1000 m, and transmitting only when on the surface, ideally synchronised with passage of a satellite. To help acquire data and locations during short tag exposure, arrangements can be made to transmit with less than the normal minimum interval of 60 s (Taillade, 1992).

Satellite tracking has also proved useful for determining migration routes and important feeding areas for species of conservation concern, including albatrosses (Jouventin and Weimerskirch, 1990; Prince *et al.*, 1992), bustard (French and Goriup, 1992), cranes (Higuchi *et al.*, 1992; 1996), storks (Berthold *et al.*, 1992) and raptors (Howey, 1992; Meyburg and Lobkov, 1994; Meyburg *et al.*, 1995; Fuller *et al.*, 1995; Brodeur *et al.*, 1996; Meyburg and Meyburg 1998). There are also opportunities to combine PTTs with other technologies, for example with use of double-labelled water and leg-mounted activity recorders to study links between flight duration and energy expenditure of wandering albatrosses (Arnould *et al.*, 1996). There is current interest in using PTTs to transmit digitised vocalisations, and even images from miniature cameras (Seegar *et al.*, 1996).

In the mid 1990s, an enterprise called Cape Aerospace in South Africa proposed to deliver tags down to about 10 g in size, to be tracked by Russian satellites. It seemed that locations were to be estimated by differences in arrival times of relatively simple signals at different satellites, in a reverse of the GPS system. A number of organisations paid deposits, but the enterprise appears to have been a fraud. Such an approach might be possible with such small tags, given appropriate receptors on a fleet like the 24 GPS satellites, but remains in the future.

3.2 GPS TAGS

The NAVSTAR Global Positioning System was introduced by the United States during the 1980s for military use. The principle is that some 24 satellites transmit timed signals (at 1.5 GHz) that enable a receiving unit to estimate where each satellite was when the signal was transmitted. 'Knowing' the location of each received satellite, the GPS unit can estimate its relative distance from each by the time that the signal takes to arrive. Signals from three satellites give an unambiguous location, on the assumption that the GPS unit is at sea-level. Signals from four or more satellites enable the unit to estimate a more accurate position that includes altitude. The large number of satellites means that locations can usually be estimated, unless receiving units have satellite signals obstructed by topography or vegetation, which absorbs the UHF signals better than those used for VHF radio tracking.

Commercial GPS units became available in the late 1980s for private use, and have decreased considerably in cost and size during the last 5 years, so that hand-sized units can be purchased for US$200 and watch-sized units are also available. An efficient half-wave dipole antenna at the UHF frequency concerned is only about 100 mm long, and more compact antenna designs are also used. GPS receivers are routinely used by biologists for finding their own location during radio tracking or survey work; the GPS receiving systems used for navigation in vehicles and aircraft include digital maps.

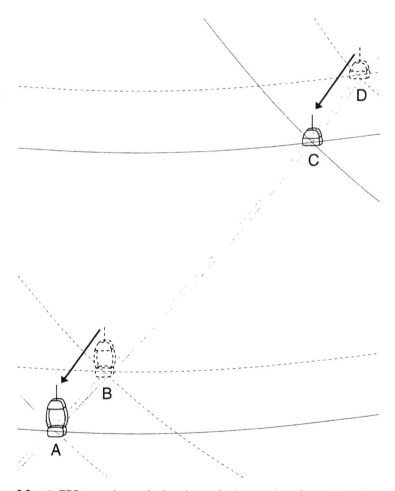

Figure 3.3. A GPS tag estimates its location at the intersection of arcs (A) estimated from the time delay in receipt of signals from different satellites. If the timing has been shifted by selective availability, as shown by the intersection of dotted lines (B), an error is introduced. However, if the true location (C) of a beacon tag is known, the deviation of the GPS location estimated for the beacon (D) at the same time can be used to estimate location (A) by differential correction, provided that the beacon is not far from the tag.

However, to prevent the system being used by other countries for precision delivery of munitions, it is degraded by 'selective availability', whereby timing signals are transmitted with an error that is varied according to a schedule controlled by the US Defense Department. There are two ways to correct the errors introduced by selective availability. One is to have US government equipment that avoids clock errors, but these receivers are not generally available to biologists. The second way is to have a base station with a precisely known location at the study area, and to use the error of locations at that station to adjust the locations registered for animals nearby. This process is known as differential correction (Fig. 3.3). Selective availability will eventually be phased out. In the meantime, differential correction is very important for obtaining adequate accuracy for location estimates from tagged animals, and also if you use GPS to record your own location during radio tracking (see section 3.4).

The first GPS tags for use on wildlife were built before commercial circuits were reduced in size, and weighed 500–1000 g. They were therefore used on animals like moose (Rempel *et al.*, 1995; Moen *et al.*, 1996). Tags are currently available from Lotek, Televilt and Telonics. They work by recording their location at pre-set intervals, and then storing the data until it can be recovered.

There are three modes of data recovery. The most primitive is to recapture the animal, or have a release on the tag that enables recovery by means of a standard VHF tag. A second approach is for the tag to download on command through a radio communication link. Data might be stored at three-hourly intervals and downloaded once a month, perhaps by approaching with a vehicle or aircraft. The third approach is to transmit the location data to the ARGOS satellite system.

In all these cases it is wise to have tags that compress the data for storage, and then transfer it rapidly. The most basic of these tags can take half an hour to download a two-month store of data at 75 baud, with a download range of only 2 km, and a need to restart if contact is broken. Keeping contact from aircraft with such tags in rough terrain provides challenging flying. With compression and an efficient data transfer, locations collected over a month can be downloaded in a minute from a 350 g tag on bighorn sheep (M. Carter, pers. comm.). Alternatively, some 20 GPS messages can be relayed to ARGOS in the 256 bits permitted per uplink (Seegar *et al.*, 1996).

Increasing the download range reduces the tag life, which is normally only about a year in any case because the receiver circuitry is quite demanding. Power requirements are least for the circuitry used in wrist watches, but the 8-channel serial operation is relatively slow and can have difficulty estimating locations if the animal is moving in dense woodland or rough terrain. The 12-channel parallel operation of other GPS cards requires more power, but for a shorter time and can provide better locations. These tags currently cost US$4000–8000 each, with ancillary equipment for data recovery and processing that can cost as much as a tag. You will also need to store large quantities of data in the same area from a beacon tag for differential correction (Moen *et al.*, 1997).

A second system, LORAN-C, is a descendant of the low frequency systems that were developed to guide bombers to targets in the Second World War. This third (C) generation LOng RANge Navigation system uses 90–110 kHz frequencies, from transmitters along the North American coast. LORAN-C can be used for animal tracking, most usefully in coastal areas (Patric *et al.*, 1988), although signals follow the earth's curvature and are detectable across the continent especially at night. An automated system in Oregon uses mobile phone technology to page transceivers, which then report their position to local receiving stations (Johnson *et al.*, 1998).

LORAN-C is scheduled to be phased out during the next decade, as is its OMEGA equivalent in Europe. They are being replaced by GPS technology, whose worldwide coverage has resulted in rapid commercial exploitation, which in turn drives the miniaturisation required for animal tags. The development of GPS boards for watch-sized personal locators is already leading to animal tags of 33 g (K. Hünerbein, (2000)).

3.3 LOCATION BY GROUND-BASED STATIONS

Fixed stations have been used mainly for presence/absence recording, especially to register the presence of animals at nests or dens, or the migrations of fish. This is the simplest way to record locations, and can be extended to record animals at several locations, or even on a grid, by switching antennas. More sophisticated fixed systems have used direction finding to estimate animal locations by triangulation or related processes. Finally, several projects have developed systems to estimate locations from time-of-arrival of tag signals at local receiving stations. A particular consideration for fixed systems with antennas on masts, whether for location finding or logging telemetry data, is to ensure protection for the electronics from atmospheric static electricity. Expensive receivers can be seriously damaged by static without actually being struck by lightning.

3.3.1 Presence/absence recording

The most frequent use of automatic stations has been to record when radio-tagged animals visit a particular site. For example, if a Yagi antenna is pointed vertically at a tree nest, and coupled to a receiver with a recorder, a steady recording peak will indicate the animal's presence in the nest, whereas a trace on the base line indicates that the animal is out of range. The earliest records, for example of seals visiting ice-flows (Siniff *et al.*, 1969), were mostly on paper chart recorders (Williams and Williams, 1970; Gilmer *et al.*, 1971), which can still be satisfactory for simple records. Figure 3.4 shows the record from a Rustrak recorder, using a large electrolytic capacitor in parallel with input from the receiver to provide a steady trace from the pulsed radio signal. The chart records a strong steady trace with the squirrel in its nest and an intermediate, irregular trace when the animal was active nearby. To avoid the problem of inaccurate timing due to variable chart speeds (Cederlund and Lemnell, 1980), which can be a particular problem when battery voltage varies with temperature in remote locations, a device can be added to provide timing marks (Gillingham and Parker, 1992).

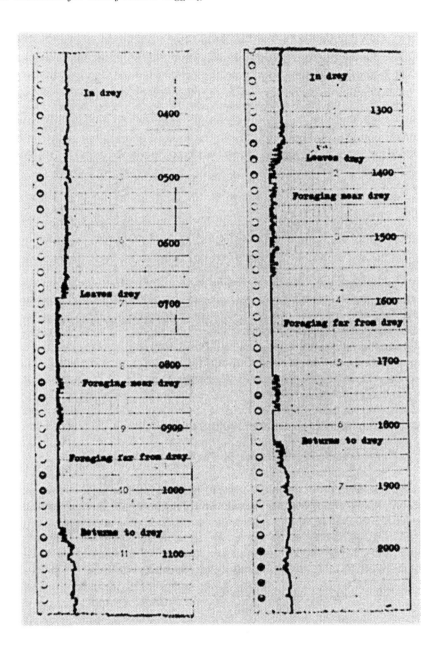

Figure 3.4.   A chart obtained on a Rustrak recorder, using a Yagi antenna pointed at a squirrel's nest. The signal was strong and steady when the squirrel was in the nest with its young, irregular when it was foraging nearby, and the trace was on the baseline when the animal was at least 50 m away.

Another approach for single-channel records has been to use an inexpensive tape recorder to record the receiver's audio output. Duration of recording can be maximised by using timers to record short samples (Downhower and Pauley, 1970; Göransson, 1980; Macdonald and Amlaner, 1980) and running the recorders at reduced speed (with normal-speed playback to aid analysis). Times of events can be estimated by counting the number of sample periods while analysing the tape, provided that a click from the machine is registered or a bleep generated at the beginning or end of each sample.

Whereas a simple tracking receiver can only be used to record one tag frequency at a time, programmable receivers can be used to record visits by tagged individuals on several different frequencies, for instance to record the passage of tagged fish (Solomon and Storeton-West, 1983). Early multichannel systems used simple electronic controllers to reset the baseline progressively across a wide chart, in synchrony with changes in receiver frequency, to create a stepped record. However, this approach required expensive chart recorders running relatively fast, and was replaced by microprocessor control of cassette data recorders and small thermal printers (Pearson, 1986). A riverside system scanned each frequency in the receiver memory every 5 minutes, for 70 days with an external car battery. Detection of a radio-tagged fish produced a printout of channel number and time, with a 15 s tape recording as a further check during the first 12 consecutive scanning periods, after which records were curtailed until loss of the signal, at which point the tag's departure was logged. Versions of this system still run, together with sensors at the recording stations to log a range of environmental variables.

Printers, chart recorders and tape recorder systems, with moving parts that may corrode in the damp and slow or even stop in cold weather, now tend to be replaced by solid-state systems (Howey *et al.*, 1984; 1988; Cooper and Charles-Dominique, 1985; Keuchle *et al.*, 1989; Nicholas *et al.*, 1992). Systems can be obtained from most equipment manufacturers, but tend to be too expensive to be used in large quantities for presence/absence recording as opposed to the collection of more sophisticated telemetry data (see section 3.5). Potential alternatives include combinations of inexpensive commercial receivers and cheap palmtop computers, provided that some form of pulse-detecting interface becomes available. Such systems could be suitable for simple presence/absence logging, which needs relatively little memory if the system is programmed merely to store the time once when a signal is acquired, and once when it is lost. However, it is important to prevent loss of data if batteries fail, either by ensuring that volatile memory is battery-backed or by using flash (EPROM) memory cards.

With suitable switching of antennas, a presence/absence system can be extended to record at a number of locations with separate antennas at many points along a river bank, or even as a more two-dimensional system with many separate logging stations (Kaneko *et al.*, 1998). To record the locations of small rodents or other terrestrial animals with small home ranges, a set of separate parallel wires can be stretched as antennas low over a study area in one direction, with another similar set at right angles to them and slightly above them. By switching between the wires, the tags approximate location can be estimated at the intersection of the two crossing wires which receive the strongest signals (Chute *et al.*, 1974; Zinnel and Tester, 1984).

3.3.2 Direction finding

The very earliest fixed stations, for example at the Cedar Creek Natural History Area in the United States (Cochran *et al.*, 1965), were intended to radio-locate a large number of tagged animals at all hours of the day and night, thus providing a great many data for home range analysis without the need for extensive mobile tracking. This sort of system is at its best in areas where few topographical errors are likely, especially in flat yet inaccessible terrain (e.g. wetlands), or where too much disturbance would be caused by mobile tracking (e.g. in very open country or small-mammal habitats).

Three types of automatic direction-finding system have been used for locating wildlife. The Cedar Creek system, which was adopted at Chizé Forest in France (Marqués, 1972; Deat *et al.*, 1980), used rotating twin Yagi antennas at the top of 20–30 m towers about 1 km apart. The Yagi direction was indicated by 360 slits in a drum that rotated past a photocell counter. The rotation speed of such systems must be relatively low, and the tag pulse rates quite fast: if signals occur at 1-second intervals, an antenna which rotates once per minute will swing through 6° between each signal. Accuracy can be improved by sweeping more slowly (4 min per revolution at Chizé), but this increases the error when each tower scans a moving animal at different times. A modification of this approach is to step a single Yagi antenna through 10° at a time, and estimate the location of the strongest signal by interpolation; a small system of this type was used to track weak tags on amphibia in a walled garden (French *et al.*, 1992).

The second type of direction-finding system replaces a stepping process by an array of fixed directional antennas oriented in different directions. A system in Illinois used six three-element Yagis pointing outwards from a mast with equal angles between them (Larkin *et al.*, 1996). The antennas were scanned electronically and tag direction estimated by interpolation. A big advantage of this approach is to dispense with moving parts, which require a great deal of servicing if used continuously. Moreover, with fast circuitry to estimate signal amplitude, the tag direction can be estimated within a single pulse, which minimises errors due to animal movements. A system of this type can give reasonable accuracy when calibrated with beacon transmitters, but was relatively inaccurate for weak tags on animals (H. McQuillen, pers. comm).

The third direction-finding approach also scans an array of fixed antennas, but in this case a ring of four or more vertical dipoles. In principle, an array of this type could be scanned in rotation to detect Doppler-effect frequency shifts, with minimal shift at the antenna closest to the tag. In reality, it is impractical to switch fast enough between antennas, and phase changes rather than frequency changes are used in a quasi-Doppler system. The system acts more like a rotating H-Adcock antenna (Chapter 2), and (like the Adcock antenna) indicates tag direction ambiguously. A commercial system of this type appeared in the early 1990s, but the low gain of dipole antennas required use of powerful tags, which also had to transmit 200 ms pulses (Angebjorn and Becker, 1992). The system was therefore only suitable for use with tags on relatively large animals, and is no longer marketed. Another quasi-Doppler system was developed for tags used in routine VHF tracking and can estimate a bearing from signals of only 10 ms in length from tags 20 kHz apart (Burchard, 1989; Schober *et al.*, 1992), although the maximum sensitivity of −125 dBm means that tags must be relatively powerful (R. Bögel, pers. comm.). A signal length of 15 ms provides scope for coding information from a sensor, with 25 ms necessary for data from two sensors and so on up to a maximum of eight sensor channels. Each antenna array can provide a bearing that is accurate to approximately 2° in good conditions. Cost (from Entwicklungsbüro Rohde) is approximately US$15 000 for each antenna array and receiver station, including battery, solar array, control system and radio-link. Software suitable for running up to 30 antenna sites costs about US$10 000.

Sophisticated software is essential to obtain good results from any automated system, at four levels. The primary software is needed to operate the hardware, at high speed in the case of an antenna-scanning system. Secondary and tertiary software provides two levels of filtering, to reject unsuitable signals. Secondary software is needed, for example, to separate the tag signal from the noise generated by a nearby electricity pylon in wet weather, perhaps on the basis of signal length or coding. Sophisticated software in Digital Signal Processing (DSP) systems may eventually provide sensitivity as good as the human ear, by separating signals very effectively from noise. Separation of signals from noise is the whole basis of signal detection, with the human ear and brain an excellent performer.

An automated direction finding system may seem very promising on paper if its primary and secondary software is adequate. However, it will estimate bearings of very variable accuracy, which tertiary software must filter. This is important for the quasi-Doppler system, because it depends on signals with a vertical polarisation and loses accuracy under conditions that affect such signals. Thus, accuracy deteriorates in woodland with many straight trees, in vertical terrain and, more seriously, if the transmitting antenna diverges appreciably from the vertical. These factors, together with movement of animals, contribute to the lower accuracy of single locations from live animals than from beacon transmitters, which must be taken into account when considering use of any automated location system. The solution with a fast-acting system is to take large samples of bearings from several stations for each location estimate, so that the software can filter those with unsatisfactory characteristics and still have a number from which to estimate median x and y coordinates.

Even good third-level software cannot deal with a particular problem affecting all direction-finding systems, namely that errors increase with distance from the antenna due to angular spread. This limits the use of direction-finding systems to animals with relatively restricted ranges, unless a large network of antennas can be installed. Tertiary software, together with the fourth level required to prevent 'drowning' in the large volumes of data that an automatic system can generate, is considered again at the end of the next section.

3.3.3 Time of arrival

GPS and LORAN are time-of-arrival (TOA) systems in which the tag estimates its position from the time taken for signals to reach it from separate sources. This requires complex, relatively large tags. Several attempts have been made to reverse this approach, and thereby to define tag location by time differences in arrival (TDOA) of a tag signal at three or more omnidirectional receiving antennas. Radio signals propagate at about 3×10^8 metres per second, or 0.3 m in a nano-second. Therefore, if a signal takes 10 ns longer to reach one of two stations, the tag is somewhere along a hyperbola that is 3 m closer to one station than the other. The time differences between these stations and a third give two further sets of hyperbolas, which intersect at the tag's position.

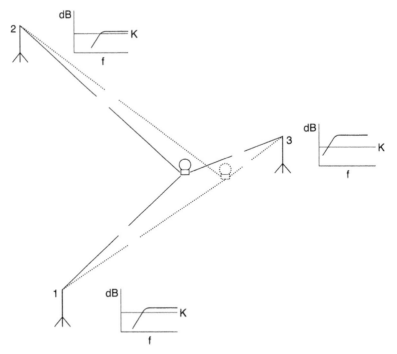

Figure 3.5. The signal-rise problem in systems that locate tags by time difference of arrival. Signal strength must reach a threshold (K) to be detected at each station. At station 3, the signal from the tag is detected earlier than would be expected without a rise-time, because proximity to the receiver causes a strong signal to cross the detection threshold earlier than at the other stations. As a result, the tag location is computed too close to station 3. Highly complex signal coding can alleviate this problem.

Needless to say, the relative positions of the receiving stations must be known precisely. Moreover, each station must detect the signal at the same point in its development. This makes tags unsuitable if their signals do not rise virtually instantaneously to peak intensity. If one receiving station is nearer to the tag than another, and therefore detects the signal closer to its onset (Fig. 3.5, station 3), then the tag's distance to that station will be underestimated. This is no trivial problem, because the rise time of typical VHF tag signals is about 1 ms, equivalent to a distance of 300 km. However, signals can be turned off rapidly, so complex coding that can help to reject indirect signals and noise (Yerbury, 1980) might also reduce the rise-time problem, by enabling several on-off times to be averaged within a signal.

In a system at Grimsö in Sweden (Lemnell *et al.*, 1983), the tags were active transponders. On activation by a powerful VHF signal from a master station, each tag sent a coded signal pulse that carried up to three channels of sensor data. Tags were on the same frequency, but were individually identifiable by being timed to transmit during a separate 2 s slot during 2-minute periods. Their VHF signals were received at the master station and at three slave stations 6–10 km apart, which immediately retransmitted a UHF signal to the master. A tag signal of 8–20 W was required for detection at these distances, so 'transmission on demand' was important for prolonging tag life. Initial tags weighed 600 g, but were later reduced to 40 g.

Another approach to the detection of signals with high timing accuracy is to use a very complex pattern of coding spread across a frequency of perhaps 1 MHz. This 'spread-spectrum' approach was used by Spencer and Savaglio (1996) to obtain an average timing precision equivalent to 13 m in wooded habitats (i.e. about 40 ns). The prototype required a minimum tag ERP of 10–15 mW, which is too demanding for the cells in the smallest tags. However, improvements in tags and detector design may eventually provide an effective TDOA system for tracking animals within the next decade. The elegance of such a system is that one adequate for small animals would scale up to larger species, with more powerful tags and greater detector spacing, without loss of accuracy that comes with increasing detection distances when direction-finding. Unfortunately, spread-spectrum tags are too heavy for small animals.

When establishing antenna sites for TDOA or direction-finding, it is important to test for local effects that may create unusual problems for accuracy. Tags with a variety of antenna orientations should be used across a matrix of test locations before masts are sited permanently (Cederlund *et al.*, 1979; Lee *et al.*, 1985; White and Garrott, 1990). Inaccurate receiving stations can then be moved. Moreover, the same approach can be used to recognise parts of a study area where location accuracy is poor, perhaps due to reflections or heavy vegetation. Where errors are consistent, tertiary software can be used to build an 'error-correction-map'.

A drawback with all automatic radio location is that it misses much information which can be gathered by tracking on foot, such as observations of whether the animal is feeding, resting, courting, etc. Moreover, fixed-station recording lacks the flexibility of a mobile tracker to increase accuracy when necessary, for instance to decide which side of a habitat boundary or in which tree an individual is feeding, or indeed whether it is in a tree or on the ground.

A final problem, that can become crucial, is how to handle the vast quantities of data that may be collected. If collected at frequent intervals, the data will fill stacks of CDs and also show strong serial correlation. Extensive top-level software will be required to extract the important elements from the data, for analyses of foraging movements, home ranges and interactions between conspecifics, different species and their environment.

3.4 CHOOSING LOCATION SYSTEMS

The following section considers the uses for various automated systems, in terms of answering biological questions. For help with the technical assessment, I am greatly indebted to Brian Cresswell, with further valuable advice from Ralf Bögel, Jim Lotimer and Shane Nelson.

The use of automated location systems is constrained by high costs and the mass of tags relative to those for mobile tracking, but also by limited accuracy. In mobile tracking, accuracy is a feature of antenna characteristics (Chapter 2) and landscape, but also depends greatly on tracking methods and distance from tags. During mobile tracking, methods and distance can be adjusted to obtain a desired tracking resolution, which sets the distance in analyses between the closest x or y coordinates. The resolution is estimated from the standard error of the tracking method. This could be 1 m for a probe antenna in grassland, 10 m for locations recorded with hand-held Yagi antennas within 100 m of tags, or 100 m for animals tracked from within 2–3 km using accurate antennas on vehicles (see Chapter 8). Tracking resolution is important both when studying how animals use areas, by estimating their home ranges, and when analysing use of resources that vary over short distances. How can you tell which tree species is being used if your tracking resolution is 100 m?

3.4.1 Resolution with PTTs and GPS tags

The resolution obtainable from GPS tags and PTTs is substantially affected by vegetation and topography. Vegetation attenuates the UHF signals used by both systems, such that a passing satellite fails to receive adequate signals from a PTT, or a GPS tag fails to acquire signals from enough of the satellites available at a given time. In experiments with GPS tags, canopy closure reduced location rates but not accuracy, whereas increase in tree density reduced both location rate and accuracy (Rempel *et al.*, 1995). Canopy closure presumably affected signal strength, whereas dense tree stands probably also caused signal reflections or interference.

Topography acts on GPS tags and PTTs in two ways. The gross affect is to block signals: animals in valley bottoms provided 86% fewer locations than those on elevated sites (Keating *et al.*, 1991). A more subtle effect is caused by tag altitude, especially for tracking by single satellites in the ARGOS system. If an increase in tag altitude (e.g. on a flying bird) reduces the distance to the satellite, an estimated location is prone to bias towards the satellite's track on the ground (Fig. 3.6). The error can be reduced if altitude data are available (Keating, 1995). The Doppler process of the ARGOS system is also affected by tag movement, giving errors in the order of 100 m for each kph of movement, and thus of around 5 km for a bird flying at 50 kph (Taillade, 1992).

Nevertheless, GPS tags can estimate 3-D locations that include their altitude if signals are received from at least four satellites, and hence make the necessary correction for elevation error. Alternatively, locations that use 2–3 satellites can be corrected to 3-D accuracy if the altitude of the tagged animal is known within 50–100 m (Moen *et al.*, 1997). It is therefore easy to imagine that GPS tags give locations with a higher precision than is recorded in the field. Even when accuracy was enhanced by differential correction, repeated re-sampling with fixed tags in uneven terrain estimated a 95% error ellipse of 5.4 ha (Carrel *et al.*, 1997), which is equivalent to a tracking resolution from one standard error (a 68% area) of about 70 m.

Software that corrects for landscape features can improve accuracy. For example, LORAN-C data from Carrel *et al.* (1997) gave accuracy ellipses of 93 ha, for a resolution of 270 m, compared with 130 m in another study where LORAN-C tags on animals could be instructed to resample if location accuracy

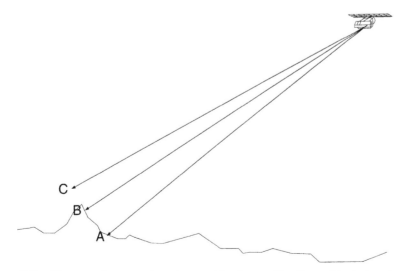

Figure 3.6. The error that can arise when tracking by satellite if tag altitude is unknown. Tags at A, B and C are equidistant from the satellite. The location of all will be given as A unless altitude data are also used in the estimate.

was inadequate (Johnson *et al.*, 1998). A tracking resolution of 20–30 m is probably practical with GPS tags in even terrain. However, this requires data to be acquired and stored in the tags with a high sampling rate, which therefore need larger batteries for frequent operation and extended downloading (Rempel and Rodgers, 1997).

Without differential correction, GPS locations had less than one tenth the accuracy of the above examples. The resolution of uncorrected GPS tags was then similar to that of PTTs in the ARGOS system. From records in the field under near-optimal conditions, Keating *et al.* (1991) gave accuracy data equivalent to resolutions of about 1200, 900 and 360 m for ARGOS accuracy classes 1, 2 and 3. With similar favourable conditions but low-power tags, on falcons in a tree-crown nest at 40°N in Kazakhstan (R.E. Kenward and E.A. Bragin, unpublished), no locations were in the most accurate classes (2 or 3), only five were in class 1 and 15 in class 0. The spread across 68% of these locations was 11 km (Fig. 3.7).

It seems reasonable to conclude that tags using differential GPS and LORAN-C can give data adequate for estimating extensive home ranges and use of coarse-grained habitats, at least for wide-ranging animals. However, only high power PTTs can be used to estimate home ranges. The low-power PTTs can be used to record long foraging movements, and wide areas visited during migration, but not for home-range data.

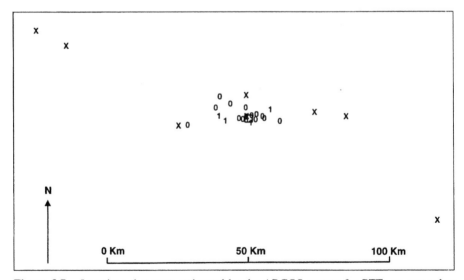

Figure 3.7. Locations that were estimated by the ARGOS system for PTTs on two saker falcon (*Falco cherrug*) chicks while they were in a tree-top nest on flat steppe-taiga in Kazakhstan. Edges of a small square are shown 1 km from repeated GPS readings at the nest. There was an 11 km spread across 68% of the locations classed 0 and 1, with greater spread across location classes A, B and Z (shown as X).

3.4.2 Automated fixed stations

An ideal automated location system would be one capable of 10 m resolution for the smallest tags now used for mobile tracking. Present radio tags with mass <1 g can be carried by many small species, and transmit at 1000–1500 ms intervals for several weeks using signals of 10 ms or less. The animals move so frequently over relatively short distances that mobile tracking of large numbers without disturbance is often impractical. However, an automated system could gather detailed data, on interactions with resources, competitors and predators. Such data are needed for modelling vertebrate food chains and webs at a woodland and farmland level, with a view to predictive conservation.

The resolution required is not available from current satellite-based systems. Nor are effective PTTs or GPS tags much under 20 g an immediate prospect, especially for species living in woodland. Automated tracking of the small tags therefore requires a direction-finding or time-of-arrival system based on local fixed stations. The main challenge for such a system is sensitivity, because current 1 g tags have an ERP of 10–100 μW. This output depends mainly on their antennas, because small cells cannot support high current drain, and boosting with large capacitors would increase tag mass unacceptably. Signals will also be weaker for creeping mammals than for birds in trees. The detection of weak tags at 50–100 m, for use with a net of receiving stations at 100–200 m intervals, needs a system with sensitivity down to −140 dBm.

Another challenge for the system is that tags for manual radio-tracking are identified by a 5–10 kHz separation of their frequencies in the 138–216 MHz band. There are advantages in being able to track manually in conjunction with an automated system, e.g. for: (i) homing on nesting or dead animals; (ii) making independent behavioural observations; (iii) locating animals that disperse beyond the system and (iv) providing data if automated tracking fails. Avoidance of signal clashes would also restrict the number of coded tags on the same frequency in a small area. Therefore, even though electronic detection of signals is simpler if signal coding is used to identify tags on a single frequency, it remains preferable for the system to support multiple tag frequencies for manual tracking.

The first time-of-arrival system that used spread-spectrum techniques to obtain a resolution around 10 m operated on a single frequency and also required a power output of 10–15 mW (Spencer and Savaglio, 1996), which prevents use of tags with small cells. Another prototype system from TICOM-Inc has detected signals with power outputs typical of standard tags (S. Nelson, pers. comm.), but used a 1 MHz bandwidth signal at 900 MHz from a tag circuit that may not easily reduce below 10–20 g. A problem with large tags is that animals capable of carrying them also tend to travel widely, and therefore need a very extensive system to track adequate samples in different habitats.

Direction finding for weak tags is practical with rotating Yagi antennas (Deat *et al.*, 1980, French *et al.*, 1992), but these are bulky and the need to service moving parts discourages use of many receiving stations. The use of fixed antenna elements in an active array is therefore more attractive for direction-finding. However, the system used in Germany is prone to considerable error if a transmitter antenna departs appreciably from the vertical (Schober *et al.*, 1992), and vertical antennas would hinder small animals. Nevertheless, other types of active antenna may avoid problems with signal polarisation (B. Cresswell, pers. comm.). Moreover advances in digital signal processing (DSP) may eventually provide sensitivity better than −160 dBm (J. Lotimer, pers. comm.) for electronic detection of the weak signals from small tags on multiple frequencies.

3.4.3 The human alternative

Automatic radio location systems with fixed local receiving stations are relatively expensive to construct and fraught with problems. It may be fair to say that none have produced the large volume of exciting research results that were probably expected. On the whole, they have failed to meet all the hardware and software requirements, although some (including the initial attempt at Cedar Creek) did collect large quantities of data.

Automated location systems also still have fundamental limitations. In particular, although carefully tuned electronic signal detectors can now compete with detection by humans, it is hard for detectors in direction finding systems to remain tightly tuned to the low bandwidth tags as temperature changes. In practice, automated signal detection for direction finding is at best about 6 dB m less sensitive than the human ear, which halves detection distances. Reduced detection distance is particularly important for tracking animals with large ranges or weak tags. Problems with detecting weak signals, and requirements for large tags in most systems, have precluded the use of automated location systems for work with small animals. This is unfortunate, because it is at the small scale that it is most easy to measure environmental characteristics in detail, and where relationships along food chains can still be investigated (large predators having been eliminated in many areas). Moreover, landscape intensification and fragmentation in developed countries increasingly puts the smaller species at risk.

Where mobile tracking is less efficient than fixed stations, a practical alternative to automation is the use of human operators. The direction-finding stations can then take accurate bearings with null-peak antenna systems (Fig. 3.8), and are easily 'programmed' to reject noise and unlikely locations! If two communicating observers plot each pair of bearings as they are taken, preferably by simply entering bearings into a microcomputer-generated map display (which also stores the data), they can immediately try again if a small mammal seems to have moved 1 km in 5 min. So, for 1–3 year projects in which there is a premium on reliable data, it is still well worth settling for a pair of stations with manually rotated twin-Yagi (null-peak) antennas. One person can operate a system if auxiliary stations are servo-driven and their signals relayed (Smith and Trevor-Deutsch, 1980; Linn and Wilcox, 1982; Spencer *et al.*, 1987).

However, beware of the extra time required for development and teething problems, and of making unreasonable demands on the system. A number of projects have tried

Figure 3.8. Twin four-element Yagi antennas used for taking null-peak bearings from a hilltop.

unsuccessfully to track foraging seabirds from the shore. For a start, it is difficult to get accurate bearings from Yagis to moving animals, such as birds in flight or bobbing on the water, because changing antenna orientation affects the signal amplitude. Moreover, the bearings from each station become very similar when birds are far out to sea, which produces large errors in the location estimates (see Chapter 8). It is also easy to forget the constraint posed by the horizon. The horizon distance in miles approximates to the square root of the observer's height above it in feet. VHF signals can be detected slightly over the horizon, by a factor of up to 1.4 in excellent propagation conditions, but more often by about 1.2. This means that a Yagi on a 100 m headland will probably not detect signals much more than 33 km out to sea, or perhaps 50 km for a bird flying at 20 m ASL, even if the tag could be heard at 200 km in line-of-sight from an aircraft.

3.5 DATA LOGGING

If tags remain comfortably within detection distance of a receiver, which could be satellite or fixed, then they can transmit sensor data to recording stations.

Otherwise, if they can be large enough, they may store the data for subsequent transmission or recovery by other means.

3.5.1 Recording stations

Sometimes tags cannot be large enough to store data. Alternatively, some types of data only have meaning when registered by a receiving site. For example simple changes in amplitude of received signals may indicate activity (Cederlund and Lemnell, 1980; Widén, 1982), or amplitude may be combined with other sensor data to distinguish different activities (Nams, 1989a). At other times, information is most relevant when compared with environmental data recorded away from the tag. For example, comparing data from beacons with data from tags sensing ambient temperature may show how an animal uses microclimates (Standora, 1977; Standora *et al.*, 1984), or deep-body temperature varies in response to environmental temperature (Mackay, 1964; McGinnis, 1967; Osgood, 1970; Swingland and Frazier, 1980). Light intensity at tags may be compared with ambient records to reveal burrow emergence and use of shade by fossorial mammals (Recht, 1992). Measuring air pressure at ground level improves the precision of estimates from altitude sensors (Bögel and Burchard, 1992), which with temperature compensation can be accurate to ±50 m up to 2500 m.

To detect signals from an animal throughout its range, a high-gain receiving antenna is needed. To cover a large area, the antenna should be mounted on a high point and preferably at least 10 m above the ground, but with the cable to the receiver as short as possible to reduce signal losses along the wire. If the antenna is to be near the centre of the study area, use an omnidirectional type. Many projects use λ/2 dipoles with ground-plane elements (Amlaner, 1980). However, a higher gain, equivalent to a Yagi with three or four elements, can be obtained with three- or four-folded dipoles in a vertical stack, provided the electrical length of the cables is carefully matched, or with a 5λ/8 over 5λ/8 over 5λ/8 collinear antenna. A more convenient alternative may be to use a three- or four-element Yagi to one side of the study area.

When siting a valuable recording station, security is often as important a consideration as optimising the signal reception. If there is no building in which to lock the receiver and logger, it may be best to put them up a tree, or to bury them under the ground in a suitable container (e.g. a plastic dustbin), taking care also to hide or camouflage the antenna and its cable. As a last-ditch deterrent, some projects put 'RADIOACTIVE', 'BIOHAZARD' or 'HIGH TENSION ELECTRICITY' warning labels on and in the equipment's container.

If only one animal is to be monitored at a time, signals can be recorded directly with a tape recorder, and the signal pulses later counted for each sampling period. This provides a cheap system that is good for detecting very weak signals, although it is time consuming to process the tapes and data are less accurate than with real-time electronic pulse-detection.

An electronic signal-processing system must be used if many frequencies are to be monitored at a time. If your receiver has an appropriate port, a high-accuracy external timer can simultaneously change the channel on the receiver and log the change on a suitable inexpensive data recorder. However, dedicated data loggers for use with frequency memory receivers are now available from several manufacturers (ATS, Lotek, Telonics, Televilt, Wildlife Materials). These are quite expensive, but the alternative of developing the equipment yourself, for example to detect signals with a phase-locked loop and feed them to a general-purpose logger or computer (Nicholas *et al.*, 1992), is only practical if you have electronic expertise available. To avoid moving parts, commercial loggers tend to use solid-state records. However, with appropriate processing before storage, very large quantities of data can be recorded on audio and videotapes (Stohr, 1989; Schober and Oehry, 1987; Schober *et al.*, 1989; Seegar *et al.*, 1996).

A number of projects have built systems for logging radio-telemetered data on microcomputers (Howey *et al.*, 1984; Cooper and Charles-Dominique, 1985; Howey *et al.*, 1988; Keuchle *et al.*, 1989; Exo *et al.*, 1992). The potential for 'intelligent' data logging with microcomputers is another reason for using remote recording rather than storage tags. For example, an event alert (EVAL) programme could be written to use signals from a simple motion sensor to determine that an animal has not moved at a time of day when it should be active, and is therefore probably dead. The system could then page a biologist through a mobile phone messaging system. EVAL programs could also be used to notify a variety of other infrequent but important events, such as dispersal, courtship and giving birth.

3.5.2 Storage tags

There are of course many animals that spend much of their time beyond the range of any fixed station. For these cases, it is invaluable to be able to log sensor data in the tag, as already described for GPS tags. Like the GPS tags, the data can either be read when the animal is recaptured, or be transmitted to a satellite or surface-based receiving station when appropriate. Early tags of this type made records on tape or film (Hill *et al.*, 1983; Kooyman *et al.*, 1983), but a solid-state approach was rapidly adopted. By the mid-1980s, tags containing sensors, analogue-to-digital conversion and logic circuitry, clock and memory, could be built for about US$450 (Robinson, 1986). Tags for storing ECGs from diving birds and mammals were available by the early 1990s (Woakes, 1992). Subsequent refinements have proved especially useful for monitoring diving parameters of marine animals, for which combinations of sensors for depth, direction and acceleration can be used to build complete dive profiles in three dimensions (R. Davis, pers. comm.; R. Wilson, pers. comm.).

Present re-transmission designs send relatively small quantities of data to ARGOS in repeated signal streams, which are triggered by a pressure sensor as the animal surfaces or by a master transmission. This is not ideal for diving mammals, because intermittent transmissions and infrequent satellite passes leads to loss of data. An alternative approach is to use location data from the satellite to find marine animals and collect data from a boat 3–4 km away (M. Sjöberg, pers. comm.). Future designs will be able to store large quantities of behavioural data, perhaps dumping these once a day to automatic receiving stations at bird or bat roost sites or near carnivore dens.

One of the most elegant methods of collecting data has been a highly streamlined tag towed on a cable behind a large pelagic fish. The tags collect data on pressure, temperature, light-level and speed for a year, before being detached by explosive action of a knife edge that severs the cable. They then float to the surface, with the battery hanging down on a wire to stabilise the tag, so that floats with the cable end at the top to serve as antenna, and 500 kb streams of data are transmitted to the ARGOS satellite system (D. Weihs and P.S. Levin, pers. comm.).

3.6 CONCLUSIONS

1. Automated systems involve: (i) tracking PTTs by satellite; (ii) GPS tags; (iii) fixed receiving stations and (iv) data-storage tags. Advanced systems often combine more than one approach.
2. The smallest PTTs are about 20 g, costing about US$3500 to run for a year or more of intermittent transmission. Resolution of Doppler-based PTT locations is in hundreds or thousands of metres. The tags are best used to record movements or relay other data from wide-ranging animals in remote areas, especially at sea or during bird migration.
3. Tags that estimate locations by signal time-of-arrival from Global Positioning System satellites each cost at least US$4000 at present and with differential correction may give location resolution of 20–30 m. Commercial GPS tags for wildlife weigh >100 g, but are becoming smaller and research prototypes have reached 33 g.
4. Fixed ground stations are often used to record animal presence or telemetry data. They can provide signal amplitude cues to activity and environmental reference data that are not available from data-storage tags, and alert biologists to rare events.
5. Storage tags do not require continuous contact with a nearby receiving station, and are therefore good for animals in remote locations, but must be removed or have radio-relay facilities to recover their data.
6. Systems have also been built for estimating locations by direction-finding with rotating or scanned antenna arrays, or from signal time-of-arrival. Quasi-Doppler direction-finding has been most successful and can be used with tags that are also suitable for mobile tracking, but is insensitive to weak signals and needs vertical transmitter antennas. The smallest tags give signals too weak for automated systems, yet larger animals move beyond receiving arrays.
7. Human operators remain most efficient for recording locations from fixed sites, but advances in software, especially for digital signal processing, may eventually enable extensive collection and use of accurate automated records from small animals.

4 Making Preparations

For many projects, the stage of planning after deciding how to find answers to the biological questions will be to buy equipment. This chapter starts by considering how best to obtain that equipment, and the software for handling and analysing data. It then examines some of the decisions and preparations that should be made before fieldwork. The construction of simple tags from transmitters and other components is covered in Chapter 5.

4.1 BUYING EQUIPMENT

There are many firms that build radio tags and receiving equipment (Appendix I). Some firms specialise in building VHF tags, others offer a more complete range of products. However, each firm tends to be best at a particular line of equipment. Some are better builders than others of robust receivers, or complex receiving systems, or UHF tags, or VHF tags. Even among firms making good VHF tags, for example, some are better than others for a particular range of animals, large or small, aquatic or avian. Consider each firm's experience of building tags for your type of animal, and above all seek advice from other users of their equipment – especially the ones with impressive publications. If you list your specifications to a number of firms by e-mail and simply accept the least expensive, you may not get the best equipment.

Once a decision on equipment has been made, for example to produce the costing for a grant application, do not forget to let the chosen manufacturer have your final order in good time. Tell the firm when you will need the tags as soon as you get the grant, and as long as possible before the next season too. Always remember that the best manufacturers are likely also to be busiest, and therefore least able to fit your order in at the last minute. Fortunately, some manufacturers recognise the problems that can arise from uncertain funding. They may let you book construction time provisionally, in cases where grants will not be allocated until shortly before tags are needed.

4.1.1 Receiving equipment

Details of the three different types of receiver have been covered in Chapter 2. Sensitivity, temperature drift and ability to track nearby tags, by use of a single gain control or an attenuator, is now much improved over some early models. The choice within each type will depend on tuning requirements, weather conditions, availability of a backlight, and whether you can use primary batteries, recharging, or both. Receivers should be able to take either a change of primary cells or an external power supply, which could be a small battery pack. A cable to run them from a vehicle battery is useful for reducing costs when using receivers with primary cells in vehicles, as well as for emergencies. If you may be operating without access to chargers, consider carefully how often and how easily you can replace batteries.

Receiving equipment will need servicing and possible emergency repair, so consider ease and reliability of service as well as purchase price. For service shipments, it is not a bad idea to buy receivers within your customs zone; also, ask other users about how reliable the supplier is with deliveries and servicing. Receiving antennas can also need replacing or repairing rapidly. Check how rapidly the supplier can provide spare parts, or a new antenna if necessary. Experienced trackers usually keep spare antennas, and spare receivers if possible.

4.1.2 Tagging equipment

The adoption of pulse generator circuits and surface-mount construction methods has raised the quality of basic transmitters in the last decade. Nevertheless, there is still appreciable variation in performance of transmitters and completed tags from different firms. Moreover, if you need the smallest possible tags, with single-stage transmitters, do not expect the same consistency and reliability as for tags with multivibrators, which are not available much below 500 mg.

Tags from different manufacturers do not differ greatly in price, so you will probably need to base a decision on the ability of tags to meet criteria for life, detection range and reliability. Be careful about making judgements based on tag life estimates, because some suppliers estimate an expected life from average tag current and specified cell capacity, while others give a more conservative estimate for 95% of the tags from tests of cell performance with the transmitters. Especially for planning survival studies (see Chapter 8), it is best to have conservative estimates of tag performance, and also to note that the impact of tags on animals may vary between manufacturers (Bro *et al.*, 1999). If in doubt, ask for data on the average current through the tag, because this is a better way of comparing life – but remember that pulse modulation by sensors may complicate the estimate.

Be wary too of range estimates. These are dependent not only on the effective power that a tag radiates (ERP), but also on factors such as terrain and the height of tags and receiving antennas above the ground. It is easy for manufacturers to give values that are sometimes true, but over-optimistic for the conditions in which you will be tracking. The best possible line-of-sight range, possible from an elevated antenna to a soaring bird, can be two orders of magnitude greater than from a hand-held antenna to a ground-creeping bird in dense woodland.

ERP depends on tag design, tuning and impedance matching to antennas. Official regulations for transmitter performance mainly concern ability to provide signals of a particular strength without causing interference to other users by also producing spurious emissions. Transmitters built to meet tight specifications, such as the EMC regulations of the European Union, may not necessarily be better performers on wildlife, but do indicate the ability of firms to meet rigorous criteria. In the absence of such type approval and without access to a spectrum analyser, there is no easy way to assess the tuning. Nevertheless, it is worth looking at the inductor coils of transmitters potted in transparent material: tags with neat, circular coils, rather than loose, splayed windings, at least indicate a meticulous approach to their construction.

Tag manufacturers should be able to specify the size, weight, voltage range, temperature drift, current, pulse width and interval of their products, perhaps even power output of specific tag types. Power output should be given as measured ERP, and not as an (optimistic) estimate based on circuit current. As a rule, temperature drift of modern tags is crucial only for specialised data logging (see Chapter 3), but a drift of more than 3 kHz between −20°C and +50°C could indicate the use of poor quality crystals.

If more than one firm can meet your tag requirements, it is probably best to shop for reliability. Good manufactures will have a three-stage test routine, for testing: (i) components; (ii) transmitters and (iii) completed tag circuits. They will test each batch of crystals and cells. Cells vary from batch to batch and are the most frequent cause of premature failures. The firm should have inspected each completed transmitter board for sound joints, and they should have tested the output of completed tags through a temperature cycle of −20°C (or 0°C for the smallest silver cells) to +50°C. These tests can not only reveal cell defects, but also unreliable solder points and damaged components.

Look for firms that can, if requested, provide performance details for each tag through a temperature cycle. They may not want to do so, because it creates extra paperwork that you do not really need, but the ability to produce the data indicates a sophisticated testing system. Also, don't be shy to ask about tests performed on cell batches and crystals.

If you order transmitters alone, to build your own tags, be sure to discuss your plans thoroughly with the manufacturer, who will want you to get the best results whatever you buy. Describe exactly what you plan to build, particularly the antenna materials and dimensions because these will affect the facilities provided to tune the tag. You may be able to get advice on where best to buy cells and other materials you need, and some firms may even give you a short lesson if you can visit.

When equipment is ready to use, test it in the field (see Chapter 7). Note that detection distances may be reduced from expected values by your particular terrain, but also by faults in an old receiving system. For instance, poor connections to the receiving antenna, usually due to a break within the antenna cable or connectors, can reduce the range to less than a quarter of the expected value. Classic problems are that a connector has broken or shorted, or that the receiver socket or its central pin has become loose, and separated from the central lead inside the receiver. Do not immediately blame the tag manufacturer if the range is far too little from *all* your tags. Do, however, return single tags that are sub-standard, and do contact the firm if you have other difficulties. If you decide to change supplier on the basis of an initial problem, without seeking an explanation, you may have the same problem with the new firm. In demanding applications, it pays to work closely with one manufacturer until any problems are solved. In the same way, let the manufacturer know if you are particularly pleased with the product, if only because this may get you a more sympathetic response when you next submit a last-minute order.

If you are using PTTs, you need to contact ARGOS at a early stage in your planning, because you have to be registered with them in order to get code numbers that a manufacturer needs for each tag. ARGOS will require a description of the project, its geographic location and the dates between which it will occur, the number of tags to be used, their manufacturer, the sensor data and its form, the transmission duty cycle and any special requirements from ARGOS, including the frequency and format in which you will require data from them. A manufacturer will help you provide this information, but you must allow plenty of time to discuss with the manufacturer, register with ARGOS and get the tags built.

Licences may well be needed for tagging, and even to visit nests of rare species. Failure to allow adequate time for this can cost a whole year of work, and thus cause outright failure of a student project. If you need to implant tags, you will in many countries need licences to comply with animal welfare legislation. In the UK at least, competence requirements include a course and an examination, so adequate forward planning is again essential.

4.1.3 Software

Obtaining analysis software (the 'fourth level' software in Chapter 3) is just as important as buying receiving equipment and tags. Without suitable software, it is very easy to 'drown' in the large quantity of data that radio-tagging can generate. The danger may seem remote at the start of a project. If you are going to be logging telemetered data, for example, there should be software sold with the logging equipment. That software will probably come as a module with capabilities apparently suited to your task, but check very carefully how much display and analysis it can provide. The software may do little more than record an estimated interval between signal pulses, but you will need to estimate mean rates after rejecting spurious records, examine how they change with time, test for differences and display the results. Check that the software can either do all you need itself, or provide files suitable for analysis in other packages.

If you will be radio tracking, there is a great temptation to delay decisions about software until the end of a project. This is a recipe for disaster, for two reasons. You may either collect data in a way that invalidates the statistics needed to answer your biological questions, or you may collect too many of the wrong sort of data and too few useful records. You are then at serious risk of joining the many radio trackers that have worked long and hard in the field without producing useful publications.

For survival analyses, it is important to estimate how many animals need to be tagged in order to get significant results (Pollock *et al.*, 1989; White and Garrott, 1990). Software is not needed to estimate sample sizes for detecting simple differences in survival. However, for more sophisticated analyses using covariates, you should probably plan the marking and survival checks with dummy data using appropriate software. A number of software packages are designed to handle survival data from radio tracking, including MICROMORT (Heisey and Fuller, 1985), SURVIV/MARK (White, 1983; White and Garrott, 1990) and SURPH (Smith *et al.*, 1994). Other packages are intended mainly for trap-mark-recapture data, so make sure you get one that is also suitable for the more detailed data available from radio tracking. The package should handle animals that are tagged at different times (staggered entry), checked frequently, have many covariates, and leave the tagged population as a result of radio failure as well as death (see Chapter 10).

Software is especially important for helping to plan the collection of location data. This is because sample sizes in robust statistical tests will not be based simply on numbers of locations collected, but on the number of animals for which you have a home-range size, an interaction score or an index of habitat use. You will need to estimate how few locations are needed to define each measure, so that you can increase the power of your analyses by increasing the sample size of home ranges or other measures. You will therefore need to run analyses from the start of the fieldwork, so that you subsequently collect the right data from each animal.

A great variety of software has been written for analysing location data. Some programs are primarily for plotting data and estimating locations from bearings, including XYLOG (Dodge and Steiner, 1986), TRIANG (White and Garrott, 1990) and LOCATE II (Nams, 1989b). Others, including MCPAAL (Stüwe and Blohowiak, 1987), TELEM-PC (Lewis and Haithcoat, 1986), TELEM88 (Coleman and Jones, 1988), HOMER (White and Garrott, 1990), WILDTRACK (see Appendix II), CALHOME (Kie *et al.*, 1996), TRACKER (Camponotus AB, 1994) and KERNELHR (Seaman *et al.*, 1998) are primarily intended to estimate home ranges. A few have gained additional useful functions, including estimation of spatio-temporal dependence between fixes by autocorrelative techniques (HOME RANGE: Ackerman *et al.*, 1990), calculation of range area increase with addition of consecutive locations, and even analyses of sociality and habitat use (RANGES IV: Kenward 1990; RANGES V: Kenward and Hodder, 1996).

The early suites of software, mainly provided at cost of discs and postage, were reviewed by Larkin and Halkin (1994). Software that is still supported and used widely in the late 1990s, reviewed by Lawson and Rogers (1997), includes two commercial packages dedicated to analysing data from radio-tagged animals, TRACKER at US$1000 and RANGES V at GB£300 (US$500). However, an alternative to dedicated analysis packages is to use a general-purpose geographic information system (GIS), for which software to analyse radio-tracking data is available as an enhancement. A useful set of animal movement analysis tools are available for the commercial ArcView GIS package (Hooge and Eichenlaub, 1997).

The choice between dedicated software or GIS enhancements depends on how much GIS capability you need for mapping, how many data you have to analyse and what other analysis functions you require. If you need sophisticated GIS functions and have relatively few animals, you may find a GIS enhancement adequate, provided that you have time to become familiar with the GIS itself. On the other hand, if you require fairly straightforward analyses of habitat use and availability for large numbers of animals, the best approach may be to use a dedicated package, especially if you can get help from a trained GIS operator to make the map. This is because GIS functions will probably require more work than the dedicated package for handling large numbers of analyses. An efficient dedicated package can estimate habitat proportions for a whole series of locations and range cores (e.g. containing 20% to 100% of locations at 5% intervals) in one run, for a large number of different animals in a format that goes straight into a spreadsheet. The dedicated package may also provide other specialised analyses, for example to study interactions between animals or to analyse dispersal and other movements in detail.

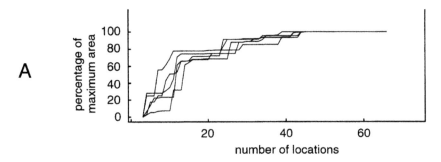

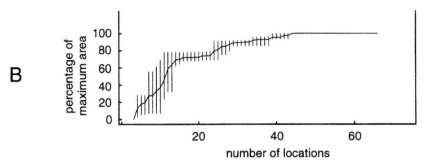

Figure 4.1. Analysis software should be able to plot how range area increases as each successive location is sampled. Plots for 66 locations recorded at two-hour intervals for adult blackbirds (*Turdus merula*) show from individual records (A), and from the mean and spread of records (B), how all five birds reached maximum range size after 45 locations.

Whatever approach you choose, make sure that the dedicated package or GIS enhancements provide the analyses you need at all stages of your project. When you are planning how often to record locations for each animal, you may wish to estimate intervals between records that minimise spatiotemporal dependence, in which case the software should include autocorrelation analysis (see Chapter 8). You may need to determine when range size no longer increases with addition of locations (Fig. 4.1), and to do so for polygons and ellipses as well as contouring techniques. To avoid running a separate analysis for each added location, it is convenient to have a system that automates incremental analysis. The suite should either be able to edit data conveniently (ideally against a map for preliminary error checking) or at least be able to import and display readily from other editing software, such as a spreadsheet. If it cannot display range outlines simultaneously for numbers of animals against a map, it should at least be able to export the edges to other packages that can, because visualising interactions between ranges and with habitats is also important (Fig. 4.2).

4.2 SOFTWARE PREPARATIONS

Another reason for you to obtain software well in advance of fieldwork is so that you have time to practice using it, and thus to make your entry of data as efficient as possible. Things that may seem trivial in terms of timesaving can accumulate to a substantially reduced effort by the end of a project. For example, you might use a GPS or other system that estimates coordinates with several decimal places, and then file them by hand. You could type those coordinates to the nearest hundredth of a metre (e.g. 108329.61,7489.20), forgetting that the resolution of your location system is no better than 100 metres. However, if practice shows that your system accepts different scales of data, you could use a scale of 100 m and enter 1083,75, thereby sparing yourself the needs to type 10 extra digits for every pair of

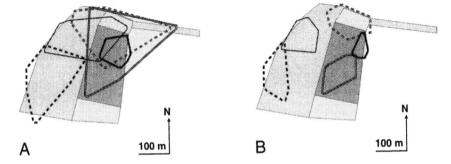

Figure 4.2.   Display of multiple ranges is important for inspecting data. Convex polygons plotted round all the locations of five adult squirrels (*Sciurus carolinensis*) in a small wood show extensive overlap (A), whereas polygons containing only the 75% of locations closest to the squirrels' nests show little overlap (B).

coordinates. Copying so many digits not only wastes time, but also increases risk of errors. It is amazing how many data sets contain hand-entered locations accurate to several decimal places.

4.2.1 Filing location data: spreadsheets or dedicated software?

The two main alternatives for filing are spreadsheets or software dedicated to display and analysis. Data may sometimes be provided directly in spreadsheet format, for example from automated systems. However, many people prefer to estimate locations by hand, in order to gain the greater detail that may be available from a paper map, or because climate or weight makes a computerised system inconvenient in the field.

If locations are collected by hand, is it better to file them in spreadsheets or dedicated software? A flexible dedicated package should be able to transfer data and results readily to and from spreadsheets, for editing or for analyses that are done best in one or other package, and you should in any case be able to display data from the start of a pilot study. For one thing, display of the data against a map will immediately indicate possible recording errors (Fig. 4.3). It will also prompt you to enter data in an appropriate format, and perhaps even save you

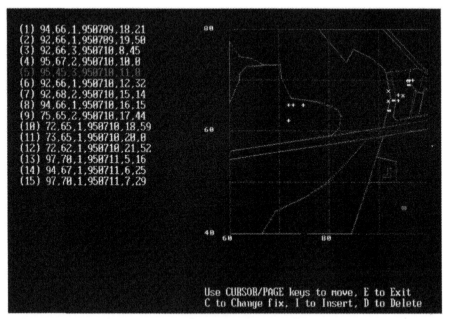

Figure 4.3. Editing the fifth fix in a file of blackbird data, which the map shows to be in an unusual open field location (small grey square, bottom right), the result of an error when copying data from a field notebook. The data in each row are the x-coordinate, y-coordinate, and four qualifying variables: an activity code, the date (YYMODD) and time (HH, MI).

time by automating the entry of dates and repeated variables. Moreover, you should fine-tune your data collection at an early stage of your study, for example by using autocorrelation analysis to estimate optimal time intervals between records or incremental analysis of range sizes. So, even if you wish to file direct to a spreadsheet, perhaps for convenience of future access, you would be wise to start importing to display and analysis software right from the start of the study.

If you file data first in a spreadsheet, remember that you will need to export it in a format acceptable to your analysis system. For example, RANGES V accepts comma-separated-values (CSV), tab-separated values (TSV) or space/break-separated values (BSV or TXT). It also exports data in column format for other systems. However, it requires numeric values (missing values should be entered as –9), and can produce unexpected results if there are empty columns.

4.2.2 Location data formats

Comprehensive analysis software will let you select particular categories of location for display and analysis from within large files that contain data from many animals. You may want to display males in one colour and females in another, or separate adults from juveniles, or overlap home ranges recorded in summer and winter, or omit locations recorded at night. To be able to make these selections, you will need to file variables that describe each pair of coordinates.

```
271 2 1 7 1995 100 64
99 65 3 950709 18 15 101 65 1 950709 19 43 104 64 1 950710 8 34 101 66 1 950710 9 35
98 62 1 950710 10 49 92 62 1 950710 11 58 98 64 1 950710 15 8 98 65 1 950710 16 11
99 63 1 950710 17 27 98 62 1 950710 18 47 97 62 1 950710 19 49 72 62 1 950710 21 41
92 61 2 950711 5 24 96 62 2 950711 6 23 100 63 2 950711 7 26 103 63 2 950711 8 56
98 63 2 950711 10 13 100 64 2 950711 11 29 99 62 1 950711 13 12 99 64 2 950711 14 14 -1 -1

369 2 2 7 1995 96 70
94 66 1 950709 18 21 92 66 1 950709 19 50 92 66 3 950710 8 45 95 67 2 950710 10 0
75 60 3 950710 11 0 92 66 1 950710 12 32 92 68 2 950710 15 14 94 66 1 950710 16 15
75 65 2 950710 17 44 72 65 1 950710 18 59 73 65 1 950710 20 0 72 62 1 950710 21 52
97 70 1 950711 5 16 94 67 1 950711 6 25 97 70 1 950711 7 29 92 64 1 950711 9 12
96 69 1 950711 10 15 96 70 1 950711 11 32 93 66 1 950711 13 15 95 67 1 950711 14 19 -1 -1

-1 -1

~ 0 3 2 4 72 104 60 70 2 10 0 20 0 0 0 0 0 0 1 10 J Y A M F
activ YYMODD HH MI 1-3 950709-950711 5-21 0-59   0
```

Figure 4.4. Location data from two blackbirds in RANGES V format. For each bird, the first row is a label (ID, age-code, sex-code, start-month, start-year and nest coordinates), and following rows contain a series of location coordinates and their qualifying variables. An appendix (starting with a tilde ~) contains a row of 20 values that define scale, labels and other properties, followed by definitions for location qualifiers.

271	*2*	*1*	*7*	*1995*	*100*	*64*	**99**	**65**	3	950709	18	15
271	*2*	*1*	*7*	*1995*	*100*	*64*	**101**	**65**	1	950709	19	43
271	*2*	*1*	*7*	*1995*	*100*	*64*	**104**	**64**	1	950710	8	34
271	*2*	*1*	*7*	*1995*	*100*	*64*	**101**	**66**	1	950710	9	35
271	*2*	*1*	*7*	*1995*	*100*	*64*	**98**	**62**	1	950710	10	49
271	*2*	*1*	*7*	*1995*	*100*	*64*	**92**	**62**	1	950710	11	58
271	*2*	*1*	*7*	*1995*	*100*	*64*	**98**	**64**	1	950710	15	8
271	*2*	*1*	*7*	*1995*	*100*	*64*	**98**	**65**	1	950710	16	11
271	*2*	*1*	*7*	*1995*	*100*	*64*	**99**	**63**	1	950710	17	27
271	*2*	*1*	*7*	*1995*	*100*	*64*	**98**	**62**	1	950710	18	47
271	*2*	*1*	*7*	*1995*	*100*	*64*	**97**	**62**	1	950710	19	49
271	*2*	*1*	*7*	*1995*	*100*	*64*	**72**	**62**	1	950710	21	41
271	*2*	*1*	*7*	*1995*	*100*	*64*	**92**	**61**	2	950711	5	24
271	*2*	*1*	*7*	*1995*	*100*	*64*	**96**	**62**	2	950711	6	23
271	*2*	*1*	*7*	*1995*	*100*	*64*	**100**	**63**	2	950711	7	26
271	*2*	*1*	*7*	*1995*	*100*	*64*	**103**	**63**	2	950711	8	56
271	*2*	*1*	*7*	*1995*	*100*	*64*	**98**	**63**	2	950711	10	13
271	*2*	*1*	*7*	*1995*	*100*	*64*	**100**	**64**	2	950711	11	29
271	*2*	*1*	*7*	*1995*	*100*	*64*	**99**	**62**	1	950711	13	12
271	*2*	*1*	*7*	*1995*	*100*	*64*	**99**	**64**	2	950711	14	14
369	*2*	*2*	*7*	*1995*	*96*	*70*	**94**	**66**	1	950709	18	21
369	*2*	*2*	*7*	*1995*	*96*	*70*	**92**	**66**	1	950709	19	50
369	*2*	*2*	*7*	*1995*	*96*	*70*	**92**	**66**	3	950710	8	45
369	*2*	*2*	*7*	*1995*	*96*	*70*	**95**	**67**	2	950710	10	0
369	*2*	*2*	*7*	*1995*	*96*	*70*	**75**	**60**	3	950710	11	0
369	*2*	*2*	*7*	*1995*	*96*	*70*	**92**	**66**	1	950710	12	32
369	*2*	*2*	*7*	*1995*	*96*	*70*	**92**	**68**	2	950710	15	14
369	*2*	*2*	*7*	*1995*	*96*	*70*	**94**	**66**	1	950710	16	15
369	*2*	*2*	*7*	*1995*	*96*	*70*	**75**	**65**	2	950710	17	44
369	*2*	*2*	*7*	*1995*	*96*	*70*	**72**	**65**	1	950710	18	59
369	*2*	*2*	*7*	*1995*	*96*	*70*	**73**	**65**	1	950710	20	0
369	*2*	*2*	*7*	*1995*	*96*	*70*	**72**	**62**	1	950710	21	52
369	*2*	*2*	*7*	*1995*	*96*	*70*	**97**	**70**	1	950711	5	16
369	*2*	*2*	*7*	*1995*	*96*	*70*	**94**	**67**	1	950711	6	25
369	*2*	*2*	*7*	*1995*	*96*	*70*	**97**	**70**	1	950711	7	29
369	*2*	*2*	*7*	*1995*	*96*	*70*	**92**	**64**	1	950711	9	12
369	*2*	*2*	*7*	*1995*	*96*	*70*	**96**	**69**	1	950711	10	15
369	*2*	*2*	*7*	*1995*	*96*	*70*	**96**	**70**	1	950711	11	32
369	*2*	*2*	*7*	*1995*	*96*	*70*	**93**	**66**	1	950711	13	15
369	*2*	*2*	*7*	*1995*	*96*	*70*	**95**	**67**	1	950711	14	19

Figure 4.5. Here, the data in Figure 4.4 are in spreadsheet format. The label is repeated for each location (first seven columns, in italics), followed by the location coordinates (bold) and qualifying variables.

These variables will fall into two main categories. One category will label the data for each individual animal within a file that may contain many individuals. If the data have been recorded as home ranges with a standard number of locations in each, it may be convenient to have range labels that separate not only an identity code, sex and age, but also the month and year in which the range was recorded. RANGES V can use up to five range label variables, and also accepts a pair of coordinates for a focal location, perhaps for a nest used at the time. Even if the data represent a continuous record for one animal, rather than separate seasonal ranges, it is often convenient to file the location and date where the animal was first marked or released. These range labels occur once at the start of a RANGES V file (Fig. 4.4) but occupy seven columns of a spreadsheet (Fig. 4.5).

The data shown in Figs 4.4 and 4.5 also contain a second category of four variables that describe each location record (called *fix qualifying variables* in RANGES V). They are a date, in the format YearMonthDay (YYMODD), a time in hours, a time in minutes and an activity code. Time variables are necessary for autocorrelation analysis, and all the variables can be used in RANGES V to select data for display and analysis. You might, for example, want only to plot the locations during hours of daylight, or when birds were not flying.

Give careful consideration to the format of qualifying variables. If your software accepts concatenated dates, the format YYMODD enables simple selection of locations with date sequences that transition across years. Thus the sequence defined by 981201–990131 selects all locations between 1 December 1998 and 31 January 1999 (note the use of 991201–20000131 for the next year). If you never need specify transitions across the years, you may prefer separate filing of YY from MODD to enables selection of sequences like 0715–0815 for mid-July to mid-August in any year, or only for 98–99. The same effect is obtained by filing YY and DD, where DD is Julian days (1 February is day 32). Separation of YY, MO and DD makes it easy to select for any month, without needing to repeat the specification for each year. However, it may then not be possible during one run to select only the days between mid-December and mid-January, unless the software lets you split and merge files easily. In RANGES V, you could rapidly create two new files, by splitting data for 16–31 December into one file and for 1–15 January into the second, and then merge them to give the required spread of dates.

Other qualifying variables may code for activity, or habitat, or social factors or accuracy of the locations, but format remains an important consideration. For example, it may be useful to have numeric codes that follow a particular sequence. If you use habitat codes 1 for grassland, 2 for crops, 3 for hedgerows and 4 for woodland, you could select 1–2 for all locations in fields, or 1–3 for all outside woodland. It can be useful to code activity in a similar way, on a scale from inactive to maximum movement.

Scaling can be useful for saving time when filing data, provided that your software supports the possibility that locations and maps have different scales (see section 4.3). You may find, however, that your data entry system only provides Universal Transverse Mercator (UTM) coordinates in metres, or only gives degrees and minutes of longitude and latitude. Some systems now offer both UTM coordinates or degrees and minutes. If not, two approaches can be used to translate longitude-latitude data to UTM coordinates. If your animals are travelling further than about 5° near the equator (or 2–3° near the poles), for example during migration, you must use a programme that converts precisely (Cooper, 1994) in order to take account of the earth's curvature.

On the other hand, for animals moving short distances, little accuracy is lost by a simpler approach. In this case, use a spreadsheet to divide minutes by 60 (and seconds by 3600 if necessary), and add to the degrees so that all units are decimal degrees. Next, subtract the smallest number of degrees of longitude from all the easting coordinates, and of latitude from all the northing coordinates. Finally, multiply the degrees of longitude by the number of metres across each degree of longitude at the latitude of your study area; similarly, multiply the degrees of latitude by the number of metres northward between degrees at the given latitude. For example, coordinates of 64° 15' E, 30° 57" N and 64° 12' E, 31° 3' N would first convert to 64.25° E, 30.95° N and 64.20° E, 31.05° N, and truncate (by subtraction of 64° E, 30° N) to 0.25° E, 0.95° N and 0.20° E, 1.05° N. If there were (for simplicity) 90 000 m across a degree of longitude and 80 000 across a degree of latitude at 64° E, 31° N, the resulting coordinates in metres would be 22500, 76000 and 18000, 84000, or 225 760 and 180 840 if you were scaling in hundreds of metres.

Even if the scale is decided for you, for example by a data-entry system that only provides UTM units, you need to think about the resolution of your tracking when you create files. This is because tracking resolution will be used during analyses (at least in RANGES) to set the size of grid cells, of boundary strips round polygons, of distance offsets from grid intersections in harmonic mean contouring and of habitat circles round locations (see Chapter 9).

You need to plan the collection and filing of data very carefully to ensure that you are efficient while missing nothing of importance. This will involve practising with the display and analysis software, using example files and dummy entry of the sort of data you intend to collect, to ensure that you record appropriate material in the field. Even if you prefer a spreadsheet for your master database, it is wise to decide in advance what columns you will use for labels and location qualifiers.

4.2.3 Telemetry data

Logging software may give you little choice of the format for telemetry data. However, there may well be other preparations needed before tracking, which will give you a chance to test the analysis facilities that you have or may need to obtain. A typical example is the need to calibrate some types of tag before they are used.

Variation between tags in the consistency of response to temperature can be expressed in terms of the slope and elevation of a calibration plot. Perfectly consistent tags would each have identical calibration plots, but this cannot be assumed. Early tags, with temperature-sensitive crystals and especially with mercury or silver batteries that were also temperature sensitive, could show appreciable variation in both the slope and the elevation of the plot. Any tags may have calibration plots that vary slightly in elevation according to whether they are warming or cooling.

It is reasonable to adjust the extent of calibration to the accuracy of the data required. Thus, to define physiological cycles from core body temperatures would require high accuracy, perhaps of ±0.1°C, of internal tags. Estimating microclimate at foraging locations, from distal sensors on external tags, could well require accuracy of only ±1°C. Detecting relatively extreme temperature changes that indicate torpor or death might require even less accuracy.

If you require high accuracy, you will always require some calibration work, unless you are confident to trust a manufacturer's guarantees of very high repeatability between tags. Your initial calibration work will be to decide whether the coefficient of variation of slopes and elevations falls within your required accuracy criteria over the likely spread of temperature. If it does not, you need to calibrate every tag individually over that temperature spread, noting the possible need to calibrate after cell voltage has stabilised and possibly in a dummy animal. If you find during initial tests that the coefficient of variation of slopes is within your criteria, but not the elevations, you may be content thereafter to calibrate at a single temperature. If variation within both slope and elevation are acceptable, you may decide against further calibration, but would be wise to check again if you later recover tags, and to check again for each new batch from the manufacturer.

A different type of calibration will develop in the future, using software to interpret animal activities not merely from signal widths and intervals, but by particular sequences of signal from posture sensors and other means of modulating pulses. Software may eventually become available to interpret sounds and even images to record automatically when animals fly, groom, feed, mate, etc.

4.3 MAPS AND GIS

If the use of software has matured during the last decade, the use of digital maps has just about started from scratch. A number of different options have become available to obtain maps suitable for a variety of different tasks. This section outlines the advantages and drawbacks to these options, which include digitising your own map, preparing a map from remote images or obtaining a map from another source.

4.3.1 Vector and raster maps

A vector map is composed of lines or curves drawn between points, to create shapes that may or may not be filled (Fig. 4.6A). The lines have no width, so shapes can be defined in detail at any scale. In contrast, a raster map is an array of identical shapes (typically square) with a defined size (Fig. 4.6B). It cannot depict detail below that size.

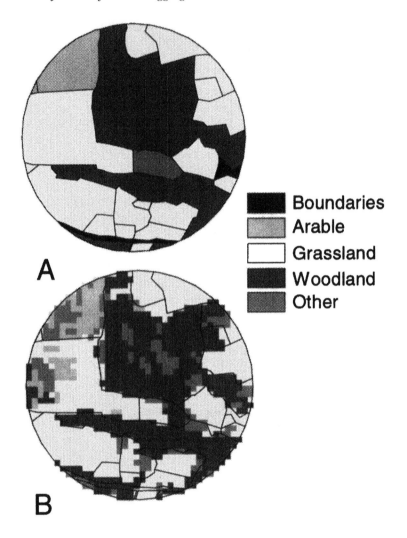

Figure 4.6. Maps of vector data (A) and rasters (B) within the same circular buffer (500 m radius). The 25 m rasters have vector lines superimposed to show how woodland and fields derived from Landsat images relate to boundaries digitised by hand (from Hodder 2000, with permission).

In principle, vector maps are simple to prepare with a digitising system. Unfilled vector maps provide a convenient two-tone background against which to display detailed patterns of movements or home-range shapes in colour. Filled vector shapes are suitable for habitat analyses, and it is relatively simple to change the attributes of field shapes as crops are changed or of forest compartments as trees age. Vector maps typically require less storage space than raster maps for a given degree of detail, but also require much computation to fill, which can make complex vector maps slower than raster maps for habitat analyses.

The computation problem can be avoided, if there is adequate file space, by converting a vector map to rasters at a scale that provides adequate detail. However, vector maps with many, complex shapes are tedious to produce, and surveying to produce them is prohibitive if there are many thousand shapes. Maps covering many square kilometres therefore tend to be prepared from remote images, which are collected in raster format.

Whatever map you use, be careful to check that the files used for the map and animal locations have the scales set correctly. Otherwise you may spend a long time wondering why one part of the combined image is far away in a corner, or seems to be absent altogether.

4.3.2 Acquiring vector maps

The ability to change the attributes of different shapes makes vector maps convenient for land management. They can therefore sometimes be obtained from farms or forest companies. If so, try to get them in a format that you can edit or add to, because they may not give all the detail you require. It is also often possible to obtain vector versions of maps that contain major landscape features and even fields from a national mapping authority. Be cautious of these, however, because a great deal of work with a sophisticated GIS may be required to create and label polygons from lines that were not originally intended for that purpose.

If you decide to digitise a map, do so with as much detail as you can manage, because it is hard to increase detail at a later stage. However, note that you will probably not manage to digitise more than 300–500 fairly simple shapes, such as field boundaries, in a day. Superhuman dedication is required to continue such work for much more than a week. Moreover, the survey for such a map may itself require careful planning, even if it is only to enter crop details of fields for which you already have an outline map.

The choice of digitising software is important. Use a package that is simple to learn and can easily export your shapes, probably as a DXF (Data eXchange Format) file because that early format from Autocad is widely supported. RANGES V can be used to digitise maps, but edits them less easily than in general-purpose GIS packages. However, it imports DXF files and treats any shape labels as habitat types. Analyses of resource use can therefore be based on maps prepared in other packages. Whichever software you use, be careful to get a digitising tablet that is accurate enough for mapping; the least expensive tablets may warp shapes.

If you digitise a vector map in ARCVIEW or ARC/INFO and want to export complete polygon shapes, you must digitise as complete shapes (clockwise, and with any holes joined anticlockwise to edges for RANGES V). This is because you may need additional software to export polygons built by ESRI routines. Lines are exported as DXF files, with a label for each habitat type. From

ARC/INFO, to which files can be imported from the more widely-used ARCVIEW, the command sequence for exporting export .DXF files is:

1. Digitise complete polygons clockwise to create *myshape*.shp
2. Convert the *myshape* to an ArcInfo coverage using: SHAPEARC <infile> <cover>
3. Create line topology using: BUILD <cover> line
4. Create .DXF file using: ARCDXF *myshape*.dxf <cover>

If you only use the 'build <cover>' default command at stage 3, ARC/INFO will export your polygons after splitting then between nodes that it assigns. If you have problems exporting polygons, you can export a raster map from ARC/INFO in GridAscii format, which RANGES V will import (see next section).

If you need to digitise a map that has several layers, think about creating shapes that contain a single habitat no matter which layer is concerned. In other words, even though the ground layer throughout an area is grass, create two shapes if there is a canopy layer of deciduous trees in part of it. In this way, you only need to digitise the whole map once, and then change some of the polygon labels for each layer. Such shapes can also be labelled for analysis of layer combinations. For example, you could have grass-with-canopy and grass-without-canopy.

4.3.3 Acquiring raster maps

It is sometimes imagined that a remote image, for example a Landsat Thematic Mapper scene (185 × 185 km from Eros Data Center, costing US$425 if more than 10 years old, otherwise US$4500), is a habitat map. Such an image is in fact only a map of the reflectance recorded for each raster, measuring about 40 × 40m for Landsat's TM data. A suitable GIS or even a graphics package may show the scene as shades that you, from your knowledge of the area, know in some cases to represent woods, or water, and so on. The map is showing landcover, rather than true habitats, because it cannot show the shrubs under the tree canopy. However, for types of landcover that give very distinct reflectances, you might be content to digitise round the shapes to obtain a map suitable for elementary habitat analyses.

If you want a raster map that indicates a variety of landcover types, you have much more work to do. You will need software that can seek groupings of similar reflectance within the image, using statistical techniques like cluster analysis. Fortunately, the US Army's GRASS GIS provides these facilities without charge on computers with Unix-type operating systems (e.g. Linux). You will need to visit sites within each grouping of reflectances, almost certainly using GPS with differential correction, and decide what landcover the grouping reflects. This will provide landcover labels for those groupings, which can be applied to the map. At that point, you will be in a position to pick random points from the map and check the accuracy of the classification, and repeat the previous steps until you are content with the result; at best, classification accuracy is typically 80–90%.

If you use a satellite image, be careful to get one without clouds. An image from one season may provide up to about 10 landcover types in a particular scene.

Comparative work, using images from different seasons, may provide another 5–10 landcover types. For example, woodland may be separable into coniferous, which gives the same reflectance year-round, and deciduous, which changes reflectance in winter. If you want your classification to apply for different scenes, you will have to do further work to correct for changes in ambient light when each scene was imaged.

If you are very lucky, someone else may already have prepared a suitable landcover map. The UK Institute of Terrestrial Ecology prepared a map of great Britain with 25 landcover types in the early 1990s (Fuller *et al.*, 1994), and is repeating the process 10 years later. An initial classification and network of ground-truth locations simplifies repetition. With rasters that are 25 m square, the map provides scanty definition for features like roads and rivers, but is ideal for work with large areas of habitat, and thus for study of wide-ranging species such as large raptors (Kenward *et al.*, in press). Imaging systems with higher resolution should make such maps suitable for a wider range of species.

Careful planning is needed to use large raster maps. Some general purpose GIS only handle maps in memory, and therefore cannot work with large raster maps. The most efficient format for rapid analysis uses one byte to provide a label of 0–255 for each raster, so a file of 100 × 100 km with rasters 25 m square requires 16 Mb for the raster array. A saving for storage can be made by file compression, for example by using every two bytes to run-length-encode up to 255 consecutive identical rasters within a row. However, rapid analysis requires uncompressed maps, so that land cover can be found at any location by random access.

RANGES V can use byte array maps of any size, because it does not try to take the whole file into memory. It imports raster maps from ARC/INFO in GridAscii format, which are large because they use space-separated ASCII codes for the rasters. RANGES V therefore converts the maps to byte arrays, which are half to a third the size of the GridAscii files, and adds labels so that colours and names or values can be attributed to raster codes after import. Ascii habitat codes must not be >99, and the GridAscii files are smallest and easiest to label if you use consecutive numbers from 1 (i.e. 1,2,3,4 . . . not 1,7,23 . . .), keeping 0 for unclassified. Raster maps are often much larger than vector maps, especially if a fine scale is used to retain detail from an original vector map. However, raster maps are processed relatively fast in RANGES V, because computing is simpler for raster rectangles than for complex vector shapes. Moreover, the analysis output is spreadsheet-ready. You can also export home range outlines from RANGES V as DXF files for use in ARCVIEW or ARC/INFO, but must then learn Arc Macro Language and to handle the less convenient data output from each operation.

4.3.4 Maps in the field

Although programs like LOCATE II are available for plotting data on maps in the field, biologists still often work with maps on paper or other fabric. This is partly because fabric maps can be used in rain, after being dropped and without power supplies, and partly because computer screens have limited resolution. The number of dots in just one inch on a good quality map is comparable with the number of pixels to span the whole screen of current computers. High resolution is necessary to estimate accurately the location of sites from which you take bearings, and then to determine whether an estimated tag location may require an immediate revision, either to correct a mistake (e.g. rabbits in lakes) or to determine the habitat really being used by an animal close to a habitat boundary.

Computer maps can be enlarged to give the required accuracy, at least with a vector map, but that can make it difficult to keep several bearing sites and the tag position on the screen at the same time. At present, computer maps used in the field are most suitable for small areas, or for large areas with little detail. Nevertheless, improvements in technology will help to realise the advantages of computer maps, such as the scope for: (i) interfacing with GPS receivers and electronic compasses to plot bearing sites and bearings automatically; (ii) estimating accuracy criteria so that inadequate tag locations can be re-estimated immediately; (iii) convenient filing of locations and (iv) real-time analyses.

If you are using fabric maps, you can prolong their life by covering them with plastic sheeting. Bearings and other notes can be drawn with marker pens and washed off later with a suitable solvent. Large maps can be cut to fit in loose-leaf books of plastic folders. If you are working in a small area, you may fasten the map to the cover of a field note-book, or to a board or both sides of a board for a larger map, and cover with adhesive plastic film to prolong its life.

4.4 CONCLUSIONS

1. Each equipment manufacturer tends to excel in a different product. It is important to seek recommendations from other users, and to place orders in good time: the best suppliers tend to be busy. Early planning is essential when registering to use the ARGOS satellite system and when obtaining licences to mark animals.

2. Tag reliability is especially important for survival studies. Transmission life estimates for 95% of tags, rather than the best ones, should be sought from manufacturers, with other data that provide evidence of extensive tag testing. When buying receivers, considerations should include duration of power supplies, ease of changing them, performance in poor weather and scope for servicing in emergency.

3. Software is available for analysing data on activities, survival and locations. Its use must be planned from the start of a study, to ensure that appropriate data are collected and that they can be processed rapidly without 'drowning'.

4. For location data, decisions must be made about whether to use dedicated software or a spreadsheet for recording data. If a spreadsheet is used, import to chosen analysis software should be planned and tested, checking date formats and conversion of degrees to UTM or other rectangular coordinate systems if necessary.

5. Analyses of resource use can be completed in dedicated software, but maps for analysis of resource-use are most easily prepared in other geographic information systems. Porting maps to dedicated software may simplify extensive analyses, but some analyses may require transfer of range data to the GIS. Ease of data transfer should be checked in advance.
6. Vector maps of habitat shapes are more compact and scalable than raster maps, and most suitable as outline backgrounds when collecting and editing data, but are tedious to prepare over large areas and often slower in analyses than raster maps. If GIS software does not export vector maps, rasters can often be exported after vector-to-raster conversion.

5 Making Tags

In many projects, it is convenient to ask tag manufacturers to put their expertise into building suitable tags. However, completing your own tags can save weight, be economical and avoid delays when re-using tags or changing a design, provided that you are technically competent and keep good contact with a friendly manufacturer for advice. Even if you would rather spend your time in other ways, you would still be wise to read the first two sections of this chapter, because it considers details of tag structure that are relevant when buying completed tags. The third section contains instructions for completing a number of widely used tag designs.

5.1 TAG COMPONENTS

If you have opted to save costs and gain flexibility by buying transmitters, the following three sections help you to prepare power sources and antennas, and then to complete the tags by encapsulating all these parts in shapes suitable for attachment to animals. Provision is made for the circuit to be broken, after assembly and testing, if the tag is not for immediate use.

5.1.1 Cell lead attachment

Since it is easy to overheat cells by soldering leads directly to them, it is usually best to obtain cells with solder tags. If small cells are overheated locally, in the area where a lead is attached, they may be damaged such that their capacity diminishes and the tag's life is thereby reduced. More severe overheating can lead to total failure of the cell, due to the establishment of internal short circuits, or even to the cell exploding. However, leads can be soldered to even the smallest cells using flux-cooled soldering. You will need fine solder, an iron with a small bit, insulating tape, a pair of snipe-nosed pliers, tinned copper wire of an appropriate thickness for the leads, and a few ml of 10% phosphoric acid. You should wear protective goggles.

You also need a pair of heavy-duty pliers set up in a vice as a third hand. These pliers are used as a heat-sink, to prevent overheating the cells. To heat-sink button cells without short-circuiting them between the jaws of the pliers, one of the opposing jaws should be wrapped in insulating tape. It is best to attach the first lead to the curved side of the cell casing, which is the anode of silver-oxide cells. Put the cell in the pliers with its casing gripped against their uninsulated face (Fig. 5.1A), being careful that the opposing electrode does not also contact an uninsulated part of the pliers while you do this. Then place a 1 mm diameter drop of 10% phosphoric acid on the cell's surface where the lead will be attached, as far as possible from the seals between anode and cathode. Use a capillary tube or a loop of wire dipped in the acid (Fig. 5.1B), which serves as a flux and to prevent local overheating. Hold the lead with one hand against the cell's surface in this spot of flux. The iron is held in the other hand, wiped clean on the soldering stand's damp sponge and tinned with a very slight excess of solder, then immediately wiped quickly against the lead and the cell surface where these meet. The iron should be used with a rubbing rather than a dabbing action, and should not contact the cell for more than half a second. There will be a brief hiss as flux evaporates, removing excess heat from the cell surface as latent heat of evaporation. You should immediately test the cell surface temperature and reject it if greater than 60°C. The soldered joint to small silver cells should be no more than about 2 mm in diameter, perhaps somewhat larger to ensure firm lead attachment to large cells. The joint should be thoroughly wiped to remove any traces of phosphoric acid, which might eventually corrode the wire lead.

In joining the second lead to the central electrode (Fig. 5.1C), again take care that the cell is not tilted such that the edge of the casing or the first lead makes a short circuit against bare metal. If the first lead was for any reason soldered to the flat base of the casing, rather than to its curved side, beware of any projecting point of solder which might pierce the insulating tape to cause a short circuit while the other lead is being attached.

An alternative lead attachment technique for very small cells is the use of conducting paint or epoxy. The tinned lead wire is wound in a tight flat spiral, to

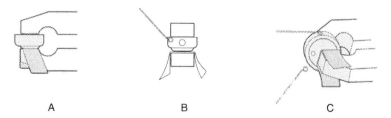

A   B   C

Figure 5.1. Flux-cooled soldering for small cells. The cell is gripped in pliers, with the negative terminal against a layer of insulating tape (A). A fine wire loop puts a drop of 10% phosphoric acid on the casing (B), and the positive lead is soldered. The process is repeated for the negative lead (C).

ensure a large area of contact with the cleaned cell surface, and adhesive or paint is smeared on both surfaces before they are brought together. The joint can then be clamped to harden in an insulated holder, protected if necessary from glue or paint with a sheet of polythene or other non-stick plastic, and taking care that the lead from a central electrode is not pressed against the cell's outer casing. This is a slower technique than flux-cooled soldering, and tends to produce a weaker join, with a greater weight of wire and attachment materials, but is less likely to damage the cell.

Small button cells can sometimes be obtained with solder tags already attached by spot-welding. However, spot-welding often reduces the capacity of the smallest cells. Skilful flux-cooled soldering may well be the best way to attach leads to button cells. After some practice, cells can be soldered directly to transmitter leads with no appreciable warming of the cell, even without pliers as a heat-sink.

5.1.2 Diode isolation

Sometimes two cells must be used in parallel to meet tag shape or life requirements. For instance, a bird might be able to carry a 30 g tag, but not of the length obtained with an AA cell. The best option might then be to use two 1/2AA cells in parallel. However, simply connecting two cells in parallel would not necessarily double the life of the tag. If the cells discharge unevenly, such that one is exhausted well before the other, the second cell will then discharge quickly through the circuit formed by the exhausted cell. This can be prevented by isolating each cell with a germanium diode, as in Fig. 5.2. Silicon diodes should not be used because they have a higher voltage drop (0.7 V) than germanium diodes (0.3 V). The voltage drop can be further reduced by using two diodes in parallel. Since diodes increase tag size, and create too great a voltage drop with 1.5 V cells, they are not worth using in the smallest packages. However, they fit easily in the space between most lithium cells and should certainly be included in large tags with cells in parallel.

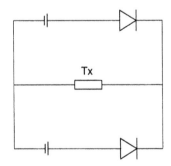

Figure 5.2. Parallel cells in transmitters should be isolated with germanium diodes.

5.1.3 Solar panels

A germanium diode should also be used to isolate the solar panels in tags that include a rechargeable cell (Fig. 5.3). Without a diode, power will be drawn from the cells by the panels in the dark. Unfortunately, even the 0.3 V voltage drop of germanium diodes is enough to require a slightly larger solar panel area than without a diode.

A particular problem with these tags is the inadequate charging of their cells during short winter days at high latitudes. Although it may not matter if a tag stops transmitting towards the end of the night, rapid discharge of the NiCad cell will minimise the chance of finding dead animals if their tags are poorly lit. To be on the safe side at high latitudes, a tag should charge enough in typical habitat on a dull day to run for at least 4 days unlit.

To protect solar panels from damage by the animal or its surroundings, they should be encapsulated with an overlying layer of hard transparent plastic. If the tags are mounted on birds, there should be enough stiff fabric or thin plastic sheeting around the tag to prevent feathers overlapping the panels.

5.1.4 Whip antennas

Transmitter whip antennas are usually single strands of nickel or high-tensile alloy, or multiple strands of stainless steel. The best materials for whip antennas on the smallest birds, bats and reptiles are new super-elastic nickel-titanium alloy wires starting at about 0.15 mm (0.006 in), which are light and highly flexible but do not kink easily. However, nickel guitar-string wire of 0.20–0.25 mm (0.008–0.010 in) diameter is also an effective antenna. NiTi wire should be this thickness for species that go in dense vegetation, because the thinner material can tangle.

Plain guitar wire in thicknesses up to 0.55 mm (0.022 in) can be used as an antenna for backpacks on larger birds, but is relatively inflexible, tends to kink and

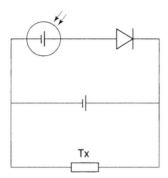

Figure 5.3. Solar-powered tags with rechargeable cells should have a germanium diode to prevent current drain through the solar panel when there is little light.

also eventually to let water into long-life tags as a result of corrosion and capillary action. A more suitable material for larger whip antennas is nylon-coated multistrand stainless steel fishing trace, the finest of which has a breaking strain of about 4.5 kg (10 lb). Fishing trace with 10–90 kg (25–200 lb) breaking strain makes good antenna material for birds that are liable to attack antennas with their beaks, such as corvids or raptors. Multistrand steel cable for bicycle gears or brakes may be used on the most powerful birds, with a coating of tough melt-lined heat-shrink tubing (e.g. SCL from Rayfast). Determined birds with powerful beaks, such as some parrots, may eventually destroy any antenna, if not the tag itself.

Whip antennas are also frequently used on collars for mammals. Small cable-tie collars use fine stainless-steel trace, either held round the cable tie with glue-lined heat-shrink tubing, or protruding as a short whip at the top. For heavy collars, of leather, machine-belting or braided nylon, stainless steel cable coated with SCL heat-shrink is appropriate. Garage-door cable is used for large mammals.

The whip antenna wire on leg-mounted or tail-mounted tags must be light yet strong, and tolerate a tremendous amount of flexing. The best material seems to be 49-strand steel cable of a type used for control cables in light aircraft. This can be coated with heat-shrink tubing to prevent 'fraying', and to reduce flexibility: if these antennas are too flexible, they whip round and snag on branches or wire fences.

Nickel wire takes solder readily, but this is not always true of alloys or multistrand stainless steel. After stripping a short length of nylon coating from the end of multistrand fishing trace, the bare wire should be dipped in 10% phosphoric acid flux. It usually then tins without too much difficulty. A strong pair of clippers is needed if these materials are to be cut without damage to the blades, especially for heavy cables. To prevent the strands being forced apart when they are cut, heat the cable at the chosen point and allow solder to flow into it. The cable then cuts cleanly and is pre-tinned.

One of the most common problems with whip antennas is breakage, through metal fatigue and/or corrosion, at the point of emergence from the potting material. There are three main ways to prevent the bending strain being concentrated at this point. The oldest approach is to set a 20–30 mm (1 in) long compression spring into the potting, so that the antenna passes through the spring as it emerges (Fig. 5.4A). Springs from ballpoint pens can be used for the smaller antennas. However, springs can be deformed by the animal, resulting in antennas losing their desired alignment. A second approach is to form a slender cone of a sealant material. Silicone sealant, such as that used for caulking baths, can been used for cones. However, it gives off acetic acid on curing, so bare wires must have a sleeve of tubing to present direct contact with the sealant (Fig. 5.4B). Hot-melt glue (e.g. T.238/12 from Itchen Adhesives), which is dispensed from a heat gun and can be shaped with moist fingers as it cools, makes flexible cones except in cold climates: this material hardens in cold temperatures. Another alternative for fine wire antennas is a thin cone of Plastidip (addresses in Appendix I). The

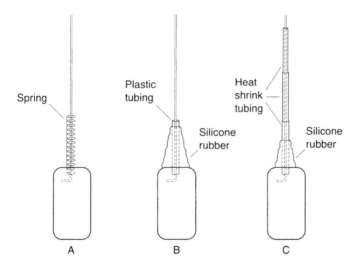

Figure 5.4. Whip antennas can be supported at their emergence with springs (A), with a cone of silicone rubber (B), or with layers of heat-shrink tubing.

third way of spreading the strain at antenna emergence points is to use concentric layers of flexible heat-shrink tubing (Fig. 5.4C). The finest tubing extends about halfway along the antenna, the next finest for one-third of the way and a cone of sealant or hot-melt glue completes the structure.

5.1.5 Loop antennas

Tuned collar loops on small mammals may be made from multistrand copper wire with a thick coating of PTFE or other very tough plastic. In one design, the collar is made as a continuous loop, with the wire folded in a stretched 'figure-8' inside a 1–1.5 cm length of heat-shrink tubing at the top of the collar (Fig. 5.5). Once slipped over the animal's head, the ends of the figure-8 are teased out from the heat-shrink to reduce the collar's circumference, before the tubing is set with a soldering iron or heated penknife blade (see Chapter 6).

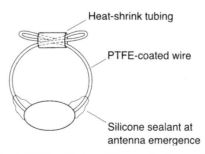

Figure 5.5. A 'sliding-8' closed-loop collar for small mammals.

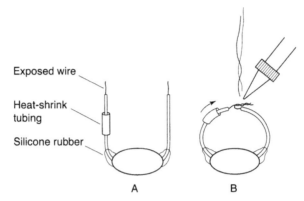

Exposed wire

Heat-shrink
tubing

Silicone rubber

A   B

Figure 5.6. An open-loop collar for small mammals. Insulation is removed from the antenna ends (A), which are then twisted together and soldered round the animal's neck (B) before being covered with heat-shrink tubing. Use a strip of paper under the wire to protect the animal while you make the join.

Small-mammal tags can also be made with an open loop of wire. With fine wire, a 1 cm length of heat-shrink tubing is slipped over the wire on one side of the collar, and the insulation on both sides is stripped to the circumference of the animal's neck (Fig. 5.6A). The wires are then twisted together and soldered, before the heat-shrink tubing is warmed over the join (Fig. 5.6B). Larger open-loop collars can be made with multistrand brass cable, such as that used for hanging pictures, and closed with a metal crimp instead of solder. The thick collar wire gives greater radiation efficiency than thin wire at a constant loop circumference, but should be covered with heat-shrink tubing or other coating to protect the animal's neck (Fig.

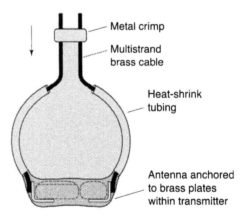

Metal crimp

Multistrand
brass cable

Heat-shrink
tubing

Antenna anchored
to brass plates
within transmitter

Figure 5.7. Crimp closure of an open-loop collar. The metal crimp is slid down to the required collar circumference, and then closed tightly. The protruding ends of the collar are trimmed, and may be soldered together beyond the crimp to maintain a good electrical connection.

5.7). On all the wire loop collars, but especially with the thicker wire, the points where the collar emerges from the potting should be protected from stress and moisture penetration with a cone of hot-melt glue. It is also important to anchor the wires thoroughly within the potting, for example by soldering to small brass strips.

Larger tuned-loop collars are best made with brass strip. Brass of 0.15 mm thickness can be used for the smaller collars, with a width of 4–8 mm, depending on the shape of the animal's neck (the wider strip being the more efficient radiator). For the largest sizes, 10 mm wide strips of brass, 0.4–1.0 mm thick, are suitable. Use thick brass for fixed circumference loops, to increase rigidity and radiation efficiency, and the thinner brass strip to give greater flexibility in collars with an adjustable circumference.

The smallest open-loop brass strip collars may be fastened by crimping, riveting or soldering the ends together, or by folding the ends over each other (Fig. 5.8A) and covering the join with stiff, adhesive-lined heat-shrink tubing. However, collars should really have their ends soldered together; otherwise the tag may transmit poorly, or fail completely, as the metal surfaces oxidise at their join. It is best to use brass nuts and bolts to join brass strips more than 4 mm wide. If the circumference is fixed, the ends are bent outwards and the nut soldered to one side, with the screw threaded to ensure correct alignment (Fig. 5.8B). Sometimes the collar must be very carefully adjusted, because the animal's head and neck circumferences are very similar. If so, solder the head of a screw inside a short brass strip on one side of the tag and drill the long strip on the other side with holes at 4–7 mm intervals (Fig. 5.8C). This keeps the inward-facing head of the screw at the side, where it is least likely to rub the animal's neck, especially if its surface is smoothed with a file.

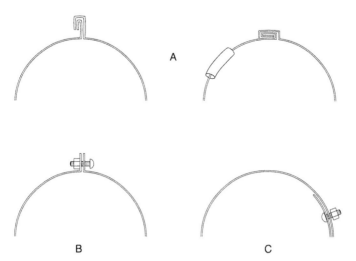

Figure 5.8. Closed-loop collars of thin brass strip can be closed by folding the strips together and covering them with heat-shrink tubing (A). Collars with a fixed circumference can be closed with a bolt at the top (B). An adjustable collar can be closed by threading a bolt, set in one brass strip at the side, through one of a series of holes in the other strip.

All holes for screws or rivets should be drilled in the collar strips and filed smooth before the brass is bent into shape. Hold the strips in a vice for drilling, not in the hand, or they may fasten to the bit as it penetrates them and cut your fingers. If the brass is first punched lightly at the point to be drilled, the bit is less likely to slide off-centre. Fasten the two sides of an open loop before you assemble the tag, using the central bolt hole if the collar has an adjustable circumference. The brass strips should be covered with heat-shrink tubing, or wrapped with 5–10 mm wide strips of fabric-backed tape to protect the animal's neck. With open-loop collars in particular, it is also best to protect the antenna emergence points with hot-melt glue.

Note that open-loop designs and brass-strip collars with variable circumferences are liable to have less efficient tuning than closed loops. They should be pre-tuned with their wires twisted together, or brass strips crimped or bolted, at the average neck circumference. Best results on animals that differ much from the average will only be obtained if the tags can re-tuned after attachment, by using a small trimming capacitor that is then covered with potting compound.

5.1.6 Coil antennas

The tightest tuning of loops is possible with those built into tags, typically on implants or where external antennas are likely to be damaged. These may be single loops round the circumference of large tags or a double loop on smaller tags. Although tuned-loop collars of fine wire are often chosen for small rodents, an alternative is to wind a length of enamelled 0.22 mm (34 s.w.g.) copper wire horizontally twice round the edges of the transmitter and cell before final potting. A subminiature trimming capacitor is used to tune this small double loop, as it would have been with the larger single collar loop. The integral antenna gives only 30–50% the range of the larger loop on a 25 g mammal, but transmission will continue if social grooming or a predator cuts the collar.

The loop is formed as a spiral on the larger implants, and may be wound on a ferritic core in the smallest implants. To keep weight to a minimum, cores should be of materials with the highest possible relative permeability to the electromagnetic flux. Any potting can affect the tuning of these antennas quite markedly, so provision should be made for a final adjustment after potting. This can be done by covering a screw slug with wax prior to potting, and leaving one end exposed so that its position in the coil can be adjusted after the potting has set. A more elegant method, however, is to use a miniature trimming capacitor, which is set in the potting with its tuning slot exposed. After tuning, the access is sealed.

5.2 TAG CONSTRUCTION

Several steps in the completion of tags from their components are common to most tag designs. These steps are considered before paying attention to individual tag designs.

5.2.1 Testing and tuning

Tag function and tuning should be tested before final potting. Potting may be done in several stages, to ensure that components are anchored together strongly and to help prevent voids (air spaces) that can later aid water penetration. However, there should be preliminary testing before it becomes too difficult to remove faulty components.

The initial test, which indicates whether the transmitter is running, can be done with an analogue microammeter (digital multimeters are not suitable). However, it will not usually reveal the presence of dry joints, in which the solder has not bonded to both metal surfaces. Dry joints are one of the most common causes of tag failure. It is particularly easy to make them when soldering transmitter leads to brass strip antennas or other large areas of metal, on which the solder may cool too quickly for bonding. These surfaces should be thoroughly tinned in advance, and you must be sure that the solder on them melts when the joint is made – holding them in pliers if necessary to protect your fingers.

Assuming that you lack a spectrum analyser, tuning involves listening to the tag signal on a receiver, without a meter in the tag circuit because this may substantially affect the tuning. A signal strength meter on the receiver can be very handy for tuning to the strongest signal, but also listen to the signal to ensure that pulses are neither becoming prolonged and 'wheepy' nor short and 'clicky'. Note that the strength of the received signal can be influenced profoundly by the relative positions of receiving and transmitting antennas, and by other objects such as yourself, especially if you touch the tag. You will probably find that the signal status changes as you tune the tag and changes again when you move your hand away. It is wise to establish standard positions for tag and receiver, probably using a short whip antenna plugged into the receiver, and to find a position for yourself by trial and error such that small movements have minimum impact on the results.

5.2.2 Switches

Apart from tags that are for immediate use, there must be some way of turning off the transmitter for storage. The simplest approach is to expose a loop in one of the cell leads as a switch. The loop is cut for storage, leaving a pair of wires that are rejoined by soldering before the tag is used. These joined wires should then be covered with a layer of potting or other sealant to prevent corrosion working along them into the tag.

For animals that may chew or otherwise damage tags, the switch wires should come to the surface where they are least likely to be attacked. On a closed loop collar, for instance, the wires are best exposed inside the loop. The same principle applies to other leads and circuit components that pass near the potting surface. Moreover, leads to brass antenna strip should be joined to the inside of the strip rather than cross its outer surface. On mammal collars, the potting tends to be chewed most at the sides, and to a lesser extent underneath, so vulnerable leads and components should be kept away from these areas if possible.

Sometimes it is impractical to pass wires through the tag's surface for switching, because there is too much risk of damage or because the transmitter and cell are completed enclosed in a moulded casing. A magnet-operated reed switch may then be used to complete the circuit. A normally-closed (NC) or changeover (CO) switch is connected so that the tag is running, and embedded near the surface of the tag. For storage, the tag can be turned off by taping a magnet to its outer surface, along the length of the switch; check with a receiver that the magnet is strong enough and in the right place. If you use these switches, be very careful to heat-shunt the leads if you solder to them close to the glass body of the switch, or else you may crack the glass. If you need small tags with neither external switch wires nor the added weight of reed switches, you will have to pot the completed circuit just before you use it.

5.2.3 Potting

Araldite epoxy resins (two-component glues) from Ciba-Geigy used to make good potting compounds for many types of tag, being compatible with electronic components, easy to apply in varying thicknesses due to their thixotrophy and setting with a degree of hardness that could be defined by the proportions mixed. Unfortunately, the manufacturer changed the formulation in a way that resulted in loss of thixotrophy and increased water penetration. Araldite is therefore no longer recommendable for radio tags. An adequate widely available replacement epoxy would be very welcome.

Dental acrylic gives an even tougher coating than epoxy adhesives. Several techniques have been developed for coating tags with acrylic, which is supplied as a liquid with a powder catalyst. The simplest method, suitable for tags of 1–2 g, is to work a thin layer of the liquid over the package, dust with powder and leave it to set. More than one layer can be used if required. A second technique is to mix a small amount of powder with the liquid acrylic, wait until it reaches a desired viscosity and then spread the resulting mix over the tag, finally dusting with more powder to speed the setting if necessary. Dental acrylic can also be mixed in a PTFE (Teflon) coated container and poured into moulds, but is less suitable than some epoxies for this purpose because it generates much heat during setting. To avoid damaging the cells or other tag components, you may need several pours if you use dental acrylic or some exothermal epoxies.

If you need to repair tags or replace cells, acrylic can be softened and removed with acetone, but this organic solvent tends to dissolve more than just the potting, as well as being a health and safety hazard. Acrylic coatings that are not too thick can be removed with wire clippers. These tools can be used for a variety of tasks, depending on their stage of wear. At first they should be kept for trimming fine copper wires during tag construction. After a time, their edges become worn and they are best set aside for cutting thicker wires, and then perhaps for cutting antennas (which really ruins the edges). At this point cutting acrylic or epoxy cannot further spoil them. As a final ignominy, they may be allocated to fieldwork: they are rather useful for mending wire-mesh traps.

Penetration of moisture is a serious problem for tags. Penetration is aided by the presence of voids in the tag, or even by small air-bubbles. However, the most common route for water entry is along antennas, either by surface corrosion of bare wires or by capillary action between a plastic coating and the potting. If the antenna lead and antenna are of different metals, electrochemical potentials help to drive the corrosion process, and the antenna may become electrically disconnected even though it remains in place. It is therefore essential to caulk these emergence points if tags are to last long in damp conditions. Water can also penetrate tags along straps, tubes and attachment cords or threads, but potting electronic components before adding attachment material reduces this problem.

In the case of implants and ingested tags, it is important not only that body fluids do not penetrate the tag, but also that the tag itself is not toxic to the animal. Although dental acrylic is relatively inert, tags should also be given an outer coat of medical grade paraffin wax or a physiologically compatible silicone compound. Tubes made of an inert plastic (e.g. polycarbonate) are suitable containers for larger implants.

A variety of other potting compounds are available. For instance, soft two-part silicone rubber may be used as a shock-absorbent filler in tags with a hard outer casing, and 3M Scotchcast is often used as a pourable potting for moulded tags. Other potting compounds should be used with care, because they may contain substances that react with component coatings, or allow the migration of charge-carrying ions.

Potting in moulds is sometimes an attractive technique for mass-producing tags. The potting compound can be mixed in a plastic bag, and squeezed out through a cut corner into plastic, silicone or epoxy moulds, with access holes left for changing cells, inserting antennas or tuning the transmitter. An even simpler system for large-mammal tags is to pot the transmitter and cell(s) in a loop formed between two strips of collar material (Fig. 5.9). The bottom of the mould (the back or front

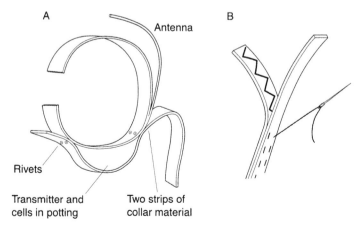

Figure 5.9. Tags for large mammals can be potted between two strips of collar material (A). The antenna can be free-standing, or fixed to the antenna with heat-shrink tubing, or sewn between two strips of the collar material (B).

of the finished tag) is formed by a plastic surface against which the strips are held. If a whip antenna is used, this passes up between the strips on the longer side of the collar. A disadvantage with moulded tags is that they usually require more potting, to ensure that the cell, transmitter and leads are adequately covered, than if the potting is spread on the tag as a thin layer. The extra weight of potting militates against using moulds for small animals.

In many tags, it is important for these tags to have cell, transmitter, antenna and attachment materials held fairly rigidly together before the outer potting is applied. Cyanoacrylate glue ('Superglue') can be used to stick parts of the smallest tags together; setting is instantaneous if sprayed with activator. However, this glue reacts to make some polymers brittle. Beware especially of its effect on cable-ties. For larger tags it is best to use small amounts of potting, which also fill up air-spaces so that the final potting can be completed well within the compound's setting time. This is especially important if the tag has been strengthened by winding with threads, for example to provide a fibre-reinforcement close to emergence of an antenna. Pre-potting is also important if sensors, such as mercury switches, are to be set in a particular position.

With any potting material, you should practise its use before applying it to tags. Note the effects on the time to start setting, the speed of setting, and with some glues the final hardness, of varying the relative amounts of the different components. With acrylic, note that setting is quite rapid if you start with much powder, but if you add more liquid and stir thoroughly as soon as the viscosity starts to increase, you can delay the setting time again. With all two-component potting compounds, it is important to mix thoroughly in order to obtain even curing and avoid uncured patches.

5.2.4 Labels

Before potting tags, transfer the transmitter's frequency label to the antenna wire, or write the frequency on the cell where it will be seen through the potting. This saves a great deal of time searching for signals later on. If tags are large enough, it may also be worth incorporating an address or 'reward' label, especially if you are studying game species. A lot of time can be wasted searching for signals from shot animals!

5.3 TAG DESIGNS

Considering the variety of power supplies, antennas and mounting techniques that are available, a vast variety of different tags can be built. Now that the general principles of tag construction have been described, it is worth covering in more detail the making and mounting of eight different designs, which can be scaled up or down to suit a very wide range of species. The mounting of these tag designs is described in Chapter 6.

5.3.1 A 1 g glue-on

The following description is for a 0.9–1.1 g tag, based on a 250 mg transmitter with a 500 mg Ag397 silver-oxide cell. The signal life is 4–8 weeks (for 10–25 ms pulses, 50/minute). The tag could weigh about 0.5 g with a 200 mg silver-oxide cell, but would then transmit for only 6–12 days. These tags are suitable for gluing directly to fur or feathers. They may also be attached with tape, tail-clips or harnesses to small birds, reptiles and amphibia.

There will be three leads emerging from the transmitter. The following description assumes that at least the positive lead can support the weight of the transmitter without bending. Start by soldering this positive lead directly to the side of the cell, which is the anode, as described in section 5.4.1. The cell's central cathode should be upwards, so that a 'switch' loop can be formed in the negative lead before this is soldered to the cell (Fig. 5.10A).

Cut the chosen length of antenna from the finest flexible alloy or guitar wire. With guitar wire, you can bend 2–3 mm at one end back to form a 'U' which will hook round the antenna lead and anchor the antenna. Super-flexible alloy cannot be bent, so the in-line join must be strong and as far as possible from where the antenna emerges from the potting. Solder the two wires. If the transmitter was not pre-tuned, give it a final adjustment now, while the cell is still active.

Put about 0.5 ml of acrylic fluid in a mixing thimble, sprinkle with powder and stir thoroughly. Holding the thickest cell lead in snipe-nosed pliers (Fig. 5.10B), use a screwdriver with a 2 mm blade to smear a thin layer over the cell and

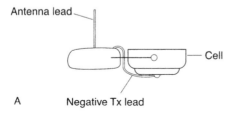

Antenna lead

Cell

A   Negative Tx lead

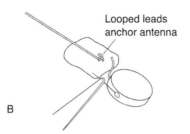

Looped leads anchor antenna

B

Figure 5.10. Making a 1.5 g tag for glue-mounting. The transmitter leads are joined directly to the cell (with flux-cooled soldering – see Figure 5.1), and the antenna held in a loop of the antenna lead for soldering. The tag can be held by the positive lead for potting.

transmitter, also working potting into the space between them. To complete the potting without getting acrylic on the pliers, move these to grip the loop in the negative lead. An artery clamp can be used to hold this loop while the potting hardens. The potting should be as thin as is consistent with complete cover. If it tends to sag appreciably, remove the excess. Any final small sag should be in the area of the antenna join and emergence.

5.3.2 An 8 g tail-mount

If built with a 1 g transmitter and 5 g 10–28 lithium/thionyl chloride cell, this tag can weigh as little as 8 g. The tag life of 7–14 months is adequate for mounting on 300 g birds (at 2.5% of body mass) that will typically moult their tail-feathers within a year. The light and flexible main antenna is attached along the shaft of one tail feather, so the tag is best for birds with relatively long tails. Tail-clips can be used on these tags, with free-standing antennas, for birds with short tails. Power output is enhanced by using an optional second 'ground-plane' antenna, about $\frac{2}{3}$ the length of the main antenna, in this case to give a total antenna length of 500 mm, which is $\lambda/4$ at 150 MHz.

To make the tag, the cell tabs are folded back on themselves, so that they do not project beyond the edge of the cell. Leave the folds slightly open at first, so that they can be used to hold the mounting ties after the transmitter leads have been soldered in place. Tin both tabs, and solder 20 mm of 0.3–0.4 mm diameter copper wire to the strip connector on the negative terminal of the cell. Tin 5 mm at the end of a 200 mm length of antenna wire, bend it at right angles and solder with another 20 mm cell lead to the connector strip on the positive terminal, using snipe-nosed pliers to hold the projecting terminal as a heat-shunt. This free-standing 'ground-plane' antenna and the 300 mm main antenna may be made of 10 kg nylon-coated multistrand fishing trace for small birds or 20–30 kg trace for larger species. Fine 49-strand aircraft cable, coated with a tough yet flexible heat-shrink polymer (e.g. Rayfast RNF 3000), is more durable as a main antenna: fishing trace usually breaks at the distal end of the supporting feather after 4–6 months, thereby reducing the detection distance. Tin the end of the main antenna and create a 'U' shape that will help to anchor it in the potting.

You have now finished soldering to the cell, so the attachment ties can be inserted and squeezed tight in each tab's fold without risk of heat damage. Fine synthetic thread for fishing nets (e.g. 30/9 Perlon) or cobblers' thread makes good ties, cut into four 30 cm lengths for each tag. Two lengths are held in the middle by each tab, so that four equal lengths hang from each end of the cell. Hold the transmitter against the side of the cell to attach the cell leads, forming a switch loop against the side of the cell. Attach the main antenna to the transmitter lead in a position where it will emerge from the potting on a level with the bottom of the tag and slightly off-centre (Fig. 5.11A), but with as little as possible of its length within 2 mm of the cell casing, to minimise detuning effects. If the tag is working, use a little cyanoacrylate glue to attach the

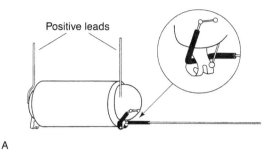

Positive leads

A

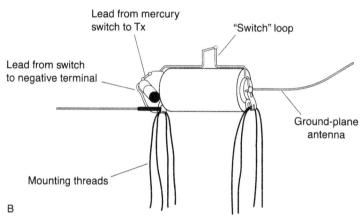

Lead from mercury
switch to Tx

"Switch" loop

Lead from switch
to negative terminal

Ground-plane
antenna

Mounting threads

B

Figure 5.11. Making a posture-sensing tag for tail-mounting. The cell leads are bent to hold the mounting threads, and the transmitter attached with its main antenna anchored behind the negative lead (A). The mounting threads may melt if heated, and are therefore inserted after all soldering to the cell leads is completed (B).

transmitter firmly to the cell. Fill around the transmitter and antenna with a small quantity of potting to hold these in place and reduce the risk of voids in awkward corners when you cover the whole tag.

Before potting, knot the eight ties loosely together. Use this knot to hold the ties out from the tag: you can then spread potting round the ties without smearing them, so that they emerge cleanly from the tag. If you start by holding the knot and positive end of the cell in one hand, the potting can be worked into the crevices round the transmitter and mercury switch, and the tag then suspended from the knot while the other end is potted. Note that unless the transmitter has been built for the specific tag design and antenna dimensions, you will probably need to leave a tuning capacitor exposed as well as the starting loop. When the potting is dry, create a 15 mm long cone of silicone or hot-melt glue around the base of the ground-plane antenna (Fig. 5.11B). After tuning and covering the tuning capacitor with a little potting, the tag can be left running for 48 hours, to check that it still runs after the potting has cooled and any initial drop in cell voltage. Then cut the starting loop to disconnect for storage.

Using battery connectors to hold the leads is an appropriate short-cut for tags that will not be exposed to water. However, for aquatic birds, the added risk of water penetration along the threads should be avoided by tying threads round the package after an initial layer of potting, by keeping antenna leads well within the potting and caulking emergence points, and preferably by eliminating the ground plane antenna.

5.3.3 A 15 g posture-sensing back-pack

Based on a 1.0–1.5 g transmitter and an 8 g 1/2AA lithium/thionyl-chloride cell, this tag will typically run for 18–36 months. Its transmitter contains extra leads, which are joined by a tilt-switch to control the pulse rate (Chapter 2.5.5). The design is used mainly for harness-mounting on birds which weigh at least 300 g, although it can also be modified for many other applications.

Start by preparing the harness tubes and antenna. Cut two straight lengths of harness tubing about 30 mm long; thin PVC tubing with a 4 mm diameter is very suitable. Halfway along one tube, cut out an elliptical slit about 10 mm long and half the depth of the tube. Tin 5 mm of the end of an antenna of 20–50 kg stainless-steel fishing trace, and coat the remainder with fine CRN (Rayfast) heat-shrink tubing. Use a heat-gun (e.g. for stripping paint) with relatively cool setting to shrink the tubing evenly onto the wire, and pinch a 2–3 mm overlap at the distal end to create a seal that prevents water access. Cover the antenna to half its length with slightly thicker adhesive-lined heat-shrink tubing, with melt-lined SCL for a third of the length if your species has a strong beak. Solder 20 mm of 0.3–0.4 mm diameter copper wire to the strip connector on the negative terminal of the cell as in the previous tag.

Glue the tubes with cyanoacrylate along the sides of the cell, about a third of the circumference apart, such that the cell can just touch a flat surface between to make the bottom of the tag. The notch in one tube should face upwards. Glue the antenna in the gap between the cell and upper surface of the other tube, for most of the length of the cell, with the soldered end towards the cell's positive lead. Use a very small quantity of inert modelling 'clay' such as Plasticine (or poster-fixing Blutak) to hold the transmitter along the cell on the same side as the antenna. Use similar material to position the tilt-switch, for example the small CM16–0 from Assemtech, beside the transmitter (Fig. 5.12A). Connect all the transmitter leads. Test the transmitter and adjust the angle of the tilt switch so that the signal is fast only when required, such as to indicate a horizontal posture during flight. When you are satisfied, use a small quantity of potting to fill the gaps around the switch, antenna and transmitter.

It is important during construction to avoid soldering near the harness tubing, which readily deforms if it is PVC, and to ensure that the wire core of the antenna remains at least 2 mm from the cell casing to minimise detuning effects. If you

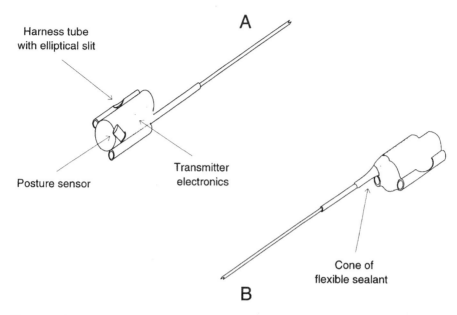

Harness tube
with elliptical slit

A

Posture sensor

Transmitter
electronics

Cone of
flexible sealant

B

Figure 5.12. A 15g posture-sensing backpack, showing the 1/2A-size lithium cell, side tubes for harness attachment, transmitter and posture sensor before potting (A), and the antenna with support cone at its base after potting (B).

propose to use very thin potting to minimise tag weight, it is wise to wind 4–5 turns of strong polyester thread around the ends of the tag to anchor harness tubes and antenna firmly in place. When potting, start by working the acrylic thoroughly into these threads and all crevices in the design, to prevent voids forming, before completing the covering. For aquatic species, it is wise to give the cell and electronics a layer of potting before attaching the harness tubes.

When you make the first tags, you will probably need at least two stages of potting to complete the covering. Start the final coating at the end of the tag furthest from the antenna and finish by holding the base of the antenna. However, note that you will probably need to leave a tuning capacitor exposed as well as the starting loop. When the potting is dry, create a 30 mm long cone of hot-melt glue to protect the emergence of the antenna (Fig. 5.12B). After tuning and covering the tuning capacitor, check that the tag runs for 48 h and disconnect for storage.

Although the tag is described with lateral harness-tubes, because this design is especially easy to attach correctly (see Chapter 6), it can also be made with the more traditional transverse tubes. In this case, the tubes are about 12 mm long and can be held in place by bending the strip connectors on the cell terminals round them, provided the species is not aquatic and *after* any soldering to the connectors. This design gives 22 g tags with 2/3AA cells, 30 g tags with AA cells and 8 g tags with 10–28 cells.

5.3.4 A 15 g necklace

This is another design with a 1/2AA cell and life of 18–36 months, suitable for birds of at least 500 g. For lighter species, it can be built with 10–28 cells at about 9 g, or with silver cells. With a whip antenna made of fishing trace, this tag has proved very satisfactory on galliformes, being quicker to attach and giving better ranges than the previous tag designs. However, birds that fly frequently (e.g. passerines) have a tendency to catch the antenna under a wing, which irritates them.

Solder a 180 mm length of 20–30 kg nylon-coated fishing trace, along with a 20 mm wire lead, to the cell's positive tab and cut off the excess of the tab. Attach the transmitter to the other tab by its negative lead, which should also form a starting loop. The main antenna is a 270 mm length of fishing trace, which is coated with two layers of heat-shrink tubing to stiffen and distance it from the cell. The inner length of tubing extends for 200 mm, and the outer for 150 mm. Bend the antenna through a right angle about 30 mm from the transmitter end, and use a dab of cyanoacrylate glue to attach this section to the side of the cell so that the main section abuts the end of the cell parallel with the ground-plane antenna (Fig. 5.13). A 3–6 mm gap should be kept between the main antenna and the ground plane: the tuning will be adversely affected if the main antenna passes too close to the ground plane. Solder the transmitter lead to the main antenna, test that it is working, and give the whole assembly an initial coat of potting.

The necklace cord is a 250–300 mm length of hollow, soft, braided artificial fibre, with 5–7 mm diameter unstretched. The 'woolly' cording often used for football-boot laces is ideal, with lesser diameter lacing for smaller tags. Do not use cotton lacing, which wears out within a few months. Slip the cord over the ground-

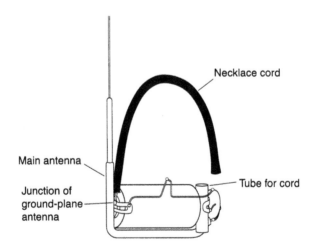

Figure 5.13.   The component layout of a 14 g necklace tag for game birds. The necklace contains a ground plane antenna, which is attached to the cell's positive terminal.

plane antenna and use a dab of glue to hold its end against the tag. At the other end, glue a 20 mm length of 4 or 6 mm diameter PVC tubing, in parallel with the antennas. Complete the potting of the tag apart from the tuning capacitor. When the potting is dry, trim any excess from the necklace tube and thread the ground-plane antenna to the position it will have on the bird become completing the tuning.

5.3.5 A 2 g collar

There are at least four ways of making a small mammal collar from a 250 mg transmitter and a small button cell (e.g. an Ag397 cell). The tag can have a closed or open tuned-loop antenna round the neck, or an integral loop encompassing the transmitter and cell alone, or a whip antenna. In the first two cases the tag is slightly heavier than the third, but has two to three times as much range. A whip antenna will give slightly lower detection distances than a tuned loop in the tag described here, which is most convenient to make, but can give more if an appreciable length emerges from the top if the collar. If you don't know whether your species will chew antennas, you can try both.

The key to this tag is a suitable strong cable-tie, with a small block to reduce bulk and weight. The Heyco #37199308 is a 3 mm width tie suitable for small collars. It is also important to per-form the tie in a way that will conform to the animal's neck, creating a curve with the package on the lower surface, and kinking the tie so that its emergence from the potting takes it up the side of the neck. You can use an artery-clamp to can hold the tie's strip in position next to the block without threading it.

Glue the transmitter and anode casing of the cell to the cable tie using small dabs of cyanoacrylate, with the cell close to the block and the lead for the main antenna at the other side of the tie. Do *not* get this glue on parts of the cable tie that will not be inside the potting, because it can make the tie brittle. If the transmitter has a stiff positive lead, you may choose to attach this first to the cell, and use an angle between these two components to help bend the tie when attaching them. Form a starter loop in the negative lead and take a thin layer of potting over the cell, transmitter and cable tie where they attach, but not over any tuning capacitor. For an aquatic species, consider pre-tuning and potting the transmitter and cell together completely before attaching to the cable tie.

Attach a fine fishing-trace antenna so that it will run round the tie to the desired circumference, and slip a similar length of adhesive-lined heat-shrink tubing over both trace and tie. After shrinking, cover the heat-shrink and antenna with potting at its emergence from the transmitter, being careful to maintain the correct curvature of the tie. If you want the antenna to leave the tubing at the top of the tag, cut the heat-shrink in half before fitting. After shrinking the section that covers the antenna, and bending the antenna to lie as required, use the second piece of tubing to help hold the antenna in that position. If the neck circumference is variable, set this piece of tubing at the longest likely circumference, and trim it back (very carefully to avoid nicking the tie) when fitting to the animal.

The tag can also be made with a fine PVC or other plastic tube slipped over the cable tie, so the tie can be slid out for replacement to enable rapid re-use on another animal. This is most worthwhile if the cell is also placed so that it can be chipped out without affecting the transmitter. However, these provisions for recycling increase the weight and depth of the tag.

Although this design is relatively easy to make, straightforward to attach and continues to transmit if the collar is severed, a tuned loop collar makes the lightest tags and can give stronger signals. For a closed-loop collar, cut a length of tough but flexible wire (e.g. PTFE-coated multistrand copper) about 30 mm longer than the maximum expected head diameter. Strip the coating from about 5 mm at each end, tin the wire and bend it into a hook. Fold the collar so that it doubles back in the 'sliding-8' format for about 12 mm close to its centre, and slip an 8–9 mm length of tight-fitting heat-shrink tubing over the doubled piece of wire (refer back to Fig. 5.5). Solder the transmitter to the cell, such that the positive lead forms a hook distal to its attachment to the cell at one end. Attach one of the antenna hooks to this hook, and the other end to the antenna lead at the other end of the package. After potting, caulk the antenna emergence points carefully with Plastidip. Tuning is very sensitive for these fine loops, and is best done with the tag on a dead animal or on a saline-filled balloon that mimics the impedance of an animal. When the loop is tuned, the signal will increase greatly in volume as it reaches resonance; the signal rate also tended to slow quite markedly on single-stage tags, but is less affected by tuning on modern transmitter circuits.

You can use similar collar wire for an open loop antenna. Cut a length about 15 mm greater than the greatest expected neck diameter, form a small tinned hook at each end and solder to the cell and transmitter, as for the closed neck loop. Then cut the wire in half, carefully adjust a wire stripper to the diameter of the conductor and strip 6–8 mm of insulation from the ends. Tin the exposed wire and bend to form a pair of hooks that interlock. To mount the tag, you will open one hook and slip a 15–20 mm length of narrow heat-shrink tubing over that wire, reform the hook and solder both together, and shrink the tubing over the bare metal. For tuning, just clip the hooks together. With the smallest cells, collars of this type can weight less than 1 g, but are very fiddly to attach.

5.3.6 A 25 g thermistor collar

An earlier version of this brass-loop collar design gave good results on 500 g squirrels, transmitting a temperature-modulated signal from a single-stage tag for 5–9 months from a lithium/copper-oxide cell. The transmitter and cell designs have been superseded. The design described here should therefore last longer, although tags on squirrels that share nests cannot be relied on to run for more than about a year without damage.

Use a metal guillotine to cut a smooth-edged strip of 0.6–0.8 mm brass about 9 mm wide. Cut a length of this strip that is about 10 mm greater than the desired internal circumference of the collar. Cut the strip in half, round all four ends, and drill holes for short brass bolts of about 3 mm ($\frac{1}{8}$ in) diameter about 4 mm from one end of each strip. Bend the brass in a right angle the same distance on the other side of the hole, insert the brass bolt and a nut and screw them tight before soldering the edges of the nut to the strip. The bolt will be removed to attach the tag, because it is easier to fasten a bolt than a nut in the field.

Shape the brass to conform with the neck, with the strips bent flat for about 10 mm at the bottom to lie along the side of a 1/2AA cell, to which they should be attached by potting. Before that attachment, however, clean and tin the end of each strip, and attach a wire that will not be covered by potting. The ideal position for the transmitter is inside the loop in the case of squirrels (Fig. 5.14), because of the tendency for squirrels to damage collars, but the shape of your transmitter may preclude that unless you have made special arrangements with the manufacturer. Attachment to the side of the cell, close to the brass strip, is the next best option.

Aim to keep all connecting leads, to the cell terminals and antenna strips, away from the outer edges of the tag, but make sure that leads do not short through the brass. In a collar for monitoring foraging by the fall in temperature outside a warm nest, the tip of the thermistor is about half-way down the side of the cell, with a thin layer of potting over its tip and the leads embedded as deeply as possible. The tip should not stand proud of the potting. When potting, remember not to cover the tuning capacitor, because careful tuning is important for all tuned-loop designs.

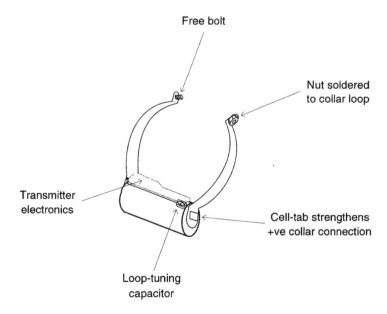

Free bolt

Nut soldered
to collar loop

Transmitter
electronics

Cell-tab strengthens
+ve collar connection

Loop-tuning
capacitor

Figure 5.14. A bolt-together collar suitable for a squirrel. The transmitter electronics and tuning capacitor are placed between the 1/2AA cell and the neck, where they cannot easily be damaged by chewing.

When the potting is dry, cover the emergence points of the brass strip with Plastidip or other sealant. Then cover the brass strips with adhesive-lined heat-shrink tubing that is also encouraged to adhere to the potting at the emergence points.

This general format can be used with smaller cells and thinner brass to build a wide range of collars with fixed diameters. Variable diameters are possible with one strip short, and a bolt soldered to its inner surface protruding outwards to engage one of several holes on a longer strip that passes right round the neck (see Fig. 5.8). However, tuning of variable-circumference collars becomes poor if they have to be adjusted far from the mean neck circumference.

5.3.7 Collars for larger mammals

Tag manufacturers are often eager to supply a standard, sealed package with a whip antenna that can be attached to a variety of collar designs. Early packages on large mammals tended to have bulky batteries to drive transmitters for several years, and bulky packages to accommodate those cells in heavily armoured packages. The ability of a pair of AA cells to drive modern VHF tags for a number of years has reduced the bulk needed for packages attached to collars, and bulky armouring is often not necessary, because many species are reluctant to let others gnaw tags close to their neck. Ask tag manufacturers and other users about the habits of your species, noting that there may be greater risk when females have cubs.

The simplest approach is to buy a tag already equipped with cells and thick SCL-covered antenna. However, it is also practical to make up such a package, using an approach broadly similar to the tag in section 5.3.3 but without tubes. If you use cells in parallel (don't forget diode isolation) the transmitter will probably fit across one end of them to reduce the depth of the package. The antenna can then pass from the transmitter along the length of the cells in the gap between them, on the side closest to the collar. Such a package can be thoroughly potted, with the antenna emergence well caulked too, before attachment to the collar.

Seek advice on the choice of collar material. Soft plastic or machine belting is sometimes used, but these materials tend sooner or later to crack. Early braided nylon strapping tended to fray and attract snags, but some modern material supplied for dog collars appears not to fray dangerously and to be very durable. Projects have found tags on such collars, closed with a standard buckle, very adequate for work on ungulates and canids, although the buckles require replacement with rivet-plates on badgers. In these cases, the package has been bound to the collar, then potted in place. The antenna has been sewn to the collar at 20 mm intervals, as far as its separation as a free-standing whip at the top, and covered to that point with tough but flexible heat-shrink. Finally, the tag itself is covered with thick melt-lined SCL. If exposed wires are used to start such tags before attachment to the collar, they should be on the surface closest to the collar.

5.3.8 A 3 g tag for fish or implanting

A radio tag developed at the Ministry of Agriculture, Fisheries and Food Fisheries Laboratory in Lowestoft is suitable for ingestion, implantation or external tagging of fish (Solomon and Storeton-West, 1983). The smallest version is based on 8 mm wide transmitters with Ag392 or Ag393 silver oxide cells in polystyrene (Sarstedt) or polycarbonate (Alge) tubes which have an external diameter of under 10 mm. These tags are suitable for external mounting on 300–500 g fish, running for up to 3 months with a 70 mAh Ag393 cell. However, much larger tags can be built along the same lines using tubes with 10 mm internal diameter for 10–28 cells or 15 mm for 1/2AA or 2/3AA cells (which may need their plastic coating removed to fit easily).

At least two tubes are needed for each tag, the blunt end being cut from one to seal the open end of the other, which is cut a little shorter than the length of the transmitter and power supply. Unless the tag is for immediate use, there should also be room for a magnet-operated reed switch. More than two tubes are needed per tag, because cutting sometimes breaks polystyrene, even if a fine hacksaw is used. However, polycarbonate tubes can be milled. The cut ends have their surfaces ground or milled to fit, and are joined with polystyrene modelling cement, which can also be used to attach a narrow longitudinal keel with mounting holes to the surface of external tags. The external tags radiate most efficiently with a whip antenna, which can be of fine PTFE-coated wire for the smallest tags and perhaps fine fishing trace for larger versions. However, there is then a higher risk of water penetration where the antenna emerges, unless some form of capacitative coupling can be used to seal off the antenna (Morris, 1992).

5.4 CONCLUSIONS

1. Constructing tags by adding cells, antennas and packaging to basic transmitters can save costs and increase flexibility, but requires time, advice and technical competence.
2. Cells should, if possible, be obtained with leads attached, because flux-cooled spot-soldering needs training. Special techniques may also be required for soldering to antenna materials.
3. Antenna emergence needs flexible support to avoid stress points, and thorough sealing to prevent water access by capillary action or along corrosion channels. Water can also penetrate along straps and harness attachment tubes unless potting is a two-stage process.
4. Readily-available epoxy adhesives are not waterproof, so potting is often with dental acrylic, which is tougher but less easy to remove for refurbishment. More flexible Plastidip may be used on small tags. There must be provision for matching antennas to circuits after potting.
5. Reed-switches, which are not entirely reliable, are usually unnecessary on self-built tags.

6. Construction is described for 1–2 g tags with silver cells, for 8 g tail-mounts with lithium cells, for backpacks, necklaces and tuned-loop collars with larger cell capacities, for large mammal collars and for implants in plastic tubes.

6 Tag Attachment

6.1 EFFECTS ON ANIMALS

Zoologists who tag animals have a moral as well as a practical obligation to ensure that there is no adverse affect on their subjects. In the long term, the moral aspect tends to override all other considerations, by motivating restrictive regulations on techniques that can cause the injury or death of animals. The need to use tags cautiously is reinforced by the current widespread public concern for animal welfare.

6.1.1 Covert effects

Effects of tagging can probably never be completely avoided. The handling alone is liable to create some stress, and there has to be an energetic cost of carrying an extra load (Pennycuick, 1989), no matter how small. However, animals may often be able to compensate for these effects so that they are effectively hidden from biologists who seek them.

When testing for differences between tagged and untagged animals, biochemical methods have been particularly revealing, especially the use of double-labelled water to record metabolic activity and hence energy expenditure. Some studies have recorded overt effects, such as reduction in flight speed or duration of pigeons, as well as energetic costs (Gessamen and Nagy, 1988; Gessamen et al., 1991), or have concentrated on detecting the energetic impact (Berteaux et al., 1996). However, other projects using double-labelled water have failed to find overt effects of tagging, indicating that animals were compensating for energetic costs. Thus, tropicbirds foraged as successfully for their young despite increased metabolic activity (Pennycuick et al., 1990). Tagged penguins used more energy than untagged birds to maintain a given swimming speed in captivity, but somehow compensated to maintain the same metabolic rate as controls in the wild (Culik and Wilson, 1991). Another covert difference after tagging was revealed by elevated steroid levels in hunting dogs (Creel et al., 1997).

For the biologist, who accepts without concern the extra cost of carrying tracking equipment, the objective is to reduce the effect on the animal to that same low level, or at least not to bias the result of the study appreciably and never to cause suffering. There should always be tests for overt effects of new tag designs and new situations. For those concerned that biologists do not always take such 'impact assessments' seriously, it is worrying that only 10% of 238 recent marked-animal studies have published evidence, in major journals in 1995, that tag impact had been considered (Murray and Fuller, in press).

6.1.2 Overt short-term effects

The overt adverse effects of tagging can be divided into long-term and short-term categories (Tester, 1971), and in each case they may be chronic or acute. Most effects are chronic and of short duration, such as a temporary reduction in foraging activity or an increase in preening or grooming at the expense of other behaviours (Boag, 1972; Bohus, 1974; Wooley and Owen, 1978; Nenno and Healy, 1979; Leuze, 1980; Sayre *et al.*, 1981; Birks and Linn, 1982; Hooge, 1991; Mikesic and Drickamer, 1992). Such effects are probably present after most tagging, even if they are not recorded, and may be as much a latent result of the capture and handling of the animal as a result of the tag's continuing presence (Côté *et al.*, 1998; Hubbard *et al.*, 1998). Chronic short-term effects may delay data recording for a few days, without being seriously detrimental to the animal. They may be reduced by attention to comparatively minor aspects of tag design, such as colour (Wilson *et al.*, 1990), or by reduction in handling time.

However, short-term effects can sometimes present acute problems for the animal. Hares, for example, seem especially prone to capture-shock (Keith *et al.*, 1968), which may be exacerbated by the subsequent presence of a tag, leading to death from chilling or predation (Mech, 1967). Other animals which normally adjust well to tagging may not do so at critical times of year: the disturbance alone may be enough to cause desertion by incubating birds (Amlaner *et al.*, 1978), or for females to abandon young broods (Horton and Causey, 1984), or to stop a dispersing animal securing a territory, or to tip an animal over the brink of starvation when food is short. Females ungulates have sometimes rejected their tagged young (Beale and Smith, 1973; Goldberg and Haas, 1978; Livesey, 1990; but cf. Coah *et al.*, 1971). Biologists should be aware of high-risk periods, and plan the tagging operations accordingly.

When there is high mortality immediately after tagging of captured animals, researchers may delay the entry point for survival estimates for a week to a month (White and Garrott, 1990). The initial reduced survival may not be primarily a result of the tags, but of the trapping and marking process, or of capture biased towards inadequate animals, which may have entered traps because they were already starving. However, discarding early deaths is not appropriate for studying survival of young animals or those being released. In these cases, the animals should be marked in time for them to adjust to the tags well before they leave the care of their parents or are released.

6.1.3 Overt long-term effects

Chronic long-term effects can result from badly fitting tags, which chafe or cut the skin, or cause feather-loss which reduces a bird's insulation and thus increases its energy expenditure (Greenwood and Sargeant, 1973). The attachment method or loading may also reduce an animal's movement efficiency or manoeuvrability (Aldridge and Brigham, 1988; Pennycuick *et al.*, 1989; Putaala *et al.*, 1997). Aquatic species that are streamlined for fast swimming are especially vulnerable to increases in drag from external tags (Mellas and Haynes, 1985; Olsen *et al.*, 1992; Wilson and Culik, 1992), but shape of collars may be important for fossorial species too (Rado and Terkel, 1989). There may be a resulting decrease in foraging performance (Massey *et al.*, 1988; Wanless *et al.*, 1988; Croll *et al.*, 1991) or growth rates (Hubbard *et al.*, 1998).

Some long-term chronic effects are slight, and may not bias the research findings. For example, a harness may reduce a bird's tendency to fly (Michener and Walcott, 1966; Ramakka, 1972; Hooge, 1991; Gessamen *et al.*, 1991), and thus perhaps reduce its success in aerial courtship but not its feeding behaviour while on the ground. With the absolute proviso that the animal is not in pain, slight long-term effects of this type may be acceptable as a last resort.

Acute long-term effects include occasional fatal complications from implants (Guynn *et al.*, 1987; Herbst, 1991). Animals may also become entangled in elastic or break-free features of tags that are designed to be shed or to accommodate growth (Weber and Meia, 1992; Kenward *et al.*, 1993b; Hill *et al.*, 1999). Predators may be attracted by conspicuous tags, for example when light flashes off solar panels (Marks and Marks, 1987; Burger *et al.*, 1991).

Sometimes the long-term or short-term effects are enough to decrease survival (Gilmer *et al.*, 1974; Erikstad, 1979; Perry, 1981; Marks and Marks, 1987; Marcström *et al.*, 1989; Sykes *et al.*, 1990; Ward and Flint, 1995) or productivity (Sibly and McCleery, 1980; Foster *et al.*, 1992; Rotella *et al.*, 1993; Pietz *et al.*, 1993; Gammonley and Kelly, 1994; Paquette *et al.*, 1997) or both (Paton *et al.*, 1991), if only for individuals that had low weight (Johnson and Berner, 1980) or were young (Cypher, 1997) or were breeding in years with a poor food supply (Marzluff *et al.*, 1997a). Again, such an effect might not influence the result of a study. For example, a tag that reduces an animal's chance of escape from a predator might not affect its range size. However, it behoves biologists to take great care, and ideally to subject themselves to independent review, in situations where research benefit (and perhaps gains for conservation) are traded against animal welfare.

6.1.4 Testing for effects

The assumptions behind tagging and considerations for testing have been thoroughly discussed by White and Garrott (1990). The most robust tests are based on study designs that compare tagged animals with properly assigned controls.

Animals not treated in the same way as those tagged may not be adequate controls, for two reasons. The worst problem is when controls are not radio tagged because they differ in some way. Such marking bias may occur because longer experience or better condition makes controls more reluctant than captured animals to enter traps. The second reason for controls being inadequate is that they were not handled in the same way. Handling may create risk for tagged animals that is not a simple product of tag presence, for example from implanting operations. Handling might even reduce risk for tagged animals, if it made a hunted species more wary of humans after tagging.

When controls can be properly assigned, because animals for tagging and comparison are equally available and handled similarly, it may be more practical to use systematic rather than random assignment. The systematic approach, of tagging every second or third capture, is appropriate if you are assigning lengthy mounting procedures to small numbers of animals each day. If you use random assignment of lengthy procedures, some days may by chance become too busy and put animals at risk because it happens to rain heavily. Other days may have no tag assignments; by Murphy's Law, these are the perfect days that are chosen to demonstrate tagging!

Testing has mainly been for differences in activity, feeding rates, productivity, survival, persistence and physical condition. Comparisons of persistence (lack of emigration or death) and condition are the least reliable. Tagged animals may suffer higher mortality than controls, but have the same recapture rate because they emigrate less (H. Smith, 1980). Animals may also compensate for foraging disadvantages by feeding for longer to maintain their weight: there is growing evidence that small-animal weight trends are generally determined by internal factors, on a seasonal basis, rather than by feeding difficulties. Individuals have been found to increase in weight immediately after tagging (Kenward, 1982a), possibly because they have responded to the presence of a tag by increasing their body reserves, in the same way that small birds respond to cold weather. Care must also be taken, if performance of adults is compared on the basis of food deliveries to young, that an untagged parent is not compensating for effects on a tagged mate (Wanless *et al.*, 1988).

The smaller the control and experimental samples, the larger the difference required for statistical significance, and thus the greater the risk of failure to detect an effect (Type I error). White and Garrott (1990) show how to calculate sample sizes for demonstration that a binary difference (e.g. survival/death) is unlikely to exceed a chosen level of acceptability. Such tests of radio impact are easiest when it is always possible to establish life or death (i.e. the probability of detection is 1). Thus, the radio tagging of a parent bird makes it relatively easy to check for survival of tagged and untagged chicks in precocial broods (Kenward *et al.*, 1993b; Korschgen *et al.*, 1996b). However, the use of bands to mark birds with and without radios may give a much lower probability of detection: hundreds of tagged and untagged goshawks were needed to show that tail-mounted radio tags did not reduce survival (Kenward *et al.*, 1999).

This means that testing for adverse effects is most practical if you work with animals that are easily seen or recaptured. If they are not easily seen, it may be possible to make comparisons using radio tags that have been shown to lack adverse effects in other studies.

Testing for adverse effects is important for improving designs. For example, the first collars in a study of snowshoe hares were too loose: they tended to catch in the animals' mouths until the design was changed (Brand *et al.*, 1975). There were similar problems with under-wing loop harnesses designed to fit growing birds, which could loosen enough to get caught in the birds' beaks until exactly the right combination of elasticity and covering was achieved (Hill *et al.*, in press). On the other hand, the recapture rate with an early squirrel collar design was lower for tagged animals than for untagged controls because the cable-tie fastener could become too tight. Squirrels with modified collars did not differ from controls in recapture rates, weight changes or breeding performance (Kenward, 1982a).

Of course, studies do not always have access to comparable controls, or samples large enough for adequate tests. It is regrettable, however, when studies take no interest in testing for tag effects despite having large samples and access to adequate controls. This not only wastes an opportunity to gain useful data, but increasingly risks the wrath of reviewers, if not the public. Even if your tags cause no marked adverse effects, you should prepare to handle an irate caller who has found one of your subjects dead or injured and blames the 'obvious' cause, the tag. Are you, and your assistants, ready to reassure such a caller – especially if you have done no tests on your tags?

6.1.5 Tag mass

Studies that used tags of two or more sizes have sometimes reported adverse effects in animals with the heavier packages, including decreased breeding success (Sibly and McCleery, 1980) or survival (Warner and Etter, 1983; Burger *et al.*, 1991; Cotter and Gratto, 1995). Theoretical considerations predict such effects (Caccamise and Hedin, 1985). However, it is important that the different tag types are allocated without bias (Hines and Zwickel, 1985; Snyder, 1985). If tags on individuals most at risk, due to age or breeding status, are incidentally a higher or lower proportion of body mass, results may be artefacts. For example, Warner and Etter (1983) tended to attach the heaviest tags to pheasants just before breeding, during which females are particularly vulnerable to nest predators.

Differences between species in how tags can be attached, and in energy required for moving, preclude a single rule about tag loading, except perhaps that small species can be expected to tolerate relatively larger tags than similar large species. Thus, flying animals may be more inconvenienced than creeping animals by tag weight, whereas swimming animals are most at risk from drag when tags are bulky. In general, however, evidence of adverse effects has tended to emerge for necklaces or collars more than about 3% of body mass and for harness-mounted tags or implants above 4–5%. Premature detachment seems likely for tags at more than

1–2% of body mass glued to feathers or fur, and if tail-feathers each carry much more than 1% (2–3% can be loaded across two or more feathers).

6.2 MINIMAL TAGGING, SOFT TAGGING

6.2.1 Tag designs

Table 6.1 provides a rough guide to the handling time and skill required for the main techniques that have been used to mount tags on terrestrial vertebrates, together with an assessment of the rate of loss and reported incidence of adverse effects for each. The most desirable techniques are those with few disadvantages (few x signs).

Table 6.1. Subjective assessments of different radio tag attachment techniques (disadvantage is greatest for techniques rated xxx).

Technique	Handling time	Skill require-ment	Rate of loss	Risk of effect	Considerations
Implant	xxx	xxx	x	xx[1]	Poor signal range; invasive.
Body-harness	xx	xxx	x	xxx[2]	Should have case-by-case tests.
Glue (fur/feather)	x	xx	xxx	xx[1]	Detachment rate varies greatly.
Skin sutures	xx	xx	xxx	xx[1]	Invasive; variable detachment.
Mammal collar	x	xx	x	x[1]	Growth of young animals.
Mammal ears	x	x	xx	xx[1]	Used mainly on young animals.
Bird necklace	x	x	x	xx[1]	Effects on flight, feeding; icing.
Bird patagium	x	xx	xx	xx[1]	Only for very large, slow birds.
Bird beak	x	xx	xx	xx[0]	Only for medium to large birds.
Bird leg	x	x	xx	x[1]	Antenna loss; not for small birds.
Bird tail	xx	xx	xx	x[1]	Moulting; not for small birds.
Scutes/horns	xx	x	x	x[0]	Few species are suitable.
Mini-harpoon		xxx	xx	xx[0]	Only for large marine animals.
Alimentary	x	x	xxx	xx[1]	Foraging behaviour.
Fish-streamer	xx	xx	xx	xxx[1]	Invasive; drag in water.

[0]Ratings with no comparisons published.
[1]Ratings based on less than 10 published comparisons.
[2]Ratings based on more than 10 published comparisons.

In general, handling time and skill requirement are quite closely related. The exception is that the time taken to shoot a harpoon-tag from a crossbow into a whale is short, but the aim must be very careful to position the tag close to the blow-hole, for maximum exposure above water, without hitting the orifice. Tags with the lowest skill requirement (Table 6.1) can be fitted from instructions, whereas training is advisable for the second category and should be obtained for implanting, harnessing and harpooning.

Techniques with appreciable risk of adverse effects and high skill requirement are best avoided, unless alternatives are either even more risky or cannot give the required duration of attachment. Thus, for birds and aquatic mammals, implants and harnesses are best avoided in favour of less durable alternatives unless long tag life is indispensable. If tags are only needed for a few weeks and there is no need of completely reliable attachment, glue-mounts may be adequate. If months of attachment are required, tail-mounts may be adequate on birds, or leg-mounts if they are to be marked before feathers are fully grown. There seem to be no published adverse effects on non-aquatic birds for either technique. However, if tags must last more than a year, or are too heavy for other attachment methods, harnesses or implants may be essential.

The best principle is to be cautious of mounting techniques that have shown adverse effects in some studies, but not necessarily to rule them out. For instance, many studies have recorded adverse effects of harnesses on feeding, flying, moulting, breeding and survival in birds. However, harnesses require much skill to attach, and it is therefore difficult to know the extent to which problems reflect inexpert attachment. In a study of spotted owls, survival varied between different teams that attached tags (Paton *et al.*, 1991). Fortunately, tail-mounts were a good alternative because the owls kept their tail-feathers for more than a year. However, other studies have attached harnesses to raptors that fly more than owls and failed to find marked adverse effects. Buzzards with harnesses survived as well as those with small tail-mounts (Kenward and Walls, 1994), and tagged falcons reared young as well as controls without backpacks (Vekasy *et al.*, 1996) except possibly in 1 year with a poor food supply.

Nevertheless, it is not at all easy to get a good fit with harnesses. Comparatively minor changes in design can affect a tendency to abrade or restrict birds (Buehler *et al.*, 1995). Even a satisfactory harness design requires changes in relative length of different straps between species, and at least two measurements may require careful tailoring to each individual bird (Kenward *et al.*, b in press). It is therefore essential, if your study requires harnesses, to precede your own tagging by practising harness attachment with someone working on a similar species or, if that is impractical, to tag some captive birds.

If any tag design is new, it is especially important to test it adequately, preferably on animals in enclosures before conducting a pilot study in the wild. Studies in advance of attaching UHF tags for tracking by satellite have been especially detailed, from original concept (Fuller *et al.*, 1984) through a line of papers involving the same authors (Pennycuick and Fuller, 1987; Obrecht *et al.*, 1988; Pennycuick *et al.*, 1989; Gessamen *et al.*, 1991; Buehler *et al.*, 1995).

6.2.2 Timing

To avoid possible chronic short-term effects, the timing of tag attachment may be important. Capture of adults may be a problem at particular stages of the reproductive cycle (Sibly and McCleery, 1980; Côté *et al.*, 1998). It may be best to tag animals before or after they require maximum effort for breeding, or during preparation for migration or torpor, or after recovery from demanding events. It may be advantageous to tag young animals while in the care of their parents, provided the process does not hinder the parent-offspring relationship, so that the young become accustomed to tags before leaving a nest or den. If animals are to be released, they should be given time to become accustomed to the tags in advance. If there is risk of a severe immediate reaction to the tag, consider keeping the animals in a dimly lit enclosure for a few hours until any initial panic reaction has subsided. Game birds, for instance, often throw themselves on their backs when first fitted with harnesses (Small and Rusch, 1985), and this must increase their susceptibility to predation if they are released immediately.

It may also be important to use the tags initially on a particular category of animal that is best able to compensate for any adverse effects. For example, it is probably better to tag adults rather than young to determine migration routes that both age groups are likely to follow (Mark Fuller, pers. comm.).

6.2.3 Sedation during tagging

Large animals generally have to be sedated during tagging, to prevent injury to the biologist or to the animals themselves. Sometimes they are caught with anaesthetic darts, which may themselves be radio tagged to help find a drugged animal or to locate lost darts (Lovett and Hill, 1977). Finding lost darts is especially important if etorphine (Imobilon) is being used, because of the high toxicity of this morphine derivative to humans who might stumble on a lost dart. If large radio-tagged animals need to be recaptured, sedative may be injected with a radio-controlled capture-collar (Mech *et al.*, 1984, 1990) which stresses the animal less than trapping (Delguidice *et al.*, 1990).

General anaesthesia is also necessary for implanting tags, whereas local anaesthetics, such as spray-on Novocaine, can be used during the attachment of tags to ears, wing patagia or fish fins. Mammals which are difficult to tag without getting bitten have been anaesthetised successfully with ether, or with more refined inhalants such as methoxyfluorane (Hardy and Taylor, 1980; Lacki *et al.*, 1989). Fluorane inhalants are also widely used for birds (Hubbard *et al.*, 1998; Schulz *et al.*, 1998). Alternatively, intramuscular agents such as phencyclidine hydrochloride and acepromazine (Mech, 1974) or ketamine hydrochloride (Cheeseman and Mallinson, 1980; Harris, 1980; Van Vuren, 1989) have given good results. Ketamine (Ketalar, Vetalar) is often used with a synergistic agent such as xylacine hydrochloride (Rompun) that reduces the quantity required and the tendency to convulse (Koehler *et al.*, 1987). The combination ketamine-medetomidine can give faster recovery times than isofluorane anaesthesia, and also requires less equipment, which makes it especially suitable for implantation in the field (Kreeger *et al.*, 1998).

Adverse effects of anaesthesia and other surgical agents have been recorded in some studies. For example, capture with xylacine hydrochloride decreased kid production and caused abandonment by female goats (Côté *et al.*, 1998). Hubbard *et al.* (1998) noted that sterilisation of instruments and implants with benzalkonium chloride (Zephiran) caused temporary loss of coordination of turkey poults after anaesthesia, but that 70% isopropyl alcohol had no such adverse effect.

If possible, however, it is wise to avoid sedatives or general anaesthetics while tagging birds and small mammals, because of the risk that the animal has not recovered completely when released. The animal may appear quite normal until it exerts itself, in flight or running up a tree, and then fall or misjudge a landing, possibly with fatal consequences. Bear in mind that humans often feel dizzy or ill for some time after being anaesthetised, despite being able to move about normally. It is quite possible that feeding or anti-predator behaviour is adversely affected for some hours or days after animals have been drugged.

6.3 ATTACHMENT TECHNIQUES

6.3.1 Glue-on

Flexible, durable glues that cause no skin reactions are required for gluing tags to mammals and birds. Fast-setting two-part epoxy adhesives (e.g. Rapid Araldite from Ciba-Geigy, Double-Bubble from Permabond) are suitable for bats (Stebbings, 1982; Jones and Morton, 1992); however, you should not use epoxies with an instruction to 'avoid skin contact'. First, experiment to see how much hardener is needed for the cured glue to remain rubbery; this usually requires appreciably more hardener than adhesive. Clip short the back fur over the scapulars, in a patch that is just large enough for the tag, so that it can attach firmly with a low tendency to wobble loose. Smear the tag lightly with the adhesive and push it onto the patch, working the surrounding hair into the adhesive. If possible, the tag should be almost covered with hair. The tag will detach some time between 10 days and several weeks later, depending on the species and the amount of hair stuck to the tag. Hair regrows fairly rapidly on the resulting bare patch. A similar approach has proved suitable for tag attachment to seals, often resulting in tracking until the next moult (Fedak *et al.*, 1983; Hammond *et al.*, 1992).

If the tag is to be glued for not much more than a week to the back of a precocial chick, cosmetic eye-lash cement can be used to fasten the down to the tag. For longer attachment to a bird's back, stick the tag first to a piece of gauze, for instance by setting the freshly potted tag down on the fabric so that the epoxy penetrates the weave before hardening. The gauze should extend beyond the tag by at least one tag-width on either side and at each end. Swab feathers in the mounting area with acetone or alcohol to degrease them, and smear cyanoacrylate glue on the underside of the tag (Green, 1988; Alexander and Cresswell, 1990). Then push the feathers gently apart, position the tag so that it sticks near their bases, and smear more glue through the gauze to contact the feathers all round. Avoid putting glue on any animal's skin unless you have a histologically compatible variety. Several recent studies have used Titan 322 epoxy (Titan

Corporation, see Appendix I) which is specially formulated for use on bird feathers (B. Cresswell, pers. comm.).

A variation on this technique is to glue the fabric base to the bird before sticking the tag to the base. Raim (1978) trimmed the spinal tract feathers in the interscapular region until only about 1.5 mm of the shaft remained, and then attached the base cloth with eyelash adhesives (the Revlon and Andrea makes were best). When the adhesive had dried, a slip of cloth sewn to the transmitter was coated with quick-drying glue, and smoothed against the base fabric with a spatula to ensure a thorough join.

6.3.2 Harnesses

Many materials have been used for bird harnesses, including neoprene, silicone rubber, PTFE and other plastic tubing, copper wires covered by these materials, clothing elastic, and cording made from nylon, terylene or other braided synthetic material. Researchers differ in their choice of material and their preferred harness design. Although there is little published information that compares materials, straps of PTFE (Teflon) ribbon (Dunstan, 1972) are probably best for long-life attachments. Teflon (from Bally Ribbon Mills and some tag suppliers) is biologically inert; it does not change with time nor cut, although free ends must be sealed to prevent unravelling. Clothing elastic has proved satisfactory for short-term attachments in several British projects (Amlaner *et al.*, 1978; Hirons and Owen, 1982; Wanless *et al.*, 1991). Monofilament nylon or fine thread should not be used, because it is liable to cut the skin at the front of the wings. Meshwork vests, which work their way under the feathers (Lawson *et al.*, 1976), are likely to be more restrictive than well-fitting straps.

Although transmitter packages have been slung under the breast of owls (Nicholls and Warner, 1968) and some other species (e.g. Siegfried *et al.*, 1977), the most popular harness format is modified from a backpack design published by Brander (1968). It makes use of a body loop which passes behind the wings from the back of the tag, and is connected ventrally to a neck loop from the front of the tag. Buehler *et al.* (1995) showed that there is less tendency to abrade the sternum if a strap along the breast is used to join the loops (Dunstan, 1972; Fig. 6.1) than if the straps are joined as an 'X' at one point under the breast. If an 'X' fitting is loose enough to avoid abrasion, there is considerable risk that the body loop will slip in front of the keel, which can greatly hinder flight.

Figure 6.1. Attachment of a side-tube back-pack. A single Teflon ribbon is knotted and smeared with glue before being pulled into the continuous side tube (A), and each end marked for length of body loop and neck loop. The body loop is threaded through the breast strap and split side tube (B) before mounting on the bird; the neck loop is then threaded beside the neck to the breast strap and back to the split tube. A crimp is squeezed to grip neck and body loop gently while both are adjusted for length, and is then tightened, the loops sewn, and the joint cemented in the split tube.

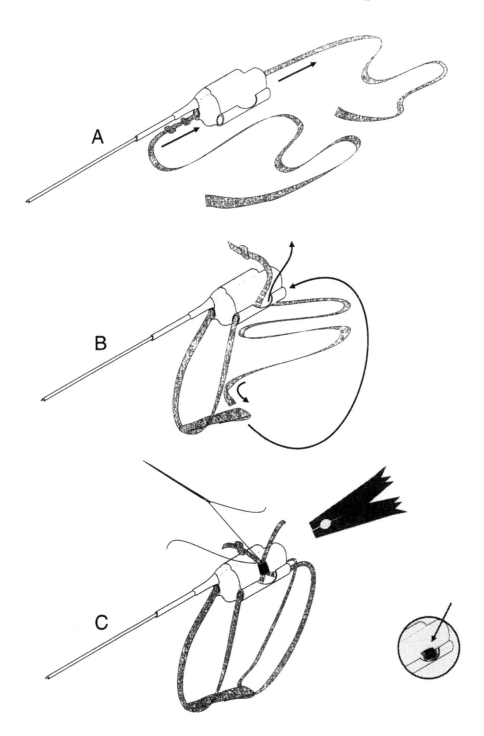

A

B

C

With a breast strap, the loop and strap lengths control the position of the tag. A relatively long breast strap puts the body loop far enough back to avoid impeding growth or movement of the flight muscles. The breast strap should meet the body loop about two thirds of the way down the keel, especially in hawks and falcons which have large flight muscles. The breast strap meets the neck loop at the point of the keel, or just behind the point, such that neither crop nor flight muscles are affected by the relatively loose loop, which lies next to the neck in the clavicular notch.

Tags have traditionally been made with transverse tubes across the front and back for fitting the harness. However, the modification in Fig. 6.1 of a side-tune design by Dwyer (1972) has several advantages (see also Fig. 5.12). The attachment of neck and body loops to the tag becomes independent of the tag length, which is convenient when using relatively long cells (Fig. 6.2). The neck and body loop can be adjusted at one point, where their compliance with predicted dimensions provides a safety check.

For raptors, the circumference of the body loop in metres is about 0.3 × the cube-root of the mass in grams of birds when leaving the nest, and the neck loop about 10% longer (Kenward *et al.*, b in press). The total length of Teflon ribbon required for a harness is just under three times the circumference of the body loop: about 0.8 m of 9 mm ribbon is suitable for a 1 kg buzzard, about 0.5 m of 6 mm ribbon for a 200 g small falcon and 1.2 m of 120 mm ribbon for a 3 kg eagle. Make two knots in the middle of the ribbon, coat them liberally with cyanoacrylate glue, and pull them into the tube that lacks a slot (Fig. 6.1A). They should be a tight fit. An easy way to pull them in, without pulling them right through the tube, is to wind the Teflon slowly round a pair of long-nose pliers at the end of the tube away from the electronics (which could be damaged by the pressure). It is a good idea to

Figure 6.2. Five saker falcons (*Falco cherrug*) equipped with 20 g backpacks with three-year life on harness of Teflon ribbon.

lightly coat 5 mm of each end of the ribbon on one surface with the glue and fold the end transversely in half to create a narrow section. This will aid threading and won't fray; when dry, cut each end at an angle to make a blunt point.

Create a breast strap about 30 mm longer than 60% of the keel length from a separate ribbon, then turn over 15 mm of each end and sew the innermost 8 mm. It is a good idea to sew double, to knot the thread at intervals on one side of the strap and to smear the knots with glue when complete. Form the body loop by threading the ribbon from the antenna end of the tag through the breast strap and into the other side tube, to come out of the central slot. Make sure the knotted side of the breast strap is away from the bird's skin. Thread a short tubular aluminium crimp on the end and tie a knot to stop the strap coming out again (Fig. 6.1B). Crimps about 0.8 mm thick with 4 mm diameter and 7 mm length can be cut off terminators for electrical wires.

Thread the other length of ribbon, which will form the neck loop, through the front of the slotted tube. Mark each end at the slot to indicate its expected length on an average bird. You can use a dab of white typing-correction fluid for this. Then unthread the neck loop. The tag is put on by holding the legs parallel with the tail so that both can be passed through the body loop. Then pass the neck loop down one side of the neck (make sure it is well in the slot next to the neck), through the breast strap, and back beside the neck into the slotted tube and finally through the crimp. As soon as the neck ribbon is just through the crimp, crush the crimp enough to grip both ends of Teflon, but not enough to prevent them sliding stiffly during loop adjustment. It is important to keep the neck loop loose, so that it does not pull the body loop forward, and then to tighten the body loop to its expected length before the neck loop.

The final adjustment of loop lengths is critical. The correct fit should be with the front of the breast strap on the anterior point of the keel, or up to 10 mm behind it. If the breast strap is too far forward, slacken the neck loop fully: you need to slide the body loop well back and reinsert it between the feathers from behind, then tighten it some more and re-set the neck loop. On a 1 kg raptor about to leave the nest, the body loop should lift about 10 mm (finger width) from the back, or slightly less for a bird that is fully grown and feathered, especially if the feathers fit tightly (as on a peregrine). The neck loop tension is correct when the transmitter can lift at the front 30% from the horizontal. When you are satisfied, crush the crimp flat, sew through the Teflon as an extra safety measure, trim the ends and record the circumference of each loop by reference to marks on the ribbons. Then glue the crimp into the slot in the tube, and finally cover with potting material (Fig. 6.1C). A leaf or slip of card between the tag and the bird's back avoids risk of adhesion to feathers. If the potting is mixed a few minutes in advance, it will cure shortly after application, so that a bird can go back into the nest without delay. With practice, a harness can be fitted in 10–15 minutes, but beginners need two to three times as long.

This method should only be used on birds that have reached full size. It is good for nestlings of raptors and other altricial birds that reach full size in the nest, provided that you can predict from previous visits when they will leave the nest (Kenward *et al.*, b in press). On species that vary little in size, training to fit tags can be very easy, using loops marked for 'one-size-fits-all'. However, on species that vary in size within and between sexes, and especially with close-fitting plumage, prolonged supervision is needed.

For species that are still growing when they leave the nest, such as many passerines, a satisfactory harness can be made with flexible cords that pass round the base of each wing without meeting ventrally (Hill *et al.*, 1999; Fig. 6.3). In this case, the elasticity of the loops makes attachment of the tag to fledglings a rapid process, as the wings are simply inserted through each loop so that the flight feathers will keep the tag in place. However, much skill is required to choose harness materials that are elastic enough to permit some growth but tensioned enough to prevent trapping the beak during preening.

Harnesses have also been used for some lizards and amphibia, and for mammals which have narrow heads relative to their necks (Fullagar, 1967; Kruuk, 1978; Mitchell-Jones *et al.*, 1984). Leather straps have proved satisfactory on many mammals. The leather should be of a type which stays the same length when wet: split rawhide has been used on badgers (Cheeseman and Mallinson, 1980), but whole chrome tanned leather, which is supple but durable, may be better. To avoid changes in harness dimensions at sea, neoprene netting has been used on seals (Broekhuizen *et al.*, 1980) and other plastic tubing on turtles (Ireland, 1980).

6.3.3 Tail-mounts

Tags may be attached to rectrices with clips (Bray and Corner, 1972), tape (Fuller and Tester, 1973), threads (Dunstan, 1973), glue (Fitzner and Fitzner, 1977) or cable-ties (Wanless *et al.*, 1989). If attached to one feather, a groove in the tag's

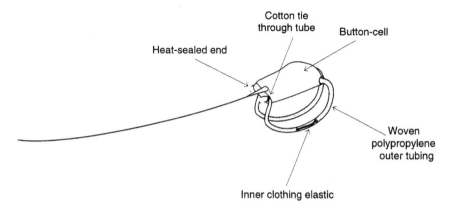

Figure 6.3. A tag with under-wing loops for small passerine birds, as described in Hill *et al.* (1999). The loops of woven polypropylene tube have limited ability to extend, and inner clothing elastic ensures a snug fit. They are glued inside the front tube, and a cotton tie through the back tube rots within a year to release the tag.

surface helps to provide a firm fit (Samuel and Fuller, 1994). Threads can be used to spread the tag's weight to a second feather and to attach the antenna along the shaft of a feather that is reasonably long.

The tags in Fig. 5.11 are suitable for attachments of this type (Kenward, 1978, 1985). Wrap the bird in a towel or custom-made straightjacket (Fuller, 1975; Maechtle, 1998), taking care to immobilise the feet. Cover the bird's head, because

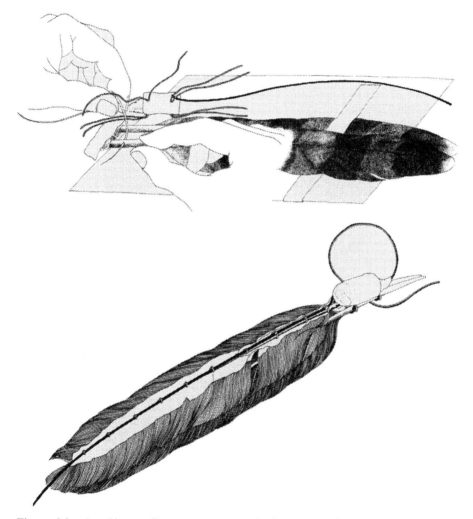

Figure 6.4. Attaching a tail-mount to a raptor. At the top, a card has been inserted under the two central rectrices, and held in place by tape across one of them. The free feather is the one to which the transmitter is being sewn. The down has been trimmed from the feather bases, and the tail coverts are held back by more tape. The lower half shows how the main antenna is tied and glued along the feather shaft to prevent breakage, with a posture sensor outlined near its exit from the tag. The ground-plane antenna curls free.

this discourages struggling, but leave the tail exposed. Slip a large filing card under the two feathers chosen for mounting, right up to their bases. It is easiest to use the two central rectrices, because these lie slightly above the others. However, if the tag must remain some time into the moult, mount it on the second and third feathers outwards from the central pair, with the main antenna on the second feather. The second pair outwards moult late in most raptors, unless they failed to moult the previous year. Push the upper tail coverts under the card on either side, and hold them in place there with a strip of insulating tape. Use another strip of tape to fasten the card across the feather that will not carry the main antenna (Fig. 6.4), and trim the down from the base of both rectrices with fine scissors.

Find the pair of ties nearest the tag's main antenna emergence, and thread one through a needle, ideally a small curved suture needle. Holding the mounting feather to prevent its base being stressed, pass the needle horizontally through the vane 3–4 cm from the feather's emergence, so that the tag is mounted on the feather without quite touching the skin. Fine pliers are useful for gripping the needle. Wind

Figure 6.5.   A posture-sensing tag fitted to the tail of a red kite (*Milvus milvus*). Note that the main antenna is fastened along the vane of one feather.

the anterior pair of ties on the same side two or three times round the same feather and glue them firmly to it. They must be firmly anchored, but need not be sewn in place. The ties on the other side of the tag are tied fairly loosely round the adjacent feather, and the knots sealed with glue. In this way, both feathers support the tag, but they can moult independently.

Use the ends trimmed off the ties, or similar thread, to bind the main antenna to the feather vane close to the transmitter. At this point, an epoxy adhesive should be used to seal the knots, to glue the tag to the main mounting feather and to fasten the main antenna for its first 1–2 cm to the vane. Make sure that there is no gap into which the bird might get its beak between the feather and the main antenna. Then use fine thread to bind the main antenna at 1–2 cm intervals along the rest of the feather shaft (Fig. 6.5), with a more flexible glue than epoxy adhesives. Evostik or Casco contact glue adheres well to the feather and to some antenna materials.

With practice, the whole mounting process can be completed in 20–25 minutes, although it may well take 40–45 minutes the first time. The bird may struggle briefly two or three times during this process. About 1% of goshawks show more severe symptoms of distress, panting very rapidly and struggling convulsively every minute or two. They die of cardiovascular collapse if the tagging process is continued. Such birds should be unwrapped without attempting to complete the mounting, and preferably released immediately.

The use of more than one feather to support a tag is a compromise between the need to ensure a firm mounting, without the feathers being stressed enough to moult prematurely, and the need to avoid hindering them spreading and moulting. Ideally, the tag should be fastened to one or both of the two central rectrices, which tend to moult together. However, for tags that are relatively large compared to the tail feathers, especially on small passerines (Johnstone, 1992), attachment plates may be needed. The tag can be mounted on a perspex plate attached with fine perspex bolts between the feather shafts to a similar plastic plate on the under side of the feathers (Fig. 6.6). Some glue may also be used, to anchor the tag really firmly, and to seal the nuts in place.

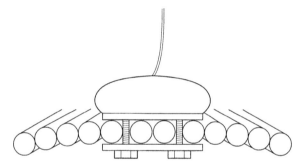

Figure 6.6. A tail clip design. The tag is mounted on a plastic base-plate on top of the tail feathers, with another plate screwed tight against the under-surface of the feathers.

When the tag is mounted dorsally on the tail, which is easiest for the biologist. it is also most accessible for the bird. However, the worst that goshawks have done is to remove part or all of the ground plane antennas. Tags without ground-plane antennas can be attached to the ventral surface of the tail where they are kept warm, dry and out of sight (Samuel and Fuller, 1994). Ventral tail-attachment sometimes works on parrots. However, it may be best to avoid ventral attachment for birds likely to copulate before they moult the tags.

6.3.4 Necklaces

The necklace design in Fig. 5.13 is among the easiest of all tags to mount. Once the necklace cord has been threaded through the tube at the other end of the tag, slip the loop over the bird's (or mammal's) neck, and knot the cord below the tube at the required length. The knot can be sealed with glue, but it is easier to re-use tags if you sew through it with thread, which can be cut later. If the free end is not too long and you bind it to prevent fraying, it can be left to hang. This avoids shortening the ground-plane antenna and simplifies re-use of the tag.

Necklaces should be loose enough to let animals swallow large food items without choking. The circumference may even be slightly greater than a bird's head, provided that the neck feathers are stiff enough to prevent the tag slipping forward from an attachment position well down the neck. It is quite important to err on the generous side: enough room for swallowing an acorn was assumed to be adequate for pheasants, until a laying female tried to swallow a large snail. In contrast, a tighter fit gave no problems on Swedish black grouse (T. Willebrand, pers. comm.), which are similar in size to pheasants but probably never eat anything much larger than birch catkins.

6.3.5 Collars

Similar considerations about swallowing apply to the attachment of collars, which need to be fairly tight to prevent them being shed or bumping against the animal's neck every time it moves. Those who study large mammals usually have rules of thumb, such as leaving room for three fingers or a hand to slide under the collar. To reduce the risk of chafing on the neck, the collar should be fastened at the side, with any metal fittings covered or at least smoothed on the inside.

Adjustable collars for small mammals are attached in several different ways. With an open wire loop, push some heat-shrink tubing down one wire before the insulation is stripped from the ends (see Fig. 5.6). After joining the bare wires, shrink the tubing over them to strengthen the loop and to give the bare wires some protection against corrosion. Slip a piece of thick paper or thin card under the tubing to insulate the animal from the heat. With closed loops, fine-nosed pliers are used to tease the double folds of wire outwards to the right circumference (Fig. 6.7), after which the heat-shrink tubing is rubbed with a portable soldering iron, or a penknife blade heated with a cigarette lighter.

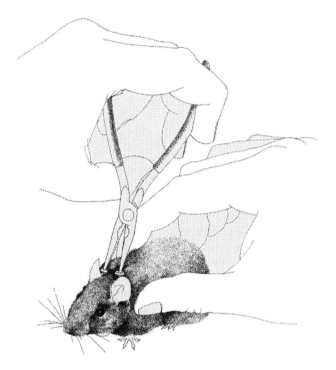

Figure 6.7. Using fine pliers to tease apart the 'sliding-8' and thus tighten a tuned loop collar on a small rodent.

Large tuned-loop collars (see Fig. 5.14) are usually closed with screws. You can avoid catching the animal's fur in the join by sliding paper or card under the collar while the screw is inserted and tightened. Squirrels and some other species can be held round the thorax while a collar is slipped over their head, using a finger on the same hand to steady the collar while the other hand inserts and tightens the screw. Squirrels that are gripped too tight during tagging may go limp, in which case they should be rested loosely with their feet on the ground until they recover. If this is not done at once, they stop breathing and die. About 10% of squirrels also roll up on the ground after release, trying to get the collar off with their hind legs. If left by themselves, however, they run off within a few minutes, and there is no evidence of this behaviour recurring.

6.3.6 Implanted and ingested tags

Tags must be coated with physiologically inert compounds before implanting, and animals anaesthetised. Inhaled methoxyfluorane and intramuscular ketamine or xylacine have been used as general anaesthetics for implanting tags in mammals, birds and reptiles (E. Smith, 1980; Smith and Worth, 1980; Eagle *et al.*, 1984; Madsen, 1984). Injected ketamine is more convenient in the field than an inhalant, but is best used in combination with xylacine or medetomidine (Koehler *et al.*, 1987; Kreeger *et al.*, 1998). Benzocaine (ethyl-*p*-aminobenzoate) has been used in water for fish (Laird and Oswald, 1975), but more popular alternatives are 2-phenoxyethanol (Johnsen, 1980; Moore *et al.*, 1992; Walker and Walker, 1992) or tricaine (ethyl-*m*-aminobenzoate-methane-suphonate) as Sandoz MS-222 (Winter *et al.*, 1978; Johnstone *et al.*, 1992). Some recent studies have found metomidate preferable (Mattson and Ripple, 1989; Kreiberg and Powell, 1991). Tricaine has also been used in water for amphibia (Stouffer *et al.*, 1983; Seitz *et al.*, 1992; Madison, 1998). Korschgen *et al.* (1984) preferred local anaesthesia with 2% lidocaine hydrochloride for abdominal implantation in diving ducks, after initial tests showed that this caused no more struggling than a general anaesthetic.

One of the earliest small mammal tags was a subcutaneous implant for *Peromyscus* (Rawson and Hartline, 1964). However, studies of sea-otters and beavers reported the expulsion of subcutaneous tags, leading to infection and death of the animals. No such problems occurred with intraperitoneal implants in these species (Garshelis and Siniff, 1983; Davis *et al.*, 1984), or in the long-term tagging of several other mammals (Melquist and Hornocker, 1979; Philo *et al.*, 1981; Eagle *et al.*, 1984; Reid *et al.*, 1986; Koehler *et al.*, 1987; van Vuren, 1989). Intraperitoneal implantation can also be recommended because tags weighing 10% of bodyweight had no effect on litter sizes in 9 g *Peromyscus* (H. Smith, 1980). If tags are sutured in place, perhaps to facilitate placement of electrodes, there is a risk of entangling the bowel (Herbst, 1991). Guynn *et al.* (1987) recommended wrapping free tags in a fold of the omentum to reduce risk of adhesions.

The peritoneal cavity is usually accessed by a mid-ventral incision to minimise bleeding. However, tags that must be sutured in place (e.g. for ECG recording) seem less prone to expulsion if the incision is slightly to one side (Folk and Folk, 1980). Healing may also be more rapid if a lateral incision is used to keep the weight of the tag off the wound, and this may be essential for species with much mid-ventral cartilage. A satisfactory longitudinal placement for snakes is a quarter of the distance from cloaca to snout (C. Amlaner, pers. comm.).

If you may want to withdraw the tag later, a small loop or socket at one end will provide something to grip and thus aid removal through a small incision. Tags are typically inserted in semi-sterile conditions, using sterile instruments and gloves, and with the animal's skin cleaned with disinfectant and alcohol but not shaved or plucked over a large area. Place tags in a surgical antiseptic solution and rinse them with sterile water before insertion. Keep animals for 1 or 2 days before release, to ensure that they do not attack the sutures or show other signs of distress. Suture material must be chosen with care for implants in fish, because healing is relatively slow (Winter *et al.*, 1978), some materials attract algae, and some fish tolerate suture materials poorly.

Sensor leads are often threaded for some distance under the skin. A flexible cannula can be used to make a channel towards the lead's point of insertion, so that the lead can be inserted along the cannula from its point and left in place as the cannula is withdrawn (Sawby and Gessamen, 1974). Similarly, flexible whip antennas can be inserted under the skin to provide a long radiator close to the animal's surface. This has been a popular approach on fish, where the radiation loss from fish to water is only 10% (Winter *et al.*, 1978). Some studies have taken whip antennas from implanted tags through the skin of birds (Korschgen *et al.*, 1996c). Subcutaneous implants with transcutaneous antennas had no adverse effect on more than 300 pheasant chicks (Riley *et al.*, 1998) but were associated with high unexplained loss of ducklings (Korschgen *et al.*, 1996b). Schulz *et al.* (1998) found more problems with migration, suture closure and leakage of subcutaneous implants than intra-abdominal implants in captive doves, but less immune response; infection was considered least likely if the skin adhered to the antenna material.

Ingested tags have been used to monitor internal temperatures (Mackay, 1964; Osgood, 1970; Swingland and Frazier, 1980), and for radio tracking species which are difficult to tag externally (Fitch and Shirer, 1971; Brown and Parker, 1976; Madsen, 1984; Oldham and Swan, 1992) or where implanting is unacceptable. The tags can be moistened and inserted into the stomach with the aid of a plunger. To avoid damaging the delicate gut of salmonids, the cylindrical tags described in section 5.3.8 are best inserted through a special tube, which presents a smooth, curved anterior surface when the convex-ended plunger is at its end (Fig. 6.8). The plunger is withdrawn when the tube's tip is in the right place, and used to push the tag down the tube and gently out into the stomach (Solomon and Storeton-West, 1983). If released in a pool with good lies, salmon usually remain there overnight, but if released in rough water between pools they often run upstream for several kilometres and through several pools before resting. The rate of tag expulsion varies from species to species: salmon seem to retain the tags for weeks, whereas sea-trout often regurgitate them within 4–7 days (Solomon and Storeton-West, 1983). To encourage tag retention by snakes, a cord was tied loosely round the abdomen just in front of the tag for a few days (Fitch and Shirer, 1971).

Fish with ingested tags are best marked with an external reward tag too, to improve the chance of tag recovery and reduce the time wasted on 'lost-tag' searches if the fish is caught by an angler or other fisherman. Labels on ears or patagia can be used to indicate that mammals and birds are tagged internally. Transmitters on ears should be attached well away from the rim, or they may easily be removed in fights and during play of young animals.

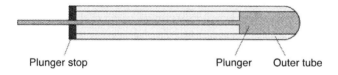

Plunger stop   Plunger   Outer tube

Figure 6.8. A plunger for inserting tags into fish stomachs. The plunger forms a smooth surface with the outer tube, and is inserted first without the tag. The plunger is then withdrawn, and used to push the tag through the tube. Redrawn, with permission, from Solomon and Storeton-West (1983).

6.4 TAG ADJUSTMENT AND DETACHMENT

Biologists working on mammal population dynamics are faced with the problem that much mortality and dispersal occurs before the animals reach full size, so that collars which are small enough to remain on juveniles will strangle them when they become adults. One solution is to use small ear tags (Serveen *et al.*, 1981) on the young animals. Even if the tags are eventually shed, they may first help to capture the full-grown animals for collar attachment. Another approach has been to use expandable collars. Expansion systems can have pleats in the collar material, held together by threads that break under pressure as the neck grows (Beale and Smith, 1973; Garcelon, 1977). Although Cochran (1980) commented that expandable collars are seldom free from problems, such as a tendency to hang low and accumulate ice masses in very cold weather (Clute and Ozoga, 1983), recent designs have proved satisfactory for bears (Strathearn *et al.*, 1984; Hellgren *et al.*, 1988) and bobcats (Jackson *et al.*, 1985).

The bobcat and later bear tags also solved another problem with collars and harnesses, in that they were designed for ultimate safe detachment. Such detachment is most easily achieved from aquatic species. For instance, if the fish tags described in Chapter 5 are made with a keel, they can be attached by a biodegradable suture which passes through the hole at one end, through the fish at the base of the dorsal fin and along a retaining plate, before passing again through the fish to the other end of the tag (Fig. 6.9). Since fish skin is delicate, it is important to sew through the musculature as well as the skin (Priede, 1980). Attachment is rapid using a suture on a ready-mounted needle (e.g. Ethicon W9415, W9418), and the tag sheds after 20–30 days. Although the sutures sometimes tear out, the injury is slight and heals in due course (Solomon and Storeton-West, 1983).

The sea is an ideal environment in which to use biodegradable materials for tag detachment. Turtle harnesses have been released by using adjacent links of aluminium and stainless steel, which form a corrosion cell (Ireland, 1980), and rubber collars on sea-otters have been released after 2 to 3 months through the corrosion of wire staples (Loughlin, 1980). Radio tags for satellite tracking have been mounted as a saddle at the base of dolphin dorsal fins, using a 6.4 mm biopsy needle to bore a hole for a retaining bolt through the fin. The stainless steel bolt, covered with a sleeve of biocompatible nylon, was inserted as the needle was withdrawn to reduce bleeding, and used with a magnesium nut which would eventually corrode to release the tag (Butler and Jennings, 1980).

Biodegradable sutures can be used to release harnesses from aquatic animals, but degrade less satisfactorily on dry land. Some projects have used elastic bands, latex harnesses (Godfrey, 1970) or other perishable materials (Karl and Clout, 1987) to release tags from birds, but this often leads to premature tag loss. However, harness straps of 4 mm wide clothing elastic were satisfactory for 3 to 5 month tags on woodcock, and had been shed by most of those recaptured a year after marking (Hirons and Owen, 1982). If double-loop harnesses are used, it is

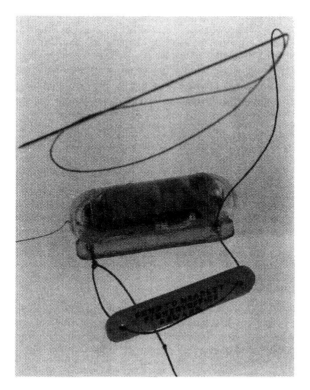

Figure 6.9. A fish tag for mounting just below the base of the dorsal fin, with suture for sewing through to the address label on the other side of the fish. (Photo by D. Solomon and T. Storeton-West.)

important that both release at the same time to free the tag cleanly. Under-wing loops held by a cotton thread through a transverse harness tube detached within a year (Hill *et al.*, 1999), and the closed tube of a side-tube tag (Fig. 6.1) could house a mechanism to release neck and body loops at the same time. Work is needed to develop compact release mechanisms that detach long-life tags reliably after a desired interval, perhaps using cell-enhanced corrosion activated by a micro-controller.

In some projects, tagged animals can be retrapped or drugged relatively easily to remove or change their tags. With this in mind, Bertram (1980) recommended that those tagging large mammals should:

1. always carry a spare tag, to replace a dying tag or to test receiving equipment;
2. routinely record the signal pulse rate of tags, to detect any changes that precede cell failure;
3. tag more than one animal in a social group, in case one tag fails, and
4. deal gently with all animals, because fear would make them harder to recapture and to observe.

These recommendations hold, in whole or in part, for all wildlife radio tagging.

6.5 CONCLUSIONS

1. Researchers should be aware that tagging is likely to affect any animal, in the short term and for as long as the tag remains attached. Theoretically, small animals can compensate for greater loading than large animals of similar shape. Energetic cost is liable to be greatest from extra loading for flying animals or from extra drag in water.

2. A slight impact may be hard to detect without rigorous tests on large samples. An impact that is only detectable for an unstudied aspect of an animal's life may be acceptable if tests with adequate samples show that it has no marked welfare implications.

3. Tests have already indicated the attachments and circumstances (species, age, season) most likely to have adverse effects. Collars have proved suitable for long-term tagging of full-grown mammals, but not for streamlined aquatic species, and even expanding collars can create risks for growing young.

4. No studies have shown adverse effects of tail-mounted tags on non-aquatic birds, but these tags last only to the moult, as do tags glued to aquatic mammals; necklaces can be suitable for birds that fly infrequently. Harnesses provide secure, non-invasive, long-life attachment with optimal transmission range, and with minimal drag if tags are low-profile and covered by feathers. However, many studies have shown adverse effects of harness-mounted tags.

5. Implants may be the only satisfactory technique for aquatic birds and mammals, for snakes, amphibia and many fish, though ingested tags can give short-term data. However, internal tags give reduced detection ranges without whip antennas, and internal or external whips on implants create risks of leakage and infection.

6. Some attachments may be fundamentally problematic, because they are invasive or create drag in water, but others only if they are used with inadequate skill. In the latter case, improved designs and training may make the difference between unacceptable effect and negligible impact. New designs should always be tested for adverse effects, as should the attachment skills of individuals when using techniques known to be problematic.

7. Methods intended to reduce risk to animals by detaching or loosening tags can themselves cause problems. Development of reliable timed-release mechanisms is desirable.

7 Radio Tracking

7.1 FIRST PRINCIPLES

Radio waves behave in much the same way as light. They are subject to reflection, refraction, diffraction, interference and polarisation, and their intensity diminishes with distance from the source. In line-of-sight transmissions, from a flying bird (or received in an aircraft), the signals roughly obey the inverse-square law: their strength is reduced by 75% if the distance is doubled. In ground-to-ground transmissions, however, the rate at which signal intensity declines may be substantially modified by absorption in vegetation, and by phase differences between direct and indirect (e.g. reflected) waves.

Radio waves can be reflected quite strongly by cliffs, hillsides, woods, buildings, and even individual rocks and trees. This is sometimes useful, because reflected waves may be detectable when direct waves are blocked. However, reflections are more often a nuisance, giving a false impression of a tag's bearing.

Both reflection and refraction are involved in the propagation of signals from underwater tags. Rays are totally reflected back from the surface if their angle of incidence is more than 6° from the vertical, and rays which pass through the surface are refracted to an increasing extent as their incident angle diverges from the vertical (Fig. 7.1). As a result, a much stronger signal is obtained at a given distance in the air straight above a tagged fish than at the same distance away at water level: elevated antennas and searches from aircraft are especially effective when tracking fish.

Diffraction, which allows electromagnetic radiation to be detected slightly off-line from the source round the edge of an impenetrable object, can give rise to interference effects around tree trunks and thus make it difficult to get accurate bearings on woodland animals. This effect, and reflection from the trees, hinders direction finding not only within woodland, but also for signals emerging from the wood into a 'Fresnel zone' which extends for about 20 wavelengths outside an abrupt woodland boundary.

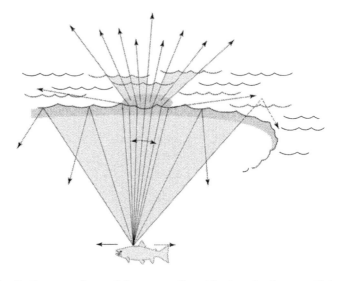

Figure 7.1. Radio waves from a fish tag are reflected back under the water if they are more than 6° from the vertical. Some waves which pass through the surface are refracted towards the horizontal, but reception is best from overhead. Redrawn, with permission, from Priede (1982).

Interference, occurring when signals reach a receiver by different path lengths and are therefore out of phase, produces very noticeable effects during radio tracking at 100–250 MHz. At 150 MHz, for instance, the wavelength is 2 m. When waves reach the receiver by pathways that differ by 2 m, they reinforce each other to produce a signal peak. On the other hand, if the pathways differ by 1 m the signal tends to a null. Peaks and nulls of this type are often noticeable when approaching tags on animals in trees. If signals on a direct path from the transmitter interact with those reflected off the ground, the signal is strong in the direction of the transmitter at one point (Fig. 7.2A), but a few paces later becomes weak in that direction (Fig. 7.2B) and may then be strongest well away from the tag's true bearing. Holding the receiving antenna with its elements vertical can help to find the true bearing again, because the vertically polarised signal is least strongly reflected from the ground. In general, however, Yagi receiving antennas should be held with their elements horizontal in woodland, even though the bearing accuracy is then somewhat reduced, because the trees produce so much reflection and diffraction interference in the vertical plane.

7.2 MAKING A START

Before starting to track fast-moving animals, it is wise to gain some experience finding stationary tags. The first step is to place one in the open, at the height and

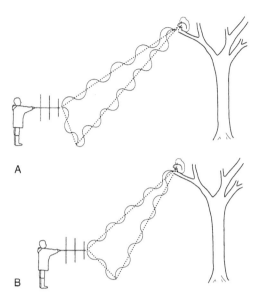

Figure 7.2. Interference from reflected waves. At A, the direct and reflected waves are in phase and produce a reinforced signal in the direction of the tag. The tracker then walks forward a couple of paces, such that the direct and reflected waves are out of phase (B). The result is a minimal signal from the tag direction, with stronger signals from other directions.

orientation it would have when a tagged animal stands on the ground. Put the tag down near a climbable, isolated tree if your study species uses these. Take your receiving equipment 50–200 m away, connect up the receiving antenna and earplug or earphones, and switch on the receiver with the gain set low. If the receiver has a separate volume control, turn this to about half the full setting. Hold the antenna where it will give the strongest signal: this is with the boom of a Yagi or the cross-piece of an H-Adcock or the plane of a vertical loop pointing towards the transmitter, but with a dipole at right angles to the direction of the transmitter. Then turn up the gain until you can easily hear the pulsed signal from the tag. If there is a lot of background hiss, try a slightly lower setting of the volume control. Leave the volume control at this 'comfortable' setting, and make any further sensitivity adjustments with the gain control.

7.2.1 Taking bearings

The polar diagram for a three-element Yagi (Fig. 2.9B) shows a marked signal peak along the line of the boom, with the shortest element (the director) towards the signal source. There is a smaller peak in the reverse direction, and relatively small peaks to the sides. Holding your Yagi with its elements vertical, turn in a circle and notice how the strong signal towards the transmitter drops off quite

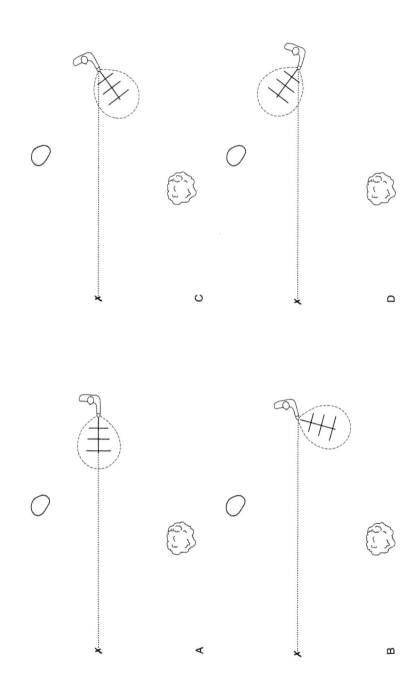

A

B

C

D

sharply as the antenna swings more than about 30° to the side, and then picks up again, but to a lesser extent, with your back to the tag. If it does not drop off when you face away from the tag, turn down the gain until there is a distinct difference between true and reverse bearings.

To take a bearing, find the peak signal position and turn down the gain until you can only just hear the tag (Fig. 7.3A). Swing the antenna slowly to the side until the signal disappears (Fig. 7.3B), and then swing back until you can just hear it again, and note a landmark along the line of the antenna boom (Fig. 7.3C). Repeat this on the other side (Fig. 7.3D). The tag bearing is the line that bisects the angle between these two directions. Now rotate the antenna boom so that the three elements are horizontal, and repeat the process. You may notice that the null directions on each side are slightly further apart than before, or perhaps slightly more difficult to find, but you should obtain a very similar bearing to the tag. Since horizontally polarised signals often give a more diffuse peak, it is usually better in open country to take bearings with the antenna vertically polarised. If you measure the bearing with a compass, hold this instrument as far as you can from the receiver and headphones, which have weak magnetic fields of their own.

The polar diagrams of loop, Adcock and dipole antennas differ from those of the Yagi in having two opposite peaks of similar strength in their reception (or transmission) pattern (Fig. 2.9A,C,D). The transmitter bearing is usually measured through one of the intervening nulls. Since the opposing peaks and nulls are symmetrical, it is harder than with a Yagi to distinguish the true bearing from the back bearing, in the reverse direction. With practice, however, this ambiguity can often be resolved.

Hold a loop antenna vertical and quite close to the front of your body. If the loop is mounted on the top of the receiver, the receiver should be horizontal at the nearest side of the loop, like the handle of a tennis racket (Fig. 7.4A). Turn in a circle, and note how there is a broad peak in the transmitter direction and in the reverse direction, with quite sharp nulls as you face at right angles to the tag. The signal should be slightly stronger when you face towards the tag, because waves are then reflected off your body onto the antenna, whereas you provide some shielding for the antenna when you face in the opposite direction. Now face in the direction of the strongest signal, looking through the loop as you hold it vertically (i.e. with the plane of the loop at right angles to the peak signal direction). Turn slightly to your left and to your right, but stop when you find the signal null: the tag is then on a bearing straight through the centre of the loop. If the loop antenna is fixed on your receiver (Fig. 7.4B), you can mark a line on the

---

Figure 7.3. Taking a bearing with a Yagi antenna. First find the direction with the strongest signal (A), making sure that it is not a reflection or back-bearing. Swing the Yagi to one side until the signal is no longer heard (B), then swing back until it is just detectable, and note a landmark in that direction (C). Repeat this procedure on the other side (D), and the tag direction is the bisector of the angle between the two landmarks.

receiver casing, through the loop axis, as a convenient sight-line for bearings. Unless the null is quite sharp, it may be best to turn down the gain until no signal is detectable at the null, and then find the directions in which the signal intensity increases markedly on either side: the bisector of these bearings points at the tag.

If the boom of an Adcock antenna is extended as a handgrip, you can use it in the same way as the loop antenna to estimate the direction of the transmitter, by turning in a circle with the elements vertical in front of you. Rather than hold the antenna across your front to find the null bearing, it may be easiest to hold it out to one side. If the antenna is well tuned, there should be a very sharp null with the boom at right angles to the line from the transmitter. If you take a line along the boom (or by lining up the vertical elements), the tag is on a bearing of 90° to this line (Fig. 7.5).

The signal received by a dipole is strongest with the antenna at right angles to the direction of the transmitter. With a short dipole in the hand, some idea of the signal direction can be obtained by using the body as a shield to help determine a tag's true bearing, as for the loop antenna. However, when frequencies around 100 MHz were being used in the UK, the 1.5 m length of a half-wave dipole prevented the body being used as an effective shield. Nevertheless, it was noticed that a dipole tilted at about 15° to the horizontal, and rotated slowly overhead (Fig. 7.6), tended to have the sharpest null with the upward pointing end of the antenna towards the tag (Parish, 1980; Fig. 7.7).

7.2.2 Estimating tag distance and position

After discovering how to determine the direction of your stationary tag, it is worth investigating the effects of tag distance and position on the signal strength. Note the gain required to just detect the signal where you are, and walk slowly towards the tag with the receiving antenna held for peak reception. The signal will increase in strength quite rapidly as you approach, and you will probably need to reduce the gain, or select signal attenuation, in order to distinguish the direction of the strongest signal. You will eventually reach a point at which you can just detect the signal with the minimum gain (and maximum attenuation). With a small-mammal collar this distance may be no more than 10 m, but it could be more than 100 m from a powerful tag.

Walk round the tag, keeping at this minimum-gain distance. Unless the tag antenna is a vertical dipole, you will notice the signal strength change, reaching a peak when you are sideways on to a loop or a horizontal whip. The signal strength will be least when you are at the axis of a tag's loop antenna, or end-on to a tag's whip antenna. If your study species generally stays on the ground, note the maximum distance at which you can hear the signal with the system set for

---

Figure 7.4.   Taking bearings with a loop antenna. The loop is held to one side to find the signal peak, and the null bearing taken looking through the loop (see text for details).

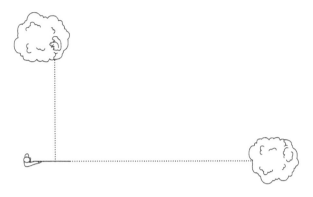

Figure 7.5. Taking a bearing with an H-Adcock which has an extended boom. The boom is used like that of a Yagi to find the centre of the null between the two signal peaks (see Figure 2.9C), and the tag bearing is then at 90° to the direction along the boom (on either side).

minimum sensitivity: if you can hear a tagged animal with this setting in the field, you will be within that distance of it.

If your study species climbs or perches in trees, put the tag 4–5 m up the nearby tree, with the same orientation as before, and again find the maximum distance at which you can hear it with minimum receiver gain. Unless your tag was originally quite high above the ground, the distance will have increased very appreciably. If you are working with animals which must not be approached within a certain distance, to avoid disturbing them or putting yourself at risk, you may wish also to note the gain setting for just detecting the signal at that distance.

If you return to the position you first used for direction-finding, you should find that the signal from the elevated tag is much stronger than when it was near the ground. Moreover, the signal strength will increase less rapidly than before if you walk towards the tag again. This effect can be useful when gathering location data on species that are sometimes in trees and sometimes on the ground. For instance, a relatively weak signal from within a squirrel's normal range followed by a rapid increase in signal strength as one approaches, is a cue that the animal is on the ground: this is often confirmed by a very marked further increase in signal strength as the disturbed squirrel runs up a tree.

As you walk towards the trial tag in the tree, you may notice the sort of peaks and nulls mentioned at the start of this chapter, resulting from interference between the direct wave and the ground wave. Another effect which sometimes makes direction-finding difficult at close range is signal swamping in the receiver. Many people unwittingly achieve this effect by turning the volume

---

Figure 7.6. Taking a bearing with a dipole antenna, which is tilted to emphasize the null towards the tag (see Fig. 6.7). Reproduced from Parish (1980), with permission.

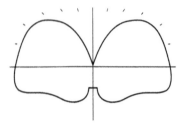

Figure 7.7. Using a horizontal dipole, the tag bearing can sometimes be discriminated from its back-bearing by tilting the antenna. With the dipole tilted at about 15° to the horizontal (Figure 7.6), the sharpest null is in the direction of the tag.

down, instead of the gain, to prevent the signal getting too strong in the headphones as they approach the transmitter. The receiver then gives much the same audio output, with the antenna in a peak direction and the automatic gain control cutting down the input, as with the antenna in a null position and no AGC action. The best remedy is not to change the volume control after the start of a tracking session.

7.2.3 Three-dimensional locations

For many species, radio tracking is essentially in one plane. In other cases, signal strength at a particular distance can indicate whether the animal is in a tree rather than on the ground, or in shallow rather than deep water, but no third-dimension bearings can be taken to show how high or deep it is. This may be either because the animal is too shy to be approached closely or because of refraction at a water surface. In some circumstances, however, animals can be tracked in three dimensions, to determine their height in trees or on cliffs or buildings.

Although loop and Adcock antennas can give elevation bearings, only a Yagi is really practical for three-dimensional tracking. If you keep the boom of a Yagi horizontal while approaching the tree that contains your stationary tag, you may notice that the signal weakens as you get very close and walk past under the tag, before peaking again as you continue on the other side. Turn round and make another approach. When you are close to the tree, sweep the Yagi through the vertical plane, with the elements first vertical and then horizontal. The signal will tend to a maximum when the antenna boom points exactly at the tag. Interference effects between the ground wave and direct wave may tend to reduce the accuracy of this elevation bearing, particularly with the Yagi elements horizontal. When you are almost under the tag, you may even get the strongest signal with the antenna pointed at the ground, probably because of reinforcement from the reflected wave (see Fig. 7.7). If the signals become confusing, try moving round the tree and checking the tag's bearing from several positions. If the animal is in a hole, you may get a much stronger signal from the side on which the hole opens.

7.3 PRACTICE TRACKING

After learning to take bearings and estimate distances to a tag in the open, try to find some tags that someone else has hidden in your study habitat. The less flat and open the habitat, and the more rapid moving your animal, the more you will benefit from the practice. Equip yourself and a companion with detailed maps of the area, and have tags hidden five or more times in the places which your study species might visit: up trees, in burrows, or in streams under a bank overhang.

This exercise can be fun, especially if you use the excuse to spend a day in the country with a friend or spouse, but do allow plenty of time. It may well take you a couple of hours to find each of the first tags. To avoid making the exercise too difficult, two or three tags should be detectable from your starting point. However, it is a good idea to have another two or three which require a search in order to detect the signal, especially if your study animals will quite frequently go out of range.

7.3.1 Triangulation and homing

When you first get a bearing to a hidden tag, do not immediately rush off in that direction. A weak signal could come from a tag on or under the ground quite a short distance away, or in an elevated position more than 1 km away, especially if you are using cars to practice tracking wide-ranging animals. Moreover, the direct route to the tag may take you through thickets or streams, which could be avoided by taking a less direct road or path. You should first take one or more cross-bearings to triangulate the tag's position, and plan your approach accordingly.

Try to take bearings from sites that will ensure reasonable accuracy. Ideally, you should be in the open, well clear of rocks, buildings, power lines, wire fencing, large trees or plantation edges, which might reflect or otherwise interfere with the signals. If you cannot avoid being surrounded by rocks or buildings, try to get on top of the highest one near you. At worst, obtain an approximate direction for the signal from several different positions, and then put the largest obstacles behind you when you take the bearing. Check the bearing from one or more nearby positions, with the elements both vertical and horizontal if you are using a Yagi. Vertical elements should give the most accurate bearing in the open, horizontal ones in woodland. If the signal is coming from a wood, you will get your most accurate bearing from at least 50 m (but preferably more than 100 m) outside it, not from among the trees.

To check that a bearing is not being affected by signal bounce, it is a good idea to repeat the bearings from 1–2 m towards the tag, and then again to one side (or both) of your position as you face it. In all cases, the estimated bearing should be the same and the nulls on each side should be equally sharp. If a null on one side is not sharp, it is probably being affected by a reflected signal; the bearing should be treated as suspect and a better position sought.

To take a second bearing, move about 200 m to the side of the first bearing. If both bearings are almost identical, the tag is a long way off, and you should arrange your route to take your next bearing at least 200 m further off the line of your first bearing when you have gone several hundred metres towards the tag (Fig. 7.8). Before you set off, make sure that you are not heading in the reverse direction to the tag! If the signal strength is not obviously weaker in the opposite direction to your bearing, try to place a large obstacle, such as a ridge, rock, building, or even a vehicle between you and the presumed position of the tag: if the signal is not appreciably weaker, and especially if it becomes stronger, suspect a back-bearing.

If the angle between the two bearings is at least 20°, and you are confident that the bearings were fairly accurate, then you have a rough idea of the tag's position. You would be wise, however, to plan another bearing from off the line of the first two as you make your approach. If you are a long way from the tag, you may be wise to triangulate it again when you are closer, preferably with 90–120° between the bearings so that you are just beyond the tag.

While you practice with one tag at a time, you may not bother to plot the bearings on a map. However, when you do plot bearings to triangulate locations, remember to use grid lines based on magnetic north as your reference. Many maps give these as well as the grid lines based on true north. If you lack such a map, you may need to correct your bearings for a difference of several degrees between true north and magnetic north. In very homogeneous habitats, such as open grassland or a large area of woodland, you may also have trouble determining your own position. Bearings are little use if you do not know their origin.

One way of finding your own position in unmapped country is to put out a grid of numbered markers before you start tracking animals (Sanderson and Sanderson, 1964). A second solution is to use a GPS receiver to determine your own position, in which case you will need to use GPS with differential correction (fig. 3.3) to obtain adequate accuracy when tracking tagged animals. In landscapes where reception of satellites leads to consistently poor GPS accuracy,

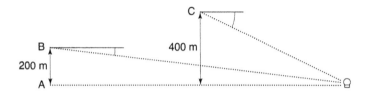

Figure 7.8. Taking bearings to triangulate a tag. The first bearing is taken from point A, after which the tracker moves across the line of the bearing to point B. There is a small angle between the bearings, so the tag is relatively far away. The tracker then moves closer to the tag, but even further away from the line of the first bearing. This produces a relatively large angle between the first bearing and the bearing at C, so that the tag's position can be estimated with enough confidence for a direct approach.

as in heavily forested terrain, a hybrid approach is to place markers using GPS, perhaps by extending equipment above the canopy to achieve adequate accuracy, and then base your tracking on the markers (Taulman *et al.*, 1998).

As you home in on a tag, and especially during the final approach, it is easiest to keep a check on the direction by swinging the antenna from side to side. This provides a sort of running average of the peak (or null) signal direction, and is especially useful in woodland or other areas where the bearing is often inaccurate. It also indicates when you start to walk past the tag, so you can adjust your course accordingly. Swing through quite a large angle, and not just where you expect to hear the tag: directions can be especially inaccurate at short range because reflections from small objects are relatively stronger. When you are very close, also swing the antenna in the vertical plane to check whether the tag is up a tree.

Once the gain has been turned down to minimum, you may have trouble deciding where a fairly loud signal is coming from, especially if you are not using a Yagi. At this point you can disconnect the antenna, turn up the gain until you can detect the signal again, and use the receiver itself to 'hunt the thimble': the signal strength increases as you move the receiver closer to the tag. If you can not hear any signal without the antenna, a straightened paperclip or other short wire can be inserted in the centre of the antenna socket to increase the gain a little.

While you practice locating hidden tags, it may be wise to attach a streamer or other visible marker to them. After all, you would probably spot the tagged animal itself, or not wish to go closer, if you were close enough to detect the tag without the receiving antenna. Nevertheless, you may sometimes have to use the receiver alone to find shed tags, or dead animals that have been cached under ground with no visible sign. If this is a frequent occurrence, you can make a loop of about 15 cm diameter in the end of a piece of 50 W coaxial cable, which is threaded through a short length of plastic piping as a handle. This can be used as a loop antenna over short distances. The peak in reception as the loop is moved to surround a tag can even be used to locate unseen tags under murky water (Solomon and Storeton-West, 1983).

7.3.2 Tags in obstructed country and towns

Triangulation from close at hand becomes practically impossible in rugged country, especially in canyons, and in their man-made equivalent, between rows of buildings. You may be able to hear a signal, but the bounce from vertical reflective surfaces makes direction-finding a nightmare. In these conditions, apply the maxim 'if in doubt, gain height'. Seek any available vantage-point above the obstructions to obtain an initial bearing, even if you have to move away from the expected tag location to find a hill, church tower or high-rise building.

If you then get a reasonable bearing on a signal, see if you can also spot another vantage-point from which to obtain a second bearing, so that triangulation provides a point from which to start a search. Without such a second bearing, you can only conduct a search along the line of the bearing that you have. If you cannot hear the signal from the second vantage-point, the tag is probably close to where you first heard it, and you can start a search there.

The search aims to identify the gap between obstacles from which you get the strongest signal, to move through that gap, then to repeat the process provided that the signal has become louder. If the signal is definitely weaker after passing a gap, you probably chose a gap that has taken you away from the tag, not towards it. Gaps in towns are often roads, so you can do this 'gap-gain' search quite effectively in a vehicle with a directional antenna or, after some practice, with a roof-mounted dipole. Whether in a vehicle or on foot, it is important to keep moving all the time, so that you are averaging signal strengths across the interference peaks and nulls caused by reflections

It is a mistake to start a gap-gain search as soon as you detect a signal, unless you have independent indications that the tag is nearby, such as lack of time for it to have moved far. The tag may be distant, in which case the signal strength will not change markedly when you go through a gap. It can then be very difficult to decide whether a change in volume is because you have really moved closer to the tag. or because of better signal reflections. Gap-gain searches are only effective when you are quite close to a tag, and ideally you need triangulation to help you reach that point.

7.3.3 Lost signals

If you cannot hear one of the hidden tags, gain height. Raising the receiving antenna only 4 m above flat ground can double the reception range. The best vantage-points are sharp hilltops or ridges, tall buildings and look-out towers (Fig. 7.9). However, even climbing onto rocks, trees or vehicle roofs can make all the difference between success and failure to detect a faint signal. If you are going to study wide-ranging animals in flat terrain, it is well worth having a hand-held mast of light tubing (e.g. plastic or aluminium), with a clip at the top for the antenna. You might also consider attaching a telescopic mast to your vehicle (section 7.5.1).

Once you have lofted yourself or the antenna, or both, hold the antenna above your head and swing it slowly round the horizon. You need only swing a loop, Adcock or dipole antenna through 180°, but a Yagi should be swung through a complete circle twice, once with its elements vertical and once with them horizontal. If you are using a Yagi with more than three elements, or a tag with a leisurely signal rate, it is especially important to swing the antenna slowly. Otherwise you may swing right past the tag bearing during the period of silence between signal pulses.

If you still cannot hear a hidden tag, then check the battery strength and antenna connections of your receiver: there should be a noticeable change in the strength of the background 'hiss' as the antenna is plugged in. Failure of antenna connections, in worn coaxial cables or inside a receiver that has a connector with a loose centre-pin, is probably the single most common cause of equipment giving poor reception. If your equipment is working properly, then the helper who hid your practice tag

Figure 7.9. A look-out tower in Kazakhstan, which was built to spot fires but also provides an ideal high point for detecting signals from distant tags.

has been careless or cruel: you can either start a mobile search (section 7.5) or give up and go home!

7.4 SIGNALS FROM TAGGED ANIMALS

7.4.1 Movement cues

When you listen to a tag on a moving animal, the signal strength often fluctuates quite noticeably. This amplitude variation results partly from changes in the orientation of the tag's antenna, which can direct peaks or nulls at the receiver, and partly from changes in the signal path, for example when the animal moves behind obstacles. The extent of these variations depends on the tag design and the habitat. Thus, there may be little variation in the signals from tags with vertical whip antennas in open country. On the other hand, signal volume can change dramatically from one moment to the next as an animal with a loop antenna forages on the woodland floor. Not only is there a null when the animal faces directly towards or away from you, but the signal also attenuates when the animal is behind large trees or in ditches. There may also be slight changes in signal frequency as the tag's antenna moves in relation to the animal or brushes against vegetation, especially if this is wet. These frequency changes are especially pronounced for UHF tags (Lawson *et al.*, 1976).

The flight of birds or bats can produce similar, if less marked, changes in signal amplitude. Accipiters (true hawks), for example, frequently change hunting perches in open country by dropping and flying inconspicuously close to the ground, before swinging up into the crown of another tree some distance away. The steady signal from a perched hawk drops to a weak, fluctuating signal as it flies near the ground, finally increasing to a steady signal again at the end of the flight. On the other hand, hawks hunting in coniferous woodland often rise to fly above the canopy between hunting perches, which produces a characteristic increase in signal volume during flight. The latter increase is most marked if the receiving antenna is horizontally polarised, and thus receives a stronger signal from the horizontal antenna on a flying hawk than from the more vertically oriented antenna on a perched hawk (Widén, 1982). Tail-mounted posture-sensors give an even more precise indication of flight, by changing to a rapid pulse rate as the tail becomes horizontal, provided that one discriminates between this rapid, fluctuating signal and the rapid, steady signal from an incubating hawk (Kenward, 1985). This is by no means the end of the story, however, because soaring in circles produces a characteristic signal too, which waxes and wanes regularly in intensity as the transmitting antenna alternately directs peaks and nulls at the receiver. The original single-stage tags also gave marked frequency fluctuations when a hawk grounded its antenna in water while bathing, but this is less detectable with modern tags. Nevertheless, a total of five behaviours can be interpreted from tags tail-mounted on live hawks: perching, soaring, other flight and, with posture sensors, feeding and incubation.

The converse of the activity cues are the indications that a tagged animal is dead. The signal then remains steady; it is often relatively weak because the corpse is lying on the ground, and it does not move between tracking periods. Tagged goshawks have a tendency to clutch the ground with their feet as they die, so that they give the same rapid signal as an incubating hawk (albeit much weaker), whereas other bird species often die on their backs.

You may well only be able to interpret the different signals if you can first watch some tagged animals. The opportunity to do this, together with the need to check that a tag has no marked adverse effect on its wearer, are two good reasons for tagging one or two captive animals before starting fieldwork.

7.4.2 Tracking moving and distant tags

Sometimes tagged animals can be detected throughout their range from one or more vantage-points. In other cases, however, they have to be followed to different parts of their range if they are to be detected. This can mean that tracking is quite difficult during the first few days of monitoring each individual. Thereafter, however, you will not only have discovered the points from which you detect signals most easily, but also have gained an understanding of the individual's habits, so that you can often predict where it will go next (Macdonald, 1978).

Before you reach this stage, you will probably face the problem of signals which go out of range while you are tracking them, especially if you are studying birds or bats which forage several kilometres from their roosts. A typical example would be for such an animal to disappear for an hour or two, and then be back at its original position when you return from a long and unsuccessful search. The best policy when tracking such animals is not to let them get too far away. It is not a good idea to sit on a nearby hilltop and wait until the signal disappears before following it, because the animal may then be far away in a well-shielded gully. Keep close to the moving tag, so that you can most easily estimate its direction as it moves away from you. If the signal fades sharply and disappears at any point, it probably means that the animal has gone over the brow of a hill: move quickly to check the next valley. If the road pattern obliges you to make detours, stop to take a quick bearing at one or two convenient high points on the way, but aim to get back close to the animal as soon as possible.

You may lose the signal in the end, especially if you are relatively inexperienced. If the animal is on a foraging trip from which it will soon return to a known site, then your best option is probably to return there and wait for its next excursion, using the intervening time to study the map and decide where it might have gone. If the animal is unlikely to return to a central place, then you should start searching for the signal from nearby high points. Aim to work along the line of the last recorded movements, although you may try first from an unusually good vantage-point off to the side or even behind you. If possible, however, stay initially within the minimum likely detection distance from the line of movement.

7.4.3 Search techniques

At some point in your search you may pick up a very faint signal. This might last for only a few moments, while transmission conditions are optimal for a very distant tag, so get the peak signal direction as quickly as you can, noting whether the signal is steady or fluctuating in volume. Record the bearing, because you may easily forget it while searching.

A fluctuating signal means that the tag is moving. In this case, the signal may be received only briefly, while the animal is in a suitable position; a typical example would be a tag on a bird that briefly flies above or between hills that usually block the signal. At such times it is very important to get a bearing to the tag as quickly as possible, but the fluctuating signal hinders the use of nulls for accurate direction-finding. The best approach is first to find the general direction of the strongest signal, swinging the antenna rapidly between this bearing and the reverse direction several times until it is certain which is the true direction rather than the back-bearing. This may be difficult if the detection is made from a site with many reflections, although checking the signal strength on both sides of an obstacle can be a good way to eliminate back-bearings.

Only after getting an unambiguous general direction is it sensible to seek a more accurate bearing, by swinging the antenna repeatedly through a wide arc until the direction for the peak signal is narrowed to a rough bearing with accuracy of 10° or so. Only if the animal stops moving is it worth trying to use nulls for an accurate bearing. Otherwise, once a rough bearing has been obtained, and no improvement is likely through a short movement to a better position, such as a hill-top, it is best to seek a cross-bearing for an approximate triangulation.

The next question is where to go for the next bearing. If the signal was weak and tended to disappear while finding a rough bearing, the animal may well be far away. It is therefore wise to select for the next bearing a top quality high point, and to select one about a quarter of the maximum detection distance from the line of the bearing. Putting this distance between the two bearings should increase the angle between them and hence improve the estimation of the tag's location. If the first bearing could have been a back-bearing, the direction to your next check-point should be perpendicular to the bearing from your initial position, but otherwise you could move at 40–60° towards the tag.

Quite often, however, the next detection attempt fails. The most common cause is that the animal is in a location that does not transmit well to the new receiving position, perhaps because a dispersal movements is taking it further away and behind obstacles. However, the tag might also have been detected initially because of a lucky antenna orientation when a predator was eating the tagged animal in a ditch, or because a broken antenna was temporarily in contact. The choice is then whether to move further towards a possibly distant tag, or back towards a possible origin for a signal that would have been too weak to detect from the second check-point. That choice will be decided by ease of movement for the tracker and the relative likelihood of dispersal, death or tag degradation.

If the signal was detected again, and the pair of bearings were likely to have been fairly accurate, then it is reasonable to make an estimate of the animal's position and immediately move close to it to take two more bearings from within the minimum distance for the required resolution. If either bearing may not have been accurate, then it is probably worth taking at least one more from across the line of the other bearing, moving towards the animal and seeking an angle of 40–60° between the bearings.

The requirement for rapid discrimination between true-bearings and back-bearings during this sort of tracking makes a unidirectional antenna desirable, which is why most biologists now use Yagi antennas. However, note that a Yagi with more than five or six elements is less than ideal, because its narrowed angle for peak reception means that rapid swinging can easily move it through the line of the signal between pulses.

It is also worth noting that frequencies may change slightly when tags are fitted to animals, due to capacitance effects. The tuning may also change after running for a few days, due to an initial fall in cell voltage, which occurs within hours for small tags or after several days for large lithium cells. It is therefore wise to record frequency and signal repeat rates after fitting tags and again after cell voltage is likely to have stabilised. That means changing all data sheets or receiver programs, so that if an animal is lost the most up-to-date records are used in the search.

7.4.4 Disappearances

If an animal with which you are familiar cannot be detected from any of your usual vantage-points within its range, and you have checked that you are receiving properly (e.g. by listening to another tag), there are three possible explanations. First, the tag may have stopped transmitting. If it was not near the end of its cell life, and you have detected no slowing of the signal pulse rate or reduced detection distance that could precede cell failure, then this is relatively unlikely. You should in any case check the second possibility, that the animal is still within its range but transmitting a very weak signal, perhaps because of antenna failure or because it has been buried by a predator. So before looking further afield, it is worth looking systematically for weak signals throughout the animal's known home range. If a broken whip antenna is suspected, visit the animal's lairs, nests or roosts at a time when it is likely to be there. It may also be worth visiting predator lairs or other possible cache sites (Fig. 7.10), and areas where you would expect poor reception from a tag on the ground. If a tagged animal lies dead in a ravine, dyke or stream bed, there may be very poor reception unless you stand at the edge.

If you still have no signal, you may decide to undertake a more extensive search away from its normal range. Whether you search on foot or by vehicle, there are several ways to increase your chance of finding the lost tag. Aim to search at a time of day when the animal's behaviour will ensure a good reception range. For instance, if your species feeds by day on the ground but spends the night in trees, search for it after dark. On the other hand, search for animals that rest in burrows when they are likely to be out. Conduct your search as soon as possible after you lose the tag, not only because the animal may get further away the longer you leave

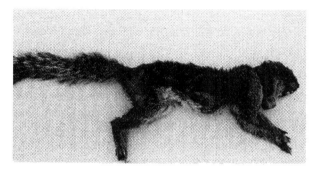

Figure 7.10. Radio tags help to find dead animals quickly, and thus to determine the causes of death. This squirrel had been cached under ground by a stoat.

it, but also because you will probably receive the signal best while it remains alive. Decide in advance on the minimum distance at which you are likely to detect the tag, and plan your search accordingly, using a map on which contour lines and other features indicate the best listening points. Do not be complacent about the distance your study species can travel. Many animals when tracked are found to travel much further than had previously been expected.

You may wish to start your search along leading features, such as hedges, ridges or lake shores that the animal may have followed. This is especially so if its last position or your previous experience with other individuals suggests that these are likely routes, but do not totally neglect to check the other directions. If you have a large area to cover, give due consideration to searching by aircraft. Large areas can usually be covered more thoroughly and much more quickly by air, especially if the terrain is rough and inaccessible. Even if you are not costing your own time, the cost of hiring a plane may not be very much greater than the high mileage cost of conducting a tight search pattern on the ground.

7.5 MOTORISED TRACKING

An important consideration for mobile searching is suppression of engine ignition noise. Take the time to check in advance that aircraft and petrol-driven cars are adequately suppressed, by running the engine at both high and low revs, with receivers and antennas in operating positions. If you make all the arrangements to use an aircraft during a window of fine weather, it is extremely irritating to install all the equipment and then discover on take-off that its electrical systems are too noisy to be usable. There can be serious noise from equipment other than the engine, including electric motors for windscreen wipers, poorly tuned radios, GPS receivers or anything else containing microprocessors. It took some time to discover that the background hiss that added after dark in a Swedish minibus came from plasma in strip-lighting. Sometimes the noise is quite weak and can be avoided by moving the tracking receiver. Alternatively, journeys can be planned for minimal use of noisy equipment. On roads or water, petrol engines can be stopped to minimise ignition noise, but the ideal ground vehicles or boats for mobile searches have diesel engines, because these produce no ignition noise.

7.5.1 Road vehicles

If you are routinely tracking over a large area in a vehicle, you will soon get to know those vantage-points which give good reception and furnish accurate bearings. If you are tracking relatively few animals, and weather conditions are usually good, you may simply climb on the vehicle roof or up to a high point near the road to enhance reception. However, if you are working with many tagged animals, or at night and in bad weather conditions, it is better to use a vehicle with an antenna which can be raised and rotated from inside. You can then sit in relative comfort to take your bearings, with easy access to maps and notebooks, without the risk of falling from an icy car roof in the dark.

Vehicles to be used as antenna mounts for mobile tracking should have easy access from the driving seat to the antenna. If the mast goes through the roof centrally just behind the front seats, drivers or passengers can rotate it. Moreover, a Yagi with five to six elements will not protrude appreciably beyond the sides of the vehicle if it needs to be turned without being raised while the vehicle is travelling. This is a convenient mast position for a Landrover (Fig. 2.12), but other vehicles, such as minibuses, have central transmission tunnels that prevent a mounting within reach of the front seats. In that case, the antenna should preferably be reached from the driver's seat by an aisle, so that a lone operator does not need to open a door in rain or freezing temperatures.

Steel tubing is suitable for a simple antenna mast (Fig. 7.11). This can be slid up and down in a second or two through a well-lubricated bush, which is mounted on ball-bearings in a short metal tube welded through the centre of the roof. The base of the mast contains a hand-grip for raising, lowering and rotating it. There should also be a device for locking the raised base to the rotatable bush, such as a pin that

Figure 7.11. A hand-raised mast mounted in a minibus. The antenna is a six-element Yagi, mounted horizontally, with a repeater compass to improve bearing accuracy.

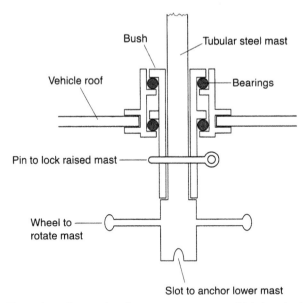

Bush

Tubular steel mast

Vehicle roof

Bearings

Pin to lock raised mast

Wheel to
rotate mast

Slot to anchor lower mast

Figure 7.12. Through-roof mounting for an antenna mast which is raised by hand. The mast must be firmly anchored in both the raised and lowered positions.

is slid through matching holes in the mast and bush (Fig. 7.12). When lowered, rotation can be prevented by cutting a vertical slot in the mast's base, which engages a pin welded across a short tube on the floor.

Lightweight pneumatic masts that are 1.5–2 m when lowered and 5–11 m when extended can be raised in 15–45 seconds by a 12 V air pump. The relatively large battery of a diesel vehicle provides adequate power, provided that several kilometres are driven between each mast operation, but reserve batteries might be required in some studies. Suitable masts, with base-plate and roof-gusset and pump, cost GB£1000–2000 (US$1600–3200) from Clark Masts (see Appendix I for address). Seals need regular lubrication and replacement at intervals to reduce leakage, which slows the elevation and prevents the mast remaining extended for long. However, if properly serviced, such masts can last for more than a decade. The gain in reception from pneumatic masts should be balanced against the time required to raise and service them, compared with simpler masts which are raised rapidly a shorter distance by hand. Tall, pneumatic masts are favoured for species that travel long distances, especially in flat country, whereas hand-raised masts have advantages for work on animals that live densely and do not move far.

The most practical antenna arrangement is a Yagi with five to six elements, with the centre of its boom fixed to the mast. Vertical antenna elements give the greatest accuracy in relatively open country, and also ensure the best reception range to tags with vertical whip antennas. For even greater range and bearing accuracy you can use a null-peak system with twin vertical Yagis, but this increases the loading and requires a stronger mast. Remember, too, that assumptions of high bearing accuracy are only justified by a relatively 'uncluttered' signal path and reception position. Moreover, vertical Yagis extend quite high above the vehicle in the lowered position, which makes them impractical in some habitats because of overhanging trees. It is therefore often most practical to use a Yagi with horizontal elements, which may extend no more than 0.5 m above the roof when lowered. An even more sophisticated approach is to use a Yagi that is stored horizontally, but has its boom rotated to bring the elements vertical when it is raised.

If you are working on a very limited budget, bearings can be taken with a compass rose fixed to the base of the mast. In this case, you should mount a compass inside the vehicle, at a point which is little influenced by the vehicle's own field (e.g. on the windscreen), and reset the compass rose to the vehicle's heading each time you stop to take a bearing. However, it is much easier and more accurate to mount a repeater compass at the top of the mast, with its display in the cab (Fig. 7.13). The compass and antenna cables should be attached to the top of each mast segment to prevent tangling, with connectors to feed them through the roof close to the mast.

Before you start tracking, you should of course zero your compass, and check that neither the compass electronics nor the vehicle lighting interferes with the receiver. Plan your route in advance so that you stop at sufficient points to cover the terrain, but travel as short a total distance as possible. During tracking, remember to rotate the antenna alternately clockwise and anticlockwise, to prevent the cables twisting up. If possible, a stop should be present to prevent over-rotation.

Figure 7.13. The display for the repeater compass of the Landrover in Figure 2.12.

It is extremely important to avoid raising masts under power lines or other wires. This mistake, which is easily made at night, not only gives erroneous bearings but can be fatal! Similarly, masts must be lowered before moving off. They should be obtained with a sensor that switches off a red light when the mast is fully lowered. However, if the mast is normally used with the engine off, it is best to have a loud buzzer wired to sound if the engine is started without the mast retracted. A warning light can be overlooked in the excitement of closing on a long-lost animal.

7.5.2 Aircraft

The ideal aircrafts for radio tracking are high-winged monoplanes, such as the Cessna 172, because the wing-struts provide excellent antenna mounts. Although loop and whip antennas can be mounted under the fuselage of other aircraft, attachment to wing struts is usually not only easier, but also keeps the antennas furthest from the de-tuning influence of the wing or fuselage surfaces. In flying a search pattern over a large area, the reception range on each side of the aircraft determines how far apart each parallel search path need be, and thus how much flying is necessary to cover the area. The greatest sideways range is obtained by mounting a vertically polarised Yagi antenna under each wing, pointing straight to the side and 15–30° downwards from the horizontal (Gilmer *et al.*, 1981). To reduce drag, and avoid any risk of over-stressing the struts, most aerial tracking uses Yagis with three to four elements. These elements should ideally be thin and on a streamlined boom: the drag on a pair of robust antennas, which are normally hand-held, is enough to slow the aircraft by up to 5%.

For optimal reception, the upper Yagi elements should be at least 15 cm below the wing surface and 30 cm forward of the leading edge (Gilmer *et al.*, 1981). Antennas can be mounted like this on a boom projecting forward from the strut, but it is difficult to fix them firmly and prevent vibration. Such vibration should be avoided, because it could eventually cause metal fatigue. Strut failure has caused a fatal crash of a light aircraft with strut-mounted military equipment (F. Anderka, pers. comm.). Although tracking antennas are relatively light, it would be wise to inspect struts carefully on aircraft used regularly for radio tracking.

A simpler attachment, which does not greatly reduce range if the antenna is mounted a bit further below the wing, involves clamping the antenna boom close to the strut (Fig. 7.14). Clamps must be fail-safe, such that neither the antenna nor clamps can detach in flight. The design in Fig. 7.15 has clips to prevent loosening of screws that hold the clamp to Cessna struts, aircraft lock-nuts to secure the section that holds the antenna boom, and the boom itself blocks any loosening of the nuts that hold this section to the strut-clamp. A layer of dense foam rubber is glued to the clamps to avoid damaging the strut, and the antenna leads are taped to the strut or wing surface to prevent possible damage due to vibration.

Figure 7.14. Side-looking Yagi antennas attached to the struts on a Cessna aircraft (left), with a close-up of the mounting bracket (right).

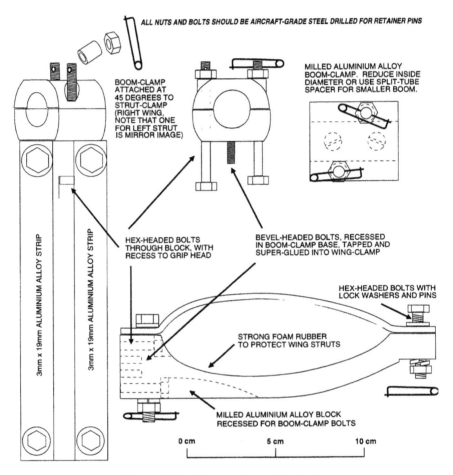

ALL NUTS AND BOLTS SHOULD BE AIRCRAFT-GRADE STEEL DRILLED FOR RETAINER PINS

BOOM-CLAMP ATTACHED AT 45 DEGREES TO STRUT-CLAMP (RIGHT WING, NOTE THAT ONE FOR LEFT STRUT IS MIRROR IMAGE)

MILLED ALUMINIUM ALLOY BOOM-CLAMP. REDUCE INSIDE DIAMETER OR USE SPLIT-TUBE SPACER FOR SMALLER BOOM.

3mm x 19mm ALUMINIUM ALLOY STRIP

3mm x 19mm ALUMINIUM ALLOY STRIP

HEX-HEADED BOLTS THROUGH BLOCK, WITH RECESS TO GRIP HEAD

BEVEL-HEADED BOLTS, RECESSED IN BOOM-CLAMP BASE, TAPPED AND SUPER-GLUED INTO WING-CLAMP

HEX-HEADED BOLTS WITH LOCK WASHERS AND PINS

STRONG FOAM RUBBER TO PROTECT WING STRUTS

MILLED ALUMINIUM ALLOY BLOCK RECESSED FOR BOOM-CLAMP BOLTS

0 cm   5 cm   10 cm

Figure 7.15. Wing-strut mounting design.

To minimise drag from the antenna leads, tape them to the training edge of the strut. Ideally, they should be fed into the cabin through an air-vent or other opening to avoid any risk of damage, but on a Cessna they reach the cabin most easily under the door. If the door edges are not rubber-clad, make sure the cables are far from a hinge, to minimise pressure that might cause internal damage. A thin strip of flat rubber cable-protector, used to keep wiring safe on office floors, can be taped to the frame as added protection.

Well-fitting, high quality headphones are essential for searching from aircraft. If your earphones do not fit well enough to reduce the engine noise, you may have to turn up the receiver volume so much that you risk damaging your ears. You should set gain for high sensitivity, and volume to a level such that a loud signal is not unpleasant. You may need to adjust the volume when in the air because of loud noises from other sources, especially near towns. Unfortunately, air-traffic control systems use frequencies close to the second harmonic of frequencies allocated for radio tagging in some countries, in which case poorly-tuned ATC equipment can be a problem near airports.

The most flexible and effective system for searching is to have a separate programmable receiver attached to each antenna. For a single operator, stereo headphones can be wired so that the output from the receiver with the Yagi on the left side goes to the left ear, and signals detected on the right side to the right ear. If there is a separate operator for each receiver, the antenna cables must be long enough to swap easily between receivers, so that you can compare the relative strength of the signal from each side by alternating their attachment to the receiver where the signal is first detected. If you are tracking animals routinely from the air, you must have a switch-box system for this, in order to avoid wear on the connectors.

If only one receiver is available, both antennas must feed that receiver, in which case destructive interference between the antennas should be minimised. It is advisable to keep their collectors close to a whole wavelength apart, with the leads the same length from each, so that signals from either side are reinforced. The leads should meet at a switch box or 'T' connector which can connect either the left, or the right, or both antennas to the receiver. To achieve maximum reinforcement of a side-on signal, especially if the antennas are not a whole wavelength apart, you can build a balun match into this box.

Well before any flying, you should check the fit of your equipment on the aircraft yourself, and ask the pilot to inspect it too. Check that you can hear a tag you have placed several hundred metres away, with and without the engine running. If you have not tracked from the air before, it is wise to put out tags in sites with normal and poor transmission characteristics (e.g. as if an animal were dead on the ground), and make a short preliminary flight to find the ranges at which you can detect them. This also provides the opportunity to check that navigation and GPS equipment are not creating unacceptable interference. You may need to move receivers around to minimise 'hiss' from the GPS.

If you are searching a wide area for several transmitters, plan your search pattern in advance. Discuss it with the pilot to ensure that turning points are easily recognised or have been programmed into the navigation GPS. There may be height restrictions in civilian airspace that make it advisable to take an alternative route in order to fly high enough for adequate detection range. It may be impossible to traverse some military zones, or at least require arrangements in advance to do so when there is no firing or parachuting in progress. The weather forecast must be given careful consideration: will the cloud base be high enough for your requirements in hilly country? Is there a convenient airfield for refuelling if strong headwinds impede your progress? Much of such planning can be left to an experienced pilot, provided you make clear exactly what you are planning. You will want to fly with an experienced pilot, because an analysis of work fatalities for biologists in developed countries might well show air-survey crashes to top the list. Remember the adage 'there are old pilots, there are bold pilots, but there are no old, bold pilots'. Ensure that you can communicate adequately with the pilot. You may need to change quickly to a communication headset. If you track frequently, it is best to obtain a headset that can switch between verbal communication and listening for tag signals.

If tags are on animals that do not move far and cannot be detected much more than about 10 km away, it is usually easiest to fly a rectangular search pattern. The parallel transects should be separated by less than twice the sideways distance at which you expect to detect a tag. If the animals have spread from one small area, as in the early stages of dispersal studies, you may wish to start near their origin and work outwards, either in all directions (Fig. 7.16A) or in the direction which the topography leads them (Fig. 7.16B). If they may be anywhere within an area, then you will probably need a regular grid search (Fig. 7.16C), although you may decide to try certain favourite sites first. For instance, when goshawk dispersal was monitored on the Baltic island of Gotland, the grid search was from north to south, but the first step in the search was usually to circumnavigate the coast from the seaward side. This was for two reasons: not only did dispersing hawks frequently settle in coastal areas, but signal transmission inland was also blocked by cliffs if hawks were on the ground near the shore.

You may have to modify the flight path in other ways to take account of local topography. For example, signals from bats roosting in caves may only be detected as you traverse the cave entrance. You may also need to fly higher above rugged terrain than the 300–1000 m above ground level (AGL) which is adequate for tracking weak tags in lowland habitats. For example, when searching the southern half of Great Britain for dispersed raptors, search paths were closer together over the uplands of Wales (Fig. 7.17). Note also that well-separated search paths made it practical to

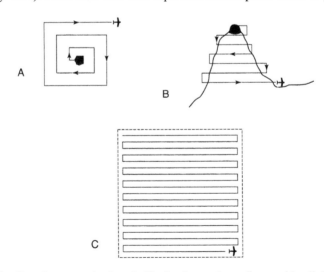

Figure 7.16. Search patterns in aircraft. If animals may have dispersed in all directions, the search may spiral outwards from a study area (A). If animals must have moved in one direction, perhaps along a peninsula, search along their likely path (B). If the tagged animals could be anywhere within a large area, then fly a grid pattern (C). In all cases, the distance between adjacent search paths is twice the minimum sideways detection distance for the tags.

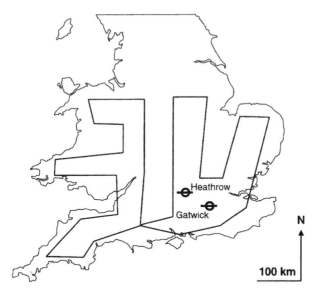

Figure 7.17. An aircraft search path that covers most of southern Britain in about 10–12 hours at 90–100 knots. The ground tracks are generally not more than 80 km apart and from the coast, which is appropriate for tags with a detection range of 40 km, except for the area around London where flying in light aircraft is restricted by Heathrow and Gatwick airports.

search a large area of Britain in 10–12 hours, because increase in tag efficiency has more than doubled the detection distances. The 10–20 km detection range of tags on Gotland in the 1970s has now increased to 40–80 km from the air when on live hawks.

Although simple receivers can be used for occasional searches for one lost tag at a time, systematic searches of this sort require programmable receivers. Switching between tag frequencies otherwise creates unacceptable fatigue and wear on the tuning knobs (not to mention your fingers!). Make sure that the receiver battery is fully charged before the flight, and enter the frequencies into the memory. The dwell time on each frequency should be set so that all frequencies are covered at least once while the aircraft's forward track is about two thirds of the sideways detection limit. Thus if the sideways detection limit is set at 4.5 km (i.e. search transects 9 km apart), a scan of all frequencies should occur every 3 km. If the maximum ground speed, which is the airspeed plus the wind speed if flying downwind, is 180 kph, then all frequencies should be scanned once per minute. This allows a dwell time of 4 s on 15 frequencies, or 10 s on 6 frequencies. Dwell times should not be reduced below about 3 s per frequency, even if tag pulse rates are quite fast. If interference occurs intermittently on particular channels, from bursts of telemetry data or voices, you may need to stop the scan briefly on those channels to obtain a clear interval.

When you think you can detect a signal, stop the scan and switch quickly to the antenna on one side and on the other: the tag is on the side with the stronger signal.

If you cannot detect the signal again, ask the pilot to circle while you listen to the antenna on the outer wing. If you still cannot detect a signal, continue flying your search grid, but make a note to give the frequency special attention during the adjacent transects on either side. When you have identified the side with the best signal, you can locate the tag by: (1) signal strength; (2) triangulation or (3) homing, depending on the accuracy required.

1. The time-saving approach when searching a wide area for many lost tags, which will then be located precisely on the ground, is to record signal strength along transects. This is best done with a data sheet on which you record each frequency detected at the top of a column, then the time and a circle of a size that indicates signal strength each time the signal is detected (Fig. 7.18). It is important that the

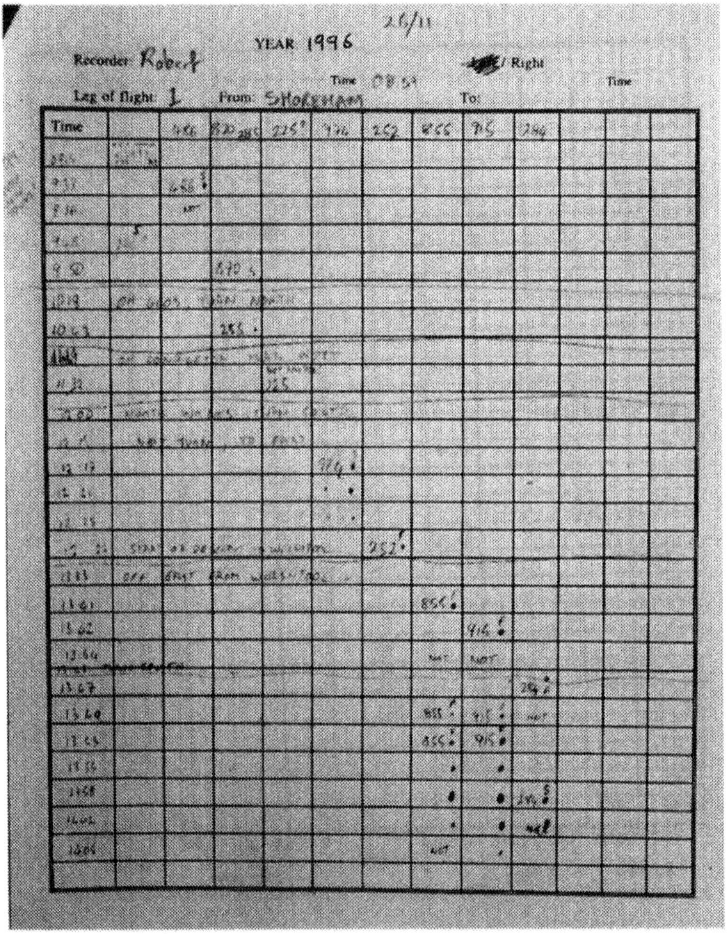

Figure 7.18. A recording sheet used in an aircraft flying from Shoreham on 26/11/1996 for the antenna on the right wing strut, for detection of different tag frequencies (in columns) at times recorded in rows, noting the strength of the detected signal from weak (dot) to strong (large circle) and whether the signal was slow (S) for a perched raptor or fast (F).

time at each turning point is recorded, so that you can later interpolate where the plane was when the signal was strongest: the tag was then at about 90° to the aircraft's heading. If you are not too busy, also note a landmark below the plane (or ask the pilot for a GPS reading) as the signal strength passes its peak. The relative strength of the signal indicates how close you are to the tag. A weak signal detectable only on one side is either (a) moderately distant from a tag transmitting poorly (e.g. from the ground) or (b) very distant and hence probably detectable on the next transect. The approximate distance can be estimated by relative signal strengths between transects. Alternatively, a strong signal equally audible on both antennas is probably beneath you.

2. If you require greater accuracy and can spare the aircraft time, ask the pilot to circle away from that side (i.e. to the left with a signal on the right). Note the headings when the signal fades and then reappears with a similar intensity: the tag bearing is very roughly at 90° to the bisector of these angles (Fig. 7.19). You

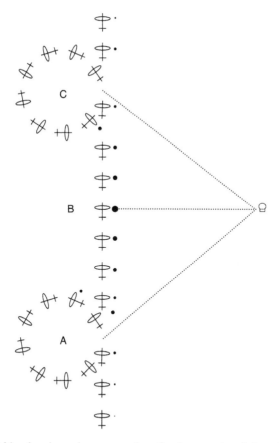

Figure 7.19. Taking bearings along a search path when an aircraft has sideways-pointing Yagi or end-fire antennas. A bearing is taken by circling when the tag is first detected (A), its bearing is at right angles to the path when the signal is strongest (B), and a third bearing is taken by circling (C) before the signal is finally lost.

can take a second rough bearing, when the signal is strongest and a third by circling again further along the same transect.

3. If you want to locate the tag more precisely, perhaps because you are dependent on aerial searches over inaccessible terrain, then you can home to it (Mech, 1983). Once you have obtained a bearing by circling, then head towards the tag. If the signal was relatively weak, you may lose it for a while, but it should reappear on one side or the other (Fig. 7.20A). Turn gently towards that side until the signal becomes equally strong from each antenna (Fig. 7.20B). You should then be heading directly towards the tag, and you may descend to 300 m AGL or even lower if the terrain permits. Keep heading towards the tag, reducing the receiver gain as you approach and using each antenna in turn to maintain the correct heading. As you pass over the tag the signal will peak and then start to fade again, especially noticeably if you are flying low. When this occurs, start a turn. If the signal is strongest on the inside wing as you turn, you have indeed gone past the tag (Fig. 7.20C). If the signal is strongest from the outside antenna (Fig. 7.20D), then you have gone through an interference peak and null instead of passing the tag: complete the turn through 360° and continue towards the tag. When you have definitely passed it, continue your turn to fly back towards the tag, listening carefully to each antenna in turn. It is unlikely that you will fly straight overhead, so the signal should be strongest on one side. Circle towards that side: if the signal remains on the inside antenna, you are circling the tag. If at any point on your turn the signal becomes stronger on the outer antenna, start circling to that side, until you are finally circling the tag at relatively low altitude.

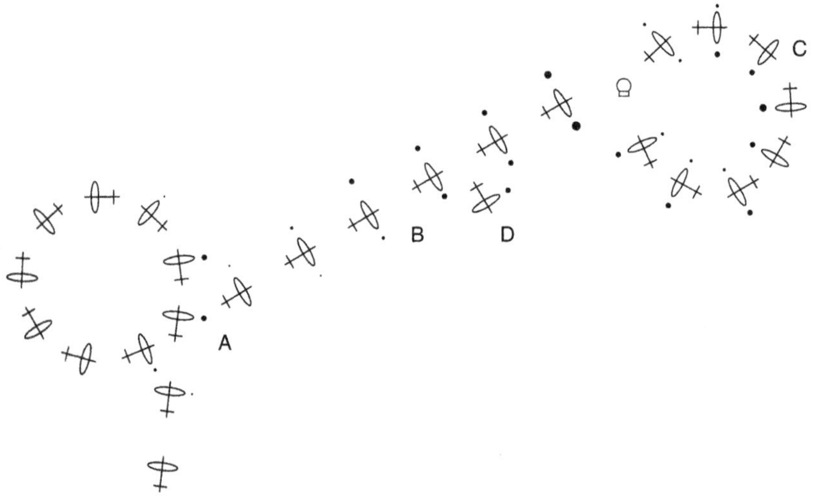

Figure 7.20. Approaching a tag in an aircraft with sideways-pointing antennas. After circling to obtain a bearing, the aircraft heads towards the tag (A), keeping the signal at equal strength in each antenna (B). When the signal fades, the plane turns. Provided the signal remains on the inside wing (C), the tag has been passed, and the plane continues towards it from a new direction. If the signal is on the outside wing during the turn (D), the turn was made too soon and the original heading must be resumed.

An alternative to the latter homing approach, which involves less turning of the aircraft and hence less discomfort, is to have a directional antenna mounted horizontally under the belly of the aircraft so that it can be rotated from the cabin during flight. In an area with a high density of nesting falcons, locations could be obtained at 5-minute intervals (Marzluff *et al.*, 1994). The system also gave higher accuracy than when tracking falcons from the ground, though partly because they were more often located at nest sites, which underestimated foraging ranges.

If you are looking for one or two animals at a time within a relatively small radius and the tags are usually detectable in line-of-sight at least 20 km away, you can use a simplified search technique that is suitable for a non-programmable receiver. This involves starting near a central point, such as an individual's last known position, and then spiralling upwards until a signal is detected or you reach, say, 5000 m (Mech, 1983). You can switch from one frequency to another, each time the aircraft's compass passes north. The higher you go, the more likely you are to achieve an uncluttered signal path from the tag. The same type of search can be fruitful for locating fish in large lakes, because the strongest signals are radiated almost vertically.

If you are tracking fish, or using an aircraft without wing-struts, you may well not be able to use a Yagi antenna system. Tags have been located fairly accurately with a vertical loop on each wing-strut, one looking forward and one to the side (Winter *et al.*, 1978). A single side-looking loop on a strut or the fuselage would probably give reasonable bearing accuracy for locating fish along a river, by flying slightly to one side and noting the position of the null between two similar peaks. Half-wave dipoles, which are best mounted horizontally and as far below any metal surface as possible, give one third to one half the range of a Yagi system, and are really only suitable for finding the general area of tags which are to be located precisely from the ground. Better gain and directionality can be obtained with an 'end-fire' array of four quarter-wave dipoles on each wing tip (Whitehouse and Steven, 1977; Whitehouse, 1980). Each antenna is one-quarter wave long and the same distance from the next, with a three-quarter wave balun match between them. The result is that signals reinforce along the line of the antennas, producing a reception peak along the wing on each side.

After making a patterned search in an aircraft, try to ground-truth a rediscovered tag as soon as possible. The ideal is to have contact with someone on the ground who can make a check on the same day as the flight, because animals which have travelled far enough to be lost may still be moving. Buzzards most distant from their nests in a particular season were also those most likely to move again before the next season. There may be occasions when an animal cannot be found from ground level, but is detected on another flight. The explanation for poor detection at ground level may be death or other factor that compromises transmission, in which case it is probably worth deviating from the grid to approach the tag and get a more accurate location from the air.

Where aircraft are being used to find specific sites, such as bird nests, bat roosts or mammal breeding dens, the added expense of using a helicopter may be worthwhile. Helicopters can home in on tags with maximum accuracy, and provide a remote landing capability which may be needed to visit distant sites and obtain data beyond that available from the tags. Helicopters may also be used to capture large mammals with radio tags.

7.5.3 Off-shore tracking

Boats for radio tracking on large lakes can be equipped with a variety of antenna systems, including loops, Adcock and Yagi antennas, on fixed or telescopic masts. Aircraft-style search patterns can be used to locate tagged fish or other animals, followed by homing to the position with the strongest, omnidirectional signal. The problem is then not so much one of finding the tag, as of determining the position of the receiving vessel. If you lack GPS, this can be done by taking bearings, with a compass or sextant, to landmarks on the shore (Winter *et al.*, 1978).

7.6 CONCLUSIONS

1. Reflection, diffraction and other interference with signals tend to hinder radio tracking. It is therefore easiest to learn to take bearings before starting to track animals. For assessing accuracy, and in some cases for safety, it is also wise to discover the distance at which tags can just be detected with minimum receiver settings.
2. Practice tracking should first use stationary tags in the open and then in the study terrain, ideally with someone hiding tags in locations suitable for the study animals. The training should involve: (i) learning to gain height in order to detect signals from distant tags; (ii) avoiding signal bounce when triangulating; (iii) homing in on tags and (iv) using probe antennas or receivers alone to find nearby tags from signal strength alone.
3. In towns, 'gap-gain' homing can be used to locate tags' by using gain in signal strength on moving through gaps between buildings.
4. Practising the techniques is important not just for routine tracking but also when animals emigrate or die. It is important also to learn the characteristic signals produced by movement of tagged animals, especially if sensors are used to indicate activity or posture. Movement cues can be studied by tagging captive animals.
5. Boats or road vehicles with mast-mounted antennas are useful for regular mobile tracking; diesel engines give no electrical interference.
6. Aircraft can save much search time; wing struts, as on Cessnas, aid antenna attachment. Priorities should be placed on flying with experienced pilots, on careful planning of routes, and on rapid ground-truthing following wide-area searches for lost tags.

8 Collecting Data

The risk of marking bias is as great in radio tagging as in other studies and there is an added risk of performance bias through the impact of the tags on the animals being studied. The great strength of radio tracking is the potential to avoid observation bias, by systematic sampling. Radio tracking is particularly effective for avoiding visibility bias when recording locations of animals that forage by night (Beyer and Haufler, 1994; Berdeen and Krementz, 1998) or in concealing habitats (Troy *et al.*, 1992). It is also valuable for checking for bias in other techniques, that may be more convenient for widespread application. Thus, radio tracking has confirmed that some visual techniques represent animal presence accurately, for example the use of faecal pellets to estimate herbivore distribution (Loft and Kie, 1988; Edge and Marcum, 1989; Forys and Humphrey, 1997). Conversely, bird singing territories do not necessarily represent home ranges (Hanski and Haila, 1988) and predation studies with artificial nests tend to overestimate the impact of visual-cued nest robbers (Willebrand and Marcström, 1988).

However, the opportunity to avoid bias is often overlooked. This chapter aims to explain why systematic sampling is so important, and to provide advice for achieving it. Some of the issues, in particular the maths involved in assessing the accuracy of locations collected manually and by automatic systems, are handled more thoroughly in the excellent review by White and Garrott (1990).

Automated radio tracking typically involves collecting locations, sometimes with qualifying data about associated activities or physiology. Manual tracking includes radio surveillance, which is here considered first, and the recording of location data. Location records require a number of decisions about sampling protocols, with accuracy an important consideration for both automated and manual approaches. The recording of survival and life-path data involves additional elements, and is therefore considered last.

8.1 RADIO SURVEILLANCE

In radio surveillance projects, the tag is used mainly to find an animal so that it can be watched. This technique has proved particularly useful for finding nocturnal animals, such as foxes or badgers, so that their behaviour can be observed with infra-red equipment or image intensifiers (Macdonald, 1978; Kruuk *et al.*, 1979). In other cases the tagging has enabled systematic collection of data on behaviours which are otherwise seen infrequently or with a strong observation bias. An example is work on the social behaviour and hunting of large predators and their prey (Seidensticker *et al.*, 1970; Mech, 1980; Bertram, 1980). The behaviour can be sampled when the observer chooses, rather than on the few occasions when lack of cover, or a particularly long chase, make it especially conspicuous. When hawks were followed continuously by radio tracking (Kenward, 1982b; Marquiss and Newton, 1982), their hunting behaviour could be quantified without the bias towards observations in open country that is likely in visual records. This type of surveillance also showed just how much bias could occur when predation was assessed by searching an area for kills instead of radio tracking the hawks (Ziesemer, 1981). Searching for kills recorded all the pale-feathered pluckings where radio-tagged hawks had killed pigeons, but only one-third of the dark-feathered pluckings of pheasants, and one eighth of the rabbit kills: the soft fur pluckings were rapidly washed down by rain, and the rabbit remains were also frequently stolen by foxes.

If animals are not to be disturbed during radio surveillance, the observer must be good not only at moving silently and inconspicuously through the countryside, but also at using characteristics of the radio signal to estimate behaviour of the tagged animal and its distance along one bearing. Posture and motion sensors on tags are extremely useful. They may indicate that the animal is lying down, or feeding, and therefore less likely to detect you. Distance estimation from signal strength is important, for example if you need to examine a kill after a predator has left. The animal will be more likely to detect you moving across its line of sight, to make a triangulation, than if you stalk it along one carefully chosen line. You should only move across the line of your approach to take a second bearing when you are close enough to be reasonably certain of the location, and can thus plan how best to use all available cover. Unless the animal is tame and unlikely to be influenced by your approach, do not simply approach until you see it, because it will most probably see your movement first. Moreover, if you get upwind of a mammal it may well respond to your scent before you have even seen it. And avoid disturbing other animals, which may either alert your subject or fall prey to it as they flee from you. To test whether your field technique disturbs the subjects, you may be able to collect before-and-after data on their movement rates or habitat choice.

Sometimes, of course, you may have to disturb an animal in order to collect data. You may, for instance, want to make measurements before much of a kill has been eaten. This requires skilful stalking, so that you do not frighten the animal too severely by suddenly appearing right in front of it (a potentially painful mistake with large carnivores). Remnants of a small kill may also be difficult to find if the predator departs unseen before you get close.

In projects where tagging is being used to find and observe social animals, such as group-living mammals or flocking birds, you may well need a visual marker to distinguish the tagged individual. Wing-tags and 'fin'-tags have been used for this purpose on birds, and reflective side-flashes or Beta-lights attached to radio tags on nocturnal mammals. The potential of radio tagging one individual in order to find flocks of animals that move around has been realised for wolves (Mech, 1983) and domestic flocks in difficult terrain (Gautestad and Mysterud, 1995). However, this 'Judas-goat' strategy could be much more widely exploited, for example to find wintering areas of migrants where many untagged individuals can also be studied (L. Peske, pers. comm.) or to keep contact with moving flocks of birds whose intake rates are required for modelling.

8.2 RECORDING LOCATIONS

If you collect locations automatically, for example by satellite or GPS tags, the number of records from each animal may be constrained mainly by the life or storage capacity of the tags. If tracking is manual, you will be constrained by the time it takes to collect each record. This puts a priority on obtaining an adequate number of records, but not more than enough, because collecting too many is liable to waste valuable time. Early studies often collected hundreds of records from each individual among a handful of tagged animals, and were then disappointed to discover that (at best) the time could been have saved by collecting many fewer records or (at worst) that too few individuals had been monitored for quantitative analyses.

8.2.1 Continuous monitoring and point sampling

Location sampling is typically done in one of two ways. Continuous monitoring involves following individuals for continuous periods, typically of 1–12 hours at a time, and either recording every discrete movement, such as each flight from tree to tree, or recording locations at regular intervals of 1–10 minutes. Point sampling, on the other hand, records locations after much longer intervals, typically of several hours or a day. The selected intervals may be entirely regular, but are more often varied, for example to ensure that daily records are sampled at different times of day.

The choice between continuous monitoring and point sampling will depend not only on the biological questions being asked, but also the stage of the project and the difficulty the observer has in moving between tagged animals. Continuous monitoring is appropriate if you need detailed data on rates of movement, perhaps from recording locations at regular short intervals. You may also need prolonged records of other behaviour, for example to estimate feeding rates or to observe the course of interactions with other individuals, and therefore combine continuous monitoring with radio surveillance. To avoid missing events while looking down at a notebook, observations may be voiced into a dictaphone or other audio-recorder.

The problem with data collected by continuous monitoring is lack of statistical independence between records from the same individual in each session. The lack of independence between habitat records is particularly obvious. For example, if an animal is in a 4 ha habitat block, and moves an average 10 m between locations, it is much more likely to remain in that habitat between moves than to change. The same argument applies to flight distances, because the tracked individual may have been especially hungry during a particular session, or hindered by bad weather.

Robust analyses therefore use a single record from each period of continuous monitoring, of mean distance per minute, or proportion of time in flight or in a particular habitat. This creates a pressure to keep monitoring periods short, so that many periods can be sampled for many individuals. Incentives for lengthening observation periods would be (i) to minimise variation between records, for example when use of a particular habitat is infrequent and (ii) when it takes a long time to move to the next individual. If animals are far apart, it may be necessary to stay with an individual all day; the data may then be separated into several time-of-day categories for analysis.

The lack of independence between locations recorded continuously is the main reason for point sampling when recording home ranges. Put simply, if you record a location every minute for an animal, at the end of a week you will have a great many records but only one home range, which may differ little from a range estimated from locations recorded every 2 hours. If your study animals occur densely, however, you might be able to obtain sample locations for another 19 during those 2 hours, to give 20 range estimates instead of just one. Moreover, you will then have data on distances between different individuals at similar times of day, which is suitable for detailed analysis of sociality. Recording at 2-hourly intervals, or less frequently, also provides point samples of habitat use and other behaviours. It can provide indices of travel rates, although these are likely to be biased compared with estimates from continuous monitoring (Laundré *et al.*, 1987; Reynolds and Laundré, 1990).

To optimise the efficiency of data collection by point sampling, you need to decide (i) what intervals to use between records and (ii) how many locations at these intervals are adequate to define a home range. Autocorrelation analysis of the spatio-temporal linkage between locations (Swihart and Slade, 1985a) can help to answer the first question; analysis of how range size increments as consecutive locations are collected helps to answer the second. These techniques are best used in a pilot study (S. Harris *et al.*, 1990), as described in the next three sections. Even when the collection of continuous records is automated, for example by the use of GPS storage tags or satellite tracking, you may still wish to use these techniques to decide how many locations to use in a standard home range.

8.2.2 Autocorrelation and timetabling

If location records are taken at minute intervals for an animal that seldom travels more than 10 m in a minute, they are unlikely to be more than 10 m apart after the first minute, 20 m after the second minute, and so on. The locations are not independent records, but are serially correlated (autocorrelated). However, if the animal travels to and fro within an area, an interval between records will eventually be reached at which the location of the first no longer has any ability to predict the distance (and hence the location) of the second. On this basis, Swihart and Slade (1985a) proposed a test of Time To Independence (TTI) with Schoener's (1981) index, $V = t^2/r^2$, where t^2 is the mean squared distance between consecutive location records and r^2 is the mean squared distance from each location to the arithmetic mean coordinates of all the locations. Swihart and Slade (1985a) showed by simulations that V follows a normal distribution with mean of 2 for distributions of independent locations. They proposed that animal locations could be considered to have negligible spatial dependence if values of V for three successive time intervals (i) along a series $i, 2i, 3i \ldots Ni$ (e.g. at 195, 200 and 205 minutes if $i = 5$ minutes) had values above the lower quartile of the distribution. The value of the quartile is about 1.6 for 10 estimates of V, 1.8 for 30 and 1.9 for 100, with slightly higher quartile values for strongly elongated ranges. A conservative test uses three consecutive values above 2.

Provided that recording intervals are short enough, the plot from an autocorrelation analysis shows an initial increase in Schoener's index with increase in sampling interval. In a classic case, there will be a point at which estimates for three successive intervals pass the threshold for negligible spatial dependence. In Fig. 8.1A, the value of V exceeds 2 consistently when the interval between samples of blackbird locations reaches 8 hours. However, the values of Schoener's index fall below this threshold for negligible dependence again after 12 and 16 hours. Moreover, the locations are very strongly spatially correlated at 24 hours, probably because the individual was revisiting areas on a daily basis. Figure 8.1B shows another individual for which V never exceeded 2, but for which the 9-hour interval for maximum spatio-temporal independence was similar to the threshold in Fig. 8.1A.

Ignoring other considerations, these figures seem to suggest that a sampling interval of about 8 hours would be appropriate for blackbirds. However, the use of such an interval would create other problems. The data were collected throughout summer days from about 6 am to 10 pm, such that three locations could be collected at 6 am, 2 pm and 10 pm. However, blackbirds were normally at roost at 10 pm, tended to be in another small area on the other side of their home-range around mid-day, and at 6 am to be close either to the roost or to the mid-day position (B.E. Kenward, R.E. Kenward and I. Hill, unpublished). It was this diurnal time-tabling, with the 10 pm and 2 pm locations far apart, that gave the 8 h 'independence' while also showing strong dependence of locations after 24 h (Fig. 8.1). This is because opposite points on the periphery of a range with radius 1 are 2 units apart, such that $t^2 = 4$ and $r^2 = 1$; adding another movement that is negligibly towards the range centre will average t^2 to 2 while r^2 remains 1, thus $V = 2$. On this basis, V is effectively an index of time to cross a home range.

A

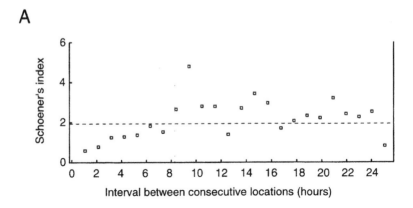

B

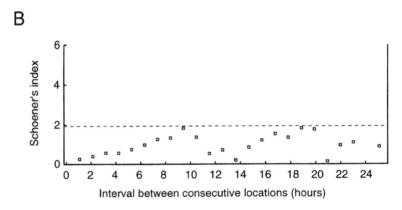

Figure 8.1. Autocorrelation analyses of timed location data from two blackbirds. In (A), Schoener's index exceeds 2 for 3 consecutive increases in interval beyond 8 hours, but data are highly correlated again when the interval is 24 hours. In (B), Schoener's index reaches maxima at 9 and 19 hours but does not exceed 2.

With locations sampled on this basis, the home-range would remain long and thin with small area. An even smaller range would result if one decided, for convenience, to sample once a day at around 2 pm. Time-tabling probably partly explains why, in several field studies, data subsampled in ways that would minimise spatial correlation have tended to under-estimate areas that used all the locations (Robertson *et al.*, 1998; Rooney *et al.*, 1998; De Solla *et al.*, 1999). Other studies have noted that, with V as an index of time to cross a home range, values of several hours seem awfully long for species that, like blackbirds, can cross their range in a minute or so (Anderson and Rongstad, 1989; Reynolds and Laundre, 1990; Minta, 1992). However, remember that the rate of travel includes searching and food handling activities, for example, as well as locomotion in a straight line. Indeed, although Swihart and Slade (1988) noted that TTI correlated with mass for a number of mammals, it was also greatly influenced by mode of foraging; for example TTI was relatively fast for central-place foragers. Even when fixes are recorded once a day, they still sometimes remain significantly autocorrelated (De Solla *et al.*, 1999), in some cases perhaps because of 24-h timetabling and in others because it really does take several days for the animal to traverse its home range.

The tendency for location records to be linked not only by speed of travel but also by timetabling probably explains why autocorrelation analysis sometimes seems to be as much a 'problem' as a 'new paradigm' (Legendre, 1993). Consider the case of an animal that takes a week to work its way across its range. Would it be better to estimate a home range quickly from two records a day during 2 weeks, or to risk that animals die or emigrate during the period needed to collect enough uncorrelated data at weekly intervals?

It now seems clear that problems can arise from slavish adherence to autocorrelation analysis. At the least, sub-sampling within field data to minimise spatial dependence may provide so few locations that range sizes are under-estimates (Rooney *et al.*, 1998; De Solla *et al.*, 1999). However, ignoring autocorrelation effects may also create difficulties. Cresswell and Smith (1992) showed with simulated data that range size was underestimated by autocorrelated bursts of observations, which would be collected in short periods of continuous monitoring instead of point sampling. Nevertheless, the effects only became severe when V was at or below 1. Sampling such that V is at least 1 may therefore provide a more practical minimal interval than requiring an interval that confers 'independence'. That would suggest sampling the blackbird data at intervals of about 3 hours.

Schoener's index can be estimated in spreadsheets if data have been collected at regular, relatively short intervals, or in HOME RANGE. RANGES V can also use locations recorded by continuous monitoring of each time an animal moved, in which case the distances and intervals are estimated on the assumption that the animal remained stationary between records. Where point sampling has been infrequent, for example with only two to four locations each day, there are few intervals in which V can be estimated. However, examination of data from early studies that used intervals of 2 to 5 hours, in recognition of a need to minimise spatial dependence before formal analysis became available, shows that V was generally at or above 1. If autocorrelation analysis is to become a useful part of radio tracking, there is a need for further work to define how the value of V relates to the optimal interval for point sampling.

Planning a protocol for location sampling involves considering not only autocorrelation, but also the time you need to move between animals' and their possible time tabling. For example, if it is difficult to sample more than six individuals three times per day, it may be better to sample 18 individuals once a day. However, if you do schedule only one or two location records per day, the risk of time tabling makes it very important to vary the sampling time between days. Otherwise, if you routinely meet at the airfield at 9 am and check animal number 43 first, you may have a good set of data for the locations of number 43 at 10 am each day; however, you would also have no knowledge of where a daily pattern of movement might have taken the animal by 5 pm. On the other hand, if your animals travel rapidly about their range and are checked at least five times daily, the daily variation likely to occur in your tracking may make it unnecessary to vary the sequence of checking. The most convenient schedule may be to sample the animals in a constant order, perhaps based on the shortest route between them, but to start with a different individual each day. This is often more practical than a random schedule, and ensures that neighbours are recorded at more or less the same time, which can be important for analysing sociality.

A location-sampling protocol may also require moderating with common sense. If you require ranges based on foraging locations, there is little point in recording half the locations in rest sites. For example, after pilot work on grey squirrels indicated that they typically foraged for a few hours in the morning and then rested before a shorter bout in the afternoon, a protocol was established to sample locations twice in the morning bout and once in the afternoon.

8.2.3 Assuming dependence among locations

Even if autocorrelation can be minimised, there remains the possibility of timetabling, and of animals having favoured sites and routes in general. It is therefore unsatisfactory to use statistical tests that assume locations from one individual to be independent observations, such as χ^2 tests or confidence limits based on numbers of locations. Robust analyses avoid potential pseudoreplication (Hurlbert, 1984) by combining the data for each animal to estimate a single measure of home-range size or structure, or of proportion of locations in habitat A, or of interaction between animals 19 and 25 (Kenward, 1992; Aebischer *et al.*, 1993b; Otis and White, 1999). Thus, sample sizes for comparisons are the number of animals for which home ranges or habitat use or social interactions have been recorded (see also Chapters 9 and 10).

8.2.4 Incremental analysis for home-range estimates

A minimum convex polygon (MCP) round all the locations (Mohr, 1947) is the most widely used home-range estimator (S. Harris *et al.*, 1990). This may be called an 'outer MCP' to avoid confusion with minimum convex polygons that include a proportion of the locations. If you plot the size of an outer MCP against the number of consecutive location records (Fig. 8.2A), the area is likely to increase rapidly as you add initial records, because the first few are unlikely to reveal the boundaries of the range that the animal is currently using. However, as you record more locations you reach a point at which new observations are adding little to the measured range size (Hayne, 1949; Stickel, 1954; Odum and Kuenzler, 1955). Beyond this point of 'sampling saturation', the animal's real range size may well continue to increase slightly, as it makes occasional excursions, but you have obtained a reasonable estimate of the range that the animal is using in the short term.

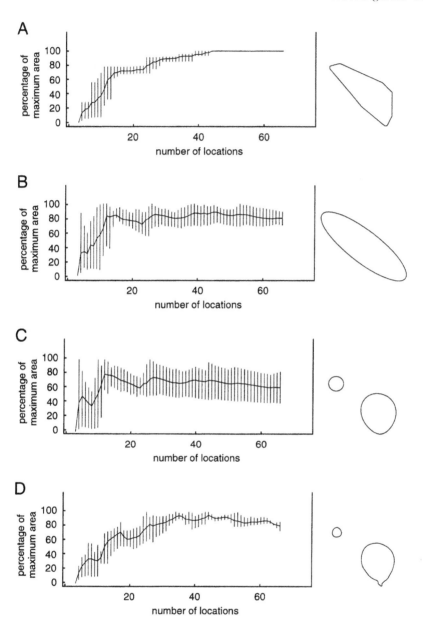

Figure 8.2. Incremental analysis of percentage maximum area against number of consecutive locations for convex polygons round all the locations (A), Jennrich-Turner ellipses plotted to include 95% of the density distribution (B), equivalent 95% kernel contours with reference smoothing (C) and equivalent 95% harmonic mean contours on a 40 × 40 grid without location centering (D). Increase in area is least variable with polygons and harmonic mean contours, but needs 35–45 locations for the mean to reach maximum size compared with 10–12 locations for ellipses and kernel contours. Data are for the 5 blackbirds in fig. 4.1.

Of course, if you track the animal for a longer time, it may disperse to a new area, or even migrate: some species are nomadic and some individuals of other species drift across the countryside, quite apart from the occasional excursions that are also likely to add to the area an animal has visited during its life. These behaviours have resulted in criticism of measuring asymptotic areas in home ranges (Gautestad and Mysterud, 1995). Nevertheless, the concept of a measurable home range remains convenient for answering many biological questions.

Although the size of an outer MCP can change only by increasing as locations are added, areas estimated by other home-range estimators can also shrink. Such shrinkage occurs when outlines include only a proportion of the locations, and records add within this core. Thus, when home range is estimated as a circle or ellipse fitted to a density distribution of the locations (e.g. Stickel, 1954; Jennrich and Turner, 1969), any locations added near the centre will reduce the spread of the distribution and the area within, say, a 95% outline (Fig 8.2B). However, shrinkage also occurs if single outlines can split into multiple outlines round separate groups of locations, for example by contouring (e.g. Dixon and Chapman, 1980; Worton, 1989) or cluster analysis (Kenward, 1987). Decreases in range area are especially likely when range cores can be defined by multiple outlines (Fig. 8.2C,D).

When ranges were estimated by outer MCPs for a number of species with different foraging behaviour (Kenward, 1982a, 1987, 1992; Parish and Kruuk, 1982; Holm, 1990), collecting two or three locations a day gave sampling saturation after about 30 records. Some 40 locations were needed for the blackbird data, which was collected at 1- to 2-hour intervals (Fig. 8.2A). Estimators that give a less detailed shape, such as the ellipses in Fig. 8.2B, and even the kernel contours in Fig. 8.2C, typically require less than half as many locations to reach a maximum size, though more locations may confer greater stability. Generally, the more detail that a range estimate can define, the greater the number of locations required for stability (Robertson *et al.*, 1998). Range analysis techniques are considered more thoroughly in Chapter 9.

8.2.5 Standard ranges and pilot studies

If analyses are based on different numbers of locations for different individuals, there is a danger that bias from autocorrelation, timetabling, seasonal changes or other factors will create artificial differences between them. The easiest way to avoid this risk is to record the same number of locations for each individual, at the same times of day. This provides not only a standard range, but is also useful in other analyses such as examining sociality based on relationships between simultaneous locations of adjacent individuals (Kenward, 1992).

A standard sample of 30 locations (from two to three daily records) has been suggested for outer MCP analyses (Kenward, 1987, and above). However, 30 locations may be too few to give stable estimates for the most detailed range estimators, especially cluster analyses (Robertson *et al.*, 1998). Other studies have recorded range area asymptotes from sampling saturation at 50–75 locations for contoured range estimates (Jaremovic and Croft, 1987; S. Harris *et al.*, 1990; Wray *et al.*, 1992b). Although sampling 30 locations with low spatial correlation gives comparability within and between species, there may be much technique-based variance if just 30 locations are used for analysis techniques that provide more detail than outer MCPs. The result would be less chance of obtaining significant results than with larger samples of locations. For this reason, it remains advisable to conduct a pilot study (S. Harris *et al.*, 1990), at least for new species and analyses.

A pilot study can be based on an initial bout of intensive tracking for an autocorrelation analysis. You could choose to track three to five animals either continuously for a day at a time or, if they are easy to track in sequence, at intervals of an hour or so, until records for about 5 days are available for each. The analysis would not be to obtain 'independent' locations but to help select an interval for efficient home-range sampling. Until more work has been done on V values that optimise sampling intervals, it is not possible to define a V threshold for efficient sampling intervals. However, based on autocorrelation analysis of the squirrel and raptor data used to define a 30-location standard for outer MCPs, it may be sensible to use a minimum sampling interval at which $V > 1$. You then need to select an interval of at least the minimum size, in which you can record locations for an adequate number of individuals in sequence.

After choosing a sampling interval, you can track a small sample of individuals to determine how many locations are needed to define a stable estimate of short-term range size, or a travel speed (Palomares and Delibes, 1991), or habitat use, or whatever must be measured to answer your biological questions. If you started by collecting data for autocorrelation analysis, you may start your incremental analysis with a plot of the initial data, sub-sampled to provide locations at the intervals chosen for the continued tracking.

If you are working primarily at one time of year, a pilot study may take only 2 to 3 weeks to develop a short-term sampling protocol. Those 2 to 3 weeks will also familiarise beginners with the necessary field techniques. However, if your work is to extend over several seasons, you may well need to spend a first year defining the times of stability and change for the study species. You may want to answer questions such as 'when are movements affected by dispersal, courtship or breeding?', 'when does performance relate most strongly to range indices?' or 'does habitat-use change even during periods when range boundaries are stable?'. The answers will help you plan a long-term protocol, in order to estimate ranges and habitat use when records are likely to be most stable and relevant.

8.3 ACCURACY

A large part of the excellent book by White and Garrott (1990) considers accuracy, including detailed suggestions for planning, testing and calibrating a radio location system of two to six fixed towers. If you are planning a fixed system, you must read that book, because your ability to answer biological questions will depend on the accuracy of the system. However, accuracy is also an important consideration if you are planning mobile tracking.

8.3.1 Accuracy in mobile systems

When two bearings are used to triangulate a tagged animal's position, errors can arise in several ways (Heezen and Tester, 1967; Springer, 1979; Macdonald and Amlaner, 1980). For a start, errors are especially likely in aforested or mountainous terrain (Hupp and Ratti, 1983; Kufeld *et al.*, 1987; Chu *et al.*, 1989), where topographical features deflect signals or cause other interference. Inaccuracy may be suspected when a bearing is taken, perhaps because a Yagi is not receiving a clear signal peak with lateral nulls. Moving a few metres forward or to one side often improves the reception in such cases. Even if such inaccuracies are not suspected, it is often wise to take a third bearing following the two required for basic triangulation.

With three bearings, an animal's location can be estimated (i) at the centre of the 'error triangle' where the three lines meet (Mech, 1983). If the triangle is too large, more bearings are taken (Fig. 8.3) until three coincide to give an acceptable error triangle, or (ii) the 'best two' from closest to the animal are used. Lenth (1981) introduced (iii) maximum likelihood estimators for deriving location from several bearings, with modifications in the Huber and Andrews estimators (Andrews *et al.*, 1972) to reduce the influence of divergent bearings. These estimators predict locations of test tags (beacons) better than the error triangle (Nams and Boutin, 1991), with the Andrews estimator most favoured because it rejects aberrant bearings and will only estimate a location if at least three are consistent (Garrott *et al.*, 1986; White and Garrott, 1990). LOCATE II, TRIANG and LOAS calculate tag locations from bearings using Lenth estimators.

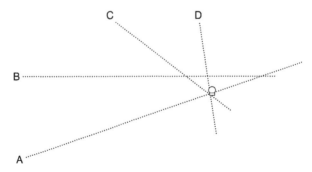

Figure 8.3. Taking repeated bearings for accurate triangulation. The angle between the first two bearings (A and B) was small, so a third bearing (C) was taken from closer to the tag. Since these three bearings did not cross at one point, a fourth bearing was taken (D) from even closer to the tag.

As well as topographical errors, which are to some extent avoidable in a mobile system, there are errors caused by the receiving system. Systematic errors can arise from misalignment of compasses in vehicles, damage to antennas, readings taken when antennas or compasses are too close to cars or other large metal objects, and other misuse of equipment. These errors too are avoidable, but even the best antenna has limited accuracy. The accuracy of the antenna in the study terrain, which can be measured as the standard deviation of repeated bearings in field tests (White and Garrott, 1990), is a fundamental variable for quantifying the error of locations.

The influence of antenna error increases with distance from the tag. For example, if the antenna gives a bearing accuracy of about 5°, the lateral error of the animal's location is about 9% (tan 5° = 0.0875) of the distance to it. It would therefore be unrealistic to give an animal's position at 1 km with a precision finer than 100 m across the line of the bearing.

When triangulating, the other important variable that combines with antenna error and distance to estimate the error of locations is the angle between bearings, as seen in an error polygon (Heezen and Tester, 1967) formed between two bearings (Fig. 8.4). The size of this polygon is minimal, at a given distance between tag and receivers, if the bearings are at 90°. With two bearings, the length of the longest diagonal of the polygon indicates the error of the location (Saltz and Alkon, 1985). With more than two consistent bearings, the Lenth estimators can be used to generate an ellipse that estimates a required probability of containing the tag. The best correlate with the distance between the true and estimated tag location is the major axis length in the ellipse (Saltz and White, 1990).

As well as the topographical and system errors, if bearings are not taken simultaneously from two (or more) sites there may be errors due to an animal's

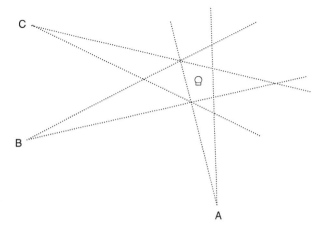

Figure 8.4. The size of error polygons increases as the angle between bearings is reduced. The error polygon between bearings A and B is the smallest.

movement. These will be especially severe if the angle between the bearings is small (Fig. 8.5), and if the animal is close and moving fast. Movement errors are particular problems when tracking birds, which can suddenly fly rapidly across country, or mammals such as canids that move continuously when foraging (Laundré and Keller, 1981). For birds, a solution is to have a motion sensor on the tag, so that you can listen for absence of movement in the time (as short as practical) between taking consecutive bearings. New pairs of bearings should be taken if the bird was heard to fly. If animals are close and moving (with a relatively constant bearing) towards or away from you, then moving rapidly yourself to obtain a cross-bearing at close to 90° will be reasonably accurate. However, moving animals are best tracked, if possible, using a 'close approach' technique of obtaining at least one bearing within the minimal detection distance. In close-approach conditions, if the tracker moves in an arc round the animal without disturbing it, continual swinging of the antenna provides many rough bearings in rapid succession. This gives a reasonably accurate location with reference to landscape features rather than by plotting.

8.3.2 Error and bias

If at all possible, you should plot bearings on a map immediately. If you merely record pairs of bearings in a notebook, without plotting them at the time, you will have no way of checking peculiar locations, and may well have to discard some of the data as unreliable. This creates a risk of bias. Consider a situation where trackers have difficulty moving about, such that locations tend to be distant and with small angles between bearings when animals are on excursions from their normal foraging areas. If the low-accuracy locations are subsequently excluded from analyses, total range areas may be substantially underestimated, thereby biasing assessment of habitats or neighbours encountered by individuals. Similarly, failure to find an animal during home-range tracking may mean that it is outside the range which you have recorded so far, possibly doing something unusual, and that you are again introducing bias. Additional bias can come from making assumptions about the habitat used when an animal is located close to a habitat boundary.

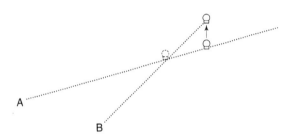

Figure 8.5. Movement error! The animal moved some distance between the taking of bearings A and B, and its triangulated position was therefore inaccurate.

Just as over-zealous autocorrelation analysis can create problems for the biologist, for example by encouraging bias from timetabling, so too can error analysis. At the least, a great deal of time may be used to investigate errors instead of answering biological questions. At worst, *post facto* exclusion of locations that fail to meet accuracy requirements can create bias as described above. The bottom line is that error is more acceptable than bias. Errors are simple 'noise' that increase sample variance and hence militate against obtaining significant results (creating Type II errors). Bias can give erroneous results. By all means seek to minimise error, but strive at all costs to avoid bias.

8.3.3 'Adequate accuracy'

The traditional approach has been to use the likely antenna accuracy in the study terrain to establish a protocol for data collection, aiming to ensure adequate accuracy for tests that would answer the biological questions. For an antenna with 5° accuracy, a typical protocol for a 10 m resolution between adjacent locations might include 'never accept a triangulation with a bearing more than 100 m from the tag', and 'always try to have an angle close to 90° between bearings'. The latter rule is less rigid in view of the finding by Chu *et al.* (1989) that accuracy of field checks to beacon tags only declined markedly for angles <20° or >160° between bearings. With access to a computer in the field, such rules may be replaced by 'do not accept locations until an ellipse with 68% probability (one standard deviation) of containing the tag is smaller than 1 ha'.

To establish a protocol, you may find it necessary to assess accuracy experimentally with beacon tags at the start of your study, especially for species not previously tracked in a particular habitat. On the other hand, if you are using standard antenna designs you may often be able to save time by using accuracy data from previous studies that tracked similar species with similar antennas in similar habitats. For studies of home-range structure, another useful rule might be to delay a search for missing tags until the end of a session, if necessary, but never to ignore them until there is evidence of death or emigration.

In habitat studies, the censoring of locations at boundaries, or their random allocation to a habitat, creates risk of bias and Type II errors, especially when recording use of habitats in fragments that are small relative to the tracking resolution (Nams, 1989b; White and Garrott, 1990; Saltz, 1994). A protocol for an adequate-accuracy approach might contain 'always check the location with a close approach if the initial plot is within the resolution distance of a habitat boundary'. Computer routines make it possible to minimise bias by assessing habitat not only at the exact location on the map but also in an ellipse, estimated to reflect the accuracy of each location (Samuel and Kenow, 1992), or in a circle, with a diameter defined by the tracking resolution (RANGES V supports all three approaches). The use of error ellipses and resolution circles to estimate habitat at locations is also favoured by limited accuracy of the raster maps that are produced by remote sensing. Nevertheless, if you put your effort into gaining accuracy data for each location and then assign habitat from a map, you are at greater risk of Type II errors than if you use adequate accuracy to record the habitat with each location in the field. In this case, use many categories rather than few to record the habitat, because it is much easier during analysis to combine categories than to split them.

8.3.4 Team tracking and difficult terrain

In terrain which is relatively flat and unobstructed, such that bearings are rarely affected by signal bounce, it may be convenient for a team of two or more people to remain at fixed sites and take synchronised bearings. Ideally, the team should be in contact and one of them should plot bearings immediately, so that likely errors can be detected and locations re-estimated. Location error will increase with distance from the fixed sites, so this technique is best reserved for situations where the terrain is too impenetrable (e.g. swampland, thicket) or the disturbance risk too high (e.g. grassland) for alternatives to be practical.

An intermediate approach that can overcome the accuracy problems may be practical where highly mobile species are to be tracked in terrain with good access, but where signal reception is poor except when bearings are taken at one or two high points. An example would be the tracking of widely foraging small birds in woodland adjacent to a tower, open hill-top or ridge. In this case, one observer on the high point may direct one or more team member in the woodland, where they may not be able to detect signals at all, to the general area of each animal so that accurate locations can be recorded. In practice, however, this approach tends to be superseded by abandonment of the high point as soon as the general areas used by each individual become known.

In some terrain, such as mountains or rainforest (and especially montane forests) with a limited network of tracks, mobile team triangulation becomes necessary. This is because the limited access makes it hard to approach animals, and signal bounce makes triangulation from fixed sites too inaccurate. In this case, the team members must plan routes carefully so that each tracker is in a good position to obtain cross-bearings on each animal in turn (F. Kerridge, pers. comm.). Tracking of this type become difficult and inefficient if the members are not in continual contact by radio or mobile phone, so that agreement on acceptable accuracy can be reached for each animal in turn. If contact is impossible, rigid routing and timing must be agreed, and revised as necessary after plotting locations each evening.

8.3.5 Accuracy in automated systems

Problems with movement errors are minimised in automated radio-location systems with fixed antennas, because bearings can be taken more or less simultaneously. Indeed, time differences of arrival are recorded for the same signal. Larger and hence more accurate antennas can be used for bearings than in a mobile system. Moreover, system accuracy is unlikely to differ between times of testing and routine operation, as it can when humans are tracking (Mills and

Knowlton, 1989). However, fixing the antennas prevents the adequate-accuracy approach that is available with mobile antennas. Accuracy changes with distance of a tag from the system. Moreover, individual antennas tend to be especially vulnerable to topographical error (signal bounce) in particular directions unless the terrain is very flat with uniform vegetation.

Errors due to topography and distance can be minimised by having a spread of many antennas across a study area. The effect of minority rogue bearings can then be minimised by using many bearings in Lenth estimators' and a tag is never far from an antenna. Locations are most accurate when a tag is close to one antenna (the equivalent of a close approach in mobile tracking), provided that the tag is not close to a line between only two antennas. There should therefore be at least three antennas, and careful placement is essential if accuracy is to be reasonably constant throughout the area covered. Accuracy is not likely to be constant if terrain is variable, in which case extensive calibration with beacons tags may be needed to map the error to be expected throughout the area. The optimal placement of up to six antennas, and many other aspects of such systems, are considered in detail by White and Garrott (1990), which is essential reading for anyone contemplating or using fixed antenna stations.

8.4 SURVIVAL DATA

Survival studies need very careful planning, with pilot work to check that tag attachment is harmless, that the tags are reliable, and that you can make allowances for dispersal. It is also important to mark adequate samples of animals. Confidence limits on survival estimates increase as the number of survivors declines: you will typically require >100 survivors for 95% confidence limits of 0.05, >50 for 0.10 and will have confidence limits around 0.15 with only 20 survivors. Keep in mind that sample sizes will decrease during the study as a result not only of death, but also of tag failure and inexplicable loss of tags. You therefore need large samples of animals, with reliable radios and tracking techniques, to detect anything other than gross differences in survival (such as those likely between released and resident animals). To avoid disappointment, you should use anticipated rates of survival and tag loss to estimate in advance how many animals must be marked to demonstrate (or refute) a given difference in survival rates (Pollock *et al.*, 1989; White and Garrott, 1990, Samuel and Fuller, 1994).

Fundamental assumptions for estimating survival rates with radio tags are: (i) that the tagged animals are representative of the study population; (ii) that the tags do not affect survival; (iii) that deaths are identified accurately and (iv) that deaths are statistically independent events. Note that if you are tagging young animals, there may be a lack of statistical independence between brood members. All the young in a brood may be exposed to the same exceptional rainstorm, or to a predator that returns repeatedly to mop them up. A correction that reduces the effective sample size may be necessary (Flint *et al.*, 1995), at least until brood break-up.

The data will be easiest to process if survival checks are close in time for all the tagged animals, i.e. spread over as few hours as possible for daily checks, and over as few days as possible for less frequent checks. Thus, if animals are checked once a week, check them all on the same day if you can, or on two consecutive days. If checks are frequent, rates can be estimated as a mean of the many short-interval samples from within each period. The variation between samples provides a basis for sensitive tests to compare rates between periods for the same set of animals, or within the same period for different groups of animals (Trent and Rongstad, 1974; Miller, 1983; Heisey and Fuller, 1985; see Chapter 10). In this case you should aim to find and check (relocate) at least 90% of the animals each time (Bunck *et al.*, 1995). On the other hand, if you are checking only at the end of long periods, and using less sensitive tests that do not assume constancy of rates (Pollock *et al.*, 1989), you should obtain highly reliable radios and aim to find all the animals each time.

It is worth making checks especially frequently just after tagging. This will provide plenty of data to test for any short-term effect of tagging, or for a tendency to trap animals that were about to die. Moreover, radios that fail due to damage from the animal often do so in the first few days after tagging. This is also a time when animals may leave the study, if they have been trapped while dispersing. If your tags give weak signals you may have many unexplained losses at this time unless you check them very often.

It is very important to detect radio failures. You should therefore record the signal pulse rate when checking each animal, and note any irregularities or changes in rate that may precede failure or normal cell expiry. Animals for which tracking ceases prematurely will then fall into two groups: those which were known to be *live* when tags failed or were shed prematurely, and those which were unaccountably *lost*. You should seek independent records of survival for each category, by seeing those that are recognisable, or by their completion of a breeding cycle or by retrapping and other recoveries. If the resighting or retrapping rates are the same for both groups, and detection of emigrants is equally likely, then it is reasonable to assume that unaccountable loss was mainly due to undetected tag failure (Kenward *et al.*, 1999). If rates are lower for *lost* than *live* animals, then some deaths may have been associated with radio failure (through impact, or destruction by a predator). If you have enough follow-up data on animals in each group, you can correct the recorded survival rates (Aebischer, 1993; Kenward, 1993, see chapter 10). Alternatively, you may be able to estimate a correction for tag failures by running an unattached sample under the conditions likely on tagged animals (Mee and Clark, 1992).

When checking that animals without mortality sensing tags are alive, it is wise not merely to rely on signal variations, but also to record location. You cannot assume that an animal is alive because the signal fluctuates: the tag, or intervening obstacles, may be blowing in the wind. The corpse of a radio-tagged buzzard that was hanging by one wing from a barbed-wire fence gave a varying pulse-rate through several checks before being discovered. Check, therefore, that the animal's location has changed since the last record.

Recording the locations gives other data that may eventually be useful even if they are not an immediate priority. How many fieldworkers say at some point in their study 'if only we had recorded those data from the start'? For example, it is not necessary to be recording home-range data to investigate habitat use. The data at locations, or in a shape round the locations, can be compared with habitat available on the map as a whole. Of course, locations recorded at weekly intervals throughout a season may eventually provide enough data to estimate a range outline, and can also be used to estimate dates when dispersal occurs. If major dispersal periods are known, field effort can be planned to minimise the risk that dispersing animals become lost.

8.5 CONTINUOUS REFINEMENT

A pilot study is by no means the only way to optimise data collection: scope for refinement can often be found throughout a study. For example, in a study aimed ultimately at modelling relationships between raptorial birds and land use, the first data were required to define when, where and how buzzards were most likely to disperse or die. Movements were recorded initially twice weekly, after marking young in the nest with long-life backpacks. This gave information of post-fledging behaviour and an initial period of dispersal (Walls and Kenward, 1994; Tyack *et al.*, 1998). When it became clear that most dispersal had finished by mid-October, checks were reduced to weekly, until late February when a further period of dispersal started and previous dispersers made philopatric movements back to natal areas. In later years checks were reduced to monthly during October to February, initially with a 30-location range record for nearby birds during this period of low movement.

With only four to five checks on location and survival in winter, time gained from fieldwork was used for data analysis. Home-range analyses indicated that ranges were strongly mononuclear, so that a small number of winter locations defined a realistic range centre. This provided: (i) the centre for a circle encompassing the average range size for that season, to estimate available habitat; (ii) coordinates for estimating how far individuals moved between seasons and hence (iii) measures for investigating whether dispersal, death, or subsequent breeding success related to the habitat. Tracking animals in this way with long-life radio tags provides 'life-path' data (Baker, 1978) that can eventually be used for individual-based modelling of how populations relate to the landscape (Kenward *et al.*, in press c). Moreover, causes of death could still be assigned into three main categories some months after birds died, by using location and X-ray evidence of fractures or shot (Simpson *et al.*, 1997).

Using radios that last for 4 to 8 years, all the data needed for age-specific survival and breeding models, and for linking these models to landscape composition, can now be gained with three checks a year. A check at the start of the breeding season records survival and breeding activity, one at the end of the breeding season records productivity (and marks any young), and a third check at a winter roost (or four to five records for adequate precision) indicates year-on-year movement. Scope for continuous refinement of data collection is an important consideration for busy supervisors, who can easily miss opportunities to help improve the efficiency of other people's fieldwork.

8.6 CONCLUSIONS

1. Radio tracking has great potential for avoiding observation bias, by systematic sampling of behaviour during radio surveillance, or of data on movements, habitat use, survival, emigration and productivity. However, it is important to develop protocols for efficient collection of unbiased data.
2. Serial correlation, timetabling and other repeated use of favoured areas militate against treating locations from the same animal as statistically independent data. Hypothesis tests should be based on individuals, rather than locations. Locations are used to estimate single indices of movement, habitat use, home range or social interaction for each individual.
3. A pilot study can be used to optimise efficiency in recording these indices, by estimating the minimal numbers of locations required from each animal, so that point sampling can collect data in a standardised format that is comparable across individuals. Autocorrelation analysis from a short period of continuous monitoring in the pilot study may help to optimise sampling intervals, provided that attention is also paid to timetabling. Incremental analysis indicates how many locations are needed to estimate a standard home range, which may require no more than 30 locations if two to four are recorded daily.
4. Accuracy of location estimates depends primarily on antenna precision, bearing distances and angles to the tag, but can be reduced by topography and animal movement. Protocols to ensure adequate accuracy for a desired tracking resolution typically require tracking from within a defined distance, but may also need close-approach techniques if animals move frequently or to resolve habitat use at boundaries. Team tracking may be necessary in difficult terrain.
5. Inaccuracy leads to error, which may be acceptable, but censoring or missing locations should be avoided because it can create bias.
6. Protocols are also necessary to avoid bias in survival estimates, with enough marked animals to test hypotheses after attrition from emigration and radio failures. Survival estimates assume absence of marking bias, lack of adverse tag effects, accurate detection of deaths and statistical independence between them. They are improved by high tag reliability; otherwise, recoveries of tags classed *live* or *lost* when tracking ceased may be used to correct survival estimates for inaccurate detection of deaths.
7. Data collection should be optimised continuously. In a study using long-life radios to analyse life paths and model population structure in relation to landscape, data collection that started by checking survival twice weekly, and recording 30-location home-ranges, was reduced to just three checks (of location, survival and productivity) in each year.

9 Behaviour and Home Ranges

Analysis techniques have been reviewed very thoroughly in the book *Analysis of Wildlife Radio-Tracking Data* in this series (White and Garrott, 1990), with another useful summary by Samuel and Fuller (1994) in the Wildlife Society's manual *Research and Management Techniques for Wildlife and Habitats*. This chapter describes different approaches to the analysis of movement records and home ranges, in enough detail for biologists to make informed choices about which to use. There are a number of mathematical expressions in the text, but please feel free to ignore them. They are there, with references to their source publications, so that those with a maths background can go deeper if they wish.

9.1 ACTIVITY AND EVENT RECORDS

A major advantage of telemetry over techniques, such as trapping or 'chance' observations, is that it records events in relation to time. Instead of registering the places visited by an animal through a series of trapping events (with the animal's foraging perhaps biased by the traps), radio tracking can show not only how often an individual visited each part of its range, but also what time of day it went there. Instead of merely describing the diet of a predator species (with possible observation bias if this involved searching for prey remains), radio surveillance can reveal the kill rate for each prey species by each individual predator.

The data from radio surveillance, telemetric monitoring and presence/absence recording tend to be continuous sequences of events along a time-base. They are most often summarised as event frequencies (e.g. beats per minute, visits per hour, kills per day), as durations between events (time submerged, time since sunrise) or as proportions of activity periods (percentage of time spent grooming, in woodland, with mother). Software is available from suppliers of logging equipment (e.g. Lotek) to graph output of data that can be transmitted by simple pulse-modulated signals.

For making comparisons between animal categories or investigating effects of covariates, a simple statistic for each animal can usually be estimated in a spreadsheet. More sophisticated procedures may be necessary, for example to estimate for each animal the strength of any tendency for events to occur in sequences (runs). Provided that the sample is one index value for each animal, and animals are not unduly influenced by each other, the data can be deemed independent for statistical tests.

Changes between periods should be examined by matched pair tests, such that sample sizes remain numbers of individuals. Trends can be examined by estimating a correlation coefficient for each individual and using the coefficients in sign-rank tests (Siegel, 1957). Analyses of variance based on a value from the same individual across several periods are pseudoreplicative (Hurlbert, 1984) and therefore not robust.

Real-time analysis of increasingly sophisticated signals from microcontrolled tags could also be used for event-alert (EVAL) systems, to inform biologists when rare events occur. Tags that detect mortality as absence of movement during a pre-set period have been available for two decades (see Chapter 2). However, the potential use of receiving systems to evaluate signal patterns and then alert biologists, by paging them through the phone system, has yet to be exploited for studying rare events. Such systems would make it much easier not only to recover fresh carcasses of radio-tagged animals, but also to record the kills of tagged predators, track dispersers and perhaps even to detect extra-pair copulations.

9.2 MOVEMENT RECORDS

Movement records from radio tags fall into two main types. These are the short-term movements that animals make daily during foraging, courting and general maintenance activity, and long-term movements that are on a seasonal or lifetime basis, such as migration and dispersal (Sanderson, 1966; Baker, 1978). Analysis of movements is typically based on distances, times, speeds and angles. The purpose of analysis can vary from checking whether the estimated speed between locations is too high for a location to be valid (White and Garrott, 1990) to sophisticated detection of dispersal.

Display of movements, perhaps with colour coding to indicate speeds or times of year, can also be useful for revealing behaviour patterns (White and Garrott, 1990; Kenward, 1992). The remaining sections of this chapter, and two of the next, involve analyses that can be seen running by downloading the RANGES V demo from *http://www.anatrack.com* The demonstration is 'live', in that it uses command files to process files of example data automatically. The site will also provide notification as modules covering survival and other analyses in previous sections become available, together with increased automation of analyses.

9.2.1 Daily movements

Movements can also be a useful index of energy expenditure, which is a fundamental measure in ecology but is generally impractical to measure directly in the field. As indirect indices of energy costs, progress has also been made in telemetric recording of respiration rate or heart rate (Gessamen, 1980; Woakes, 1992). Energy-demanding behaviour can be logged as the duration of locomotion, or the time outside a warm nest, if tags have suitable sensors (Osgood, 1970; Kenward, 1982b).

The distance moved by an animal provides another index of energy expenditure. Sometimes it is possible to record every perch position of a bird, and hence to estimate distances directly (Kenward 1982b; Robertson *et al.*, 1998). More often there may be only two to six locations registered each day, perhaps primarily for home-range estimation. Provided the locations were recorded at systematic intervals, the daily sum of the consecutive distances between them (e.g. roost–midmorning, midmorning–midday, midday–afternoon, afternoon–roost) may give a useful index for comparing the movement of different individuals over a number of days.

Nevertheless, if measurement of distances travelled is an important study aim, it is best if they were registered at biologically appropriate intervals (Samuel and Fuller, 1994) perhaps defined during a pilot study. More frequent records would be needed for central-place foragers than for transient foragers that take several days to move throughout their range. Even for transient foragers like coyotes, locations recorded at daily intervals do not correlate well with either the results of continuous tracking or of day-movement indices from small numbers of locations (Laundré *et al.*, 1987; Reynolds and Laundré, 1990). Small and Rusch (1989) found movements of grouse between daily locations were very variable, so that five-day smoothing was necessary to detect the underlying pattern.

The formatting of measurement units is likely to be important for analyses. If you lack large samples, it is wise to base comparisons on non-parametric statistical tests (e.g. Siegel, 1957) because these do not assume a particular distribution of the data. It is also wise always to plot distributions before analysis, because multiple modes may indicate a need to subdivide data before making comparisons. If you must use parametric statistics, for example in multivariate analyses, distances may best approximate a normal distribution as logarithms. If you are comparing speeds (rates of change of position) between different animals or different times of day, you will probably have measurements over a variety of time intervals (e.g. 210 m in 7 min, 495 m in 4 min, 125 m in 11 min, etc.). These can be converted into distances per second or per minute for each animal and then transformed.

Particular care is also needed in the treatment of angles, remembering for example that 2° and 358° are only 4° apart. Guidance on the treatment of circular statistics should be sought in Zar (1974) and Batschelet (1981), with a practical examples for dispersing grouse in Small and Rusch (1989). White and Garrott (1990) provide a worked example for testing whether consecutive movements recorded for a translocated grizzly bear were orientated significantly towards the area in which it was trapped before translocation, i.e. whether it was homing.

The most accurate evaluation of behaviour from radio tags may come from combining sensor data on activity with movement between consecutive locations.

In this way, Palomares and Delibes (1993) showed that mongoose walking, a fast activity, could be classified with 100% accuracy, compared with 95% for foraging and 83% for eating. Activity patterns could be determined by recording activity time in 7 days, but needed 2 weeks of movement data for similar precision (Palomares and Delibes, 1991). Thus, accuracy of classifying behaviour and of activity indices in general varies both with type of behaviour and recording method.

It is important to remember that movement records from an individual animal are not statistically independent data. Separate movements will be influenced similarly by an animal's state: in the short term perhaps by need to avoid a predator, in the medium term by hunger, and in the longer term by seasonal effects. As in all radio-tracking data, a test result from analysis of a within-animal effect, such as a correlation between rate of movement and time since previous meal, is a single index for testing whether a sample of individual animals moves faster when hungry. In the case of events governed by common local conditions, such as weather-dependent migration, the independent samples for analysis will not be individuals but a single estimate for each year (White and Garrott, 1990).

9.2.2 Seasonal movements

The separation of daily movements from seasonal movements can be important for defining whether an animal has moved at all, and if so, when the movement occurred. For instance, although a dispersal movement is sometimes easily recognised (Fig. 9.1A), at other times animals seem to drift across the countryside (Fig. 9.1B). Such drift is a relatively rare occurrence for the buzzards shown in Fig. 9.1, but frequent among foxes (Storm *et al.*, 1976) to the extent that Doncaster and Macdonald (1991) provided the concept of a 'prevailing range' in which an animal was constantly adding new areas and ceasing to visit others. Using these observations, and records from a wandering flock of sheep, Gautestad and Mysterud (1995) called the concept that animals have a definable home range a 'ghost'.

Nevertheless, although home ranges may adjust somewhat with time, radio-tracking provides the ability to define them as 'repeatedly traversed areas' (see section 9.3) even when drift occurs. That definition of home range also implicitly provides a basis for objective definition of when an animal has made a seasonal or dispersive movement.

At least five methods can show that an animal has made a major movement. The most basic method is the creation of subjective rules. For example, plots of goshawk movements round their nests showed that hawks did not return after travelling more than 1 km, so travel beyond 1 km was used to define dispersal (Kenward *et al.*, 1993a). Unfortunately, such rules may need to become complex. For instance, buzzards make excursions up to 25 km from the nest before dispersing, so in their case the rule became 'travelling more than 1 km and not returning within 3 days' (Walls and Kenward, 1994).

A

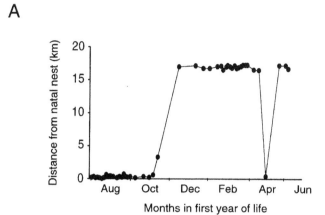

B

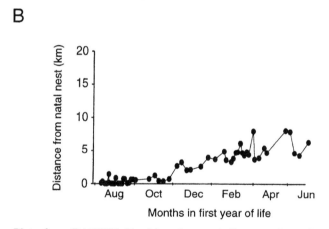

Figure 9.1. Plots from RANGES V of locations and distances from the nest of two buzzards (*Buteo buteo*) during their first winter, showing the abrupts dispersal and subsequent philopatric visit in spring that was typical of most birds (A) and the drift recorded for one of the 77 tagged birds (B).

Three other methods are based on comparing locations A before and B after a predefined time. Doncaster (1990) proposed using an index of concordance, based on Spearman rank correlation between numbers of locations in each grid cell during period A with the number of locations in the same grid cells during period B. This approach is appropriate when large numbers (hundreds) of locations have been collected. Tens of locations would be adequate for the proposal by White and Garrott (1990) to use Multiple Response Permutation Procedures (Mielke and Berry, 1982), which compare distances between locations in set A plus those in set B with the set including all *A*-to-*B* distances. A third approach would be to compare the distance between range centre estimates for periods *A* and *B* with the distribution of all distances within set *A* plus set *B*, in a *t*-test or non-parametric equivalent.

The initial approach is appropriate for estimating when a major movement occurred, provided that the subjective rules are applied without bias. The extent of movement can then be estimated by, for example, the distance between range centres. In contrast, the last three approaches indicate the extent of an animal's movement between two time periods, but not the time at which the movement occurred. However, if you want to study whether migration or dispersal is linked to sociality, disturbance, change in food availability or weather, you will need to estimate when the movement starts. The following method indicates objectively when an animal starts a dispersal or migration movement.

The principle of vectored dispersal detection (Fig. 9.2) is estimation of a probability that *n* consecutive new daily locations are beyond the distribution of all the previous *N* daily locations in one direction. The day of dispersal is based on the *n* that gives significant departure from *N* at a chosen α level. The direction (vector) and daily criteria are important, to avoid the detector being triggered by the short-term excursion movements. In principle, the time interval between consecutive records could be less than a day in animals for which a shorter time-scale would be appropriate. The initial implementation of a vectored dispersal detector, in RANGES V (Kenward and Hodder, 1996) starts with $N = 3$ locations, by estimating the arithmetic mean centre (*Ac*) for these and confidence limits for distance from the *Ac* using a selected alpha level. It then estimates the arithmetic mean for the next $n = 3$ locations, and defines the tentative 'dispersal' vector between these. Dispersal is flagged if the orthogonal distances of the *n* fixes along this vector are all outside the confidence limit for the first *N*. If not, *N* increments by 1, and the process repeats through the set of fixes.

Minor refinements in this simple detector involve setting a minimum distance for dispersal to be flagged, so that detection is reserved for when birds leave the natal area and not when they merely leave the nest, and accepting dispersal with $n < 3$ if subsequent readings were absent because the animal had disappeared. However, there remains scope for improvement. For example, the process of (correctly) rejecting dispersal when an animal makes consecutive excursions in different directions leads to delayed or missed detection if the animal makes an initial movement in one direction and then departs in the other direction. A slightly more sophisticated method, based on geometric means and additional inter-fix distances, proved no more sensitive but slower. Nevertheless, improvements based on estimating alpha levels for ellipses or contours round the first *N* locations might be worthwhile.

Note that although the last four approaches are objective in the sense of using statistical procedures, any estimates of probability rest on the assumption that locations are independent observations. The likely violation of that assumption means that test results are indices of the likelihood that movement has occurred, but not true probabilities.

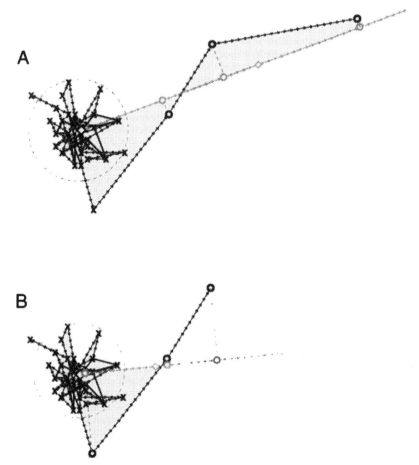

Figure 9.2. In vectored dispersal-detection, a vector (grey line) is estimated through the arithmetic mean coordinate (grey diamond) of the last three locations (o) from the arithmetic mean coordinate (black diamond) of all the previous locations (x). If orthogonal distances of the last three locations along this vector (intersection of grey perpendiculars) are beyond a circle at the 95% confidence limit for distances from the previous locations from their arithmetic mean, the animal has dispersed (A). Dispersal did not occur before the previous location, because its orthogonal distance along the relevant vector was small (B).

9.3 HOME-RANGE ANALYSIS

The concept of an animal having a 'home range' has existed for very many years (Seton, 1909). The widely-used formal definition by Burt (1943) as the 'area traversed by the individual in its normal activities of food gathering, mating and caring for young' seems to imply the exclusion of excursive movements, but conveniently parallels the concept of central place foraging for species with specific nests or dens. However, this definition makes no allowance for change with time (Cooper, 1978), and studies frequently define areas for animals that are not mating or rearing young. Moreover, it is hard to define 'normal' objectively (S. Harris *et al.*, 1990; White and Garrott, 1990), though many studies have (subjectively) used 95% of the locations or their estimated density distribution.

For purposes of analysis, the definition of a home range simply as 'an area repeatedly traversed by an animal' is convenient. This definition recognises a basis for home range in the movements pattern, which can be defined objectively. The 'repeatedly traversed area' includes excursions, where the movement is there and back, but enables a home range to be separated statistically from any unidirectional dispersal that leads from the area (see section 9.2.2) or (by reversing the dispersal detector) to the area. The repeated-traverse concept accommodates the 'prevailing range' of Doncaster and Macdonald (1991), which omits areas no longer traversed.

The home range information required to answer biological questions typically involves three 's' measures: size, shape and structure. Estimates of home-range size may be needed for management purposes, such as planning reserves, or as an indicator of differences in energy used by species or by categories within species (Schoener, 1968; Harestad and Bunnell, 1979); size may also correlate with performance (Kenward, 1985; Gatti *et al.*, 1989). Shape may be important for analysing how a home range is placed in the landscape, as one indicator of resource and security requirements (Covich, 1976), or how home ranges of conspecifics fit together, to indicate territoriality or social cohesion (Macdonald *et al.*, 1980). Structure within a home range may be required for studying how the intensity of using different areas relates to habitat and sociality (Adams and Davis, 1967), or for predicting likelihood of encounter during census work (see Chapter 10). Questions of reserve planning or general energy expenditure may require home ranges to be recorded over long periods, to minimise temporal fluctuations, whereas habitat questions may require study in seasons when particular foods are important or young being reared. Social questions may need investigation only during a short period of courtship.

Home ranges may therefore need to be defined annually, or seasonally, or for even shorter periods. However, you should if possible use a standard collection of locations, which is large enough to sample throughout the range used in that period, across all the animals. The use of a standard range, based on incremental analysis with the intended home-range estimators during a pilot study (see Chapter 8), avoids the problem of continual small increments in range size as animals change their foraging and make excursions. An alternative to a standard number of locations is to treat the number of locations as a covariate in each analysis. For parametric analyses, you should log-transform range size, for which a positive skew is likely. You should also log-transform the number of locations, because the relationship between them and range size is likely to be a power function (Loehle, 1990; Gautestad and Mysterud, 1995).

A great many home-range estimators have been published. Those that are readily available in software, and are thus used relatively frequently in publications, fall broadly into two groups. One group estimates the density of locations in ellipses or contours. The second group creates polygons that minimise the sum of link-distances between locations. There are also hybrid techniques that create tesselations or other polygons based on location density; these are difficult or slow to compute and not widely used.

The techniques within each main group are capable of defining range shape and structure with varying degrees of detail. They are illustrated here for squirrels that were radio tracked on a small island. The squirrels lived mainly in fragmented woodland, and seldom visited the beach or two industrial sites cleared for oil extraction machinery. Data are standard 30-location home ranges collected for two squirrels that lived round the edge of the oil sites. The data for one squirrel were analysed twice: once as they were recorded, and again with a single outlying fix moved to the other end of the island, where this adult male squirrel was recorded on another occasion. This illustrates the effect of a single distant outlier on the various range estimators.

As with previous parts of this chapter, don't worry about the mathematical details. You may even prefer to return to sections 9.3.1 to 9.3.4 after reading section 9.3.5. That final part uses data from recent studies to advise you on which of the estimators may best answer different biological questions in particular circumstances.

9.3.1 Location density estimators

The process of quantifying space-use patterns of animals as home ranges began well before radio-tagging, as a result of the need to summarise data from arrays of traps. Hayne (1949) suggested that an animal's (trap) range could be seen as a centre of activity, defined as the mean of the x and y coordinates, surrounded by zones of decreasing probability of trapping the animal. A way of calculating this probability distribution was provided by Dice and Clark (1953), leading to the description of ranges as probabilistic circles (Calhoun and Casby, 1958; Harrison, 1958). The basic approach was to estimate frequency of occurrence, assuming a normal distribution of independent locations, initially from one estimated range centre and later from multiple 'attraction points' (Don and Rennolls, 1983) or nodes on a grid that could be used to interpolate density contours (Dixon and Chapman, 1980; Worton, 1989). The term 'utilization distribution' was introduced to express the relationship between area estimate and frequency of occurrence (Van Winkle, 1975).

Ellipses

The probability circle was first modified to allow for the asymmetry present in ranges by estimating bivariate ellipses (Jennrich and Turner, 1969). The maths of these ellipses and their subsequent development is reviewed elegantly by White and Garrott (1990). They point out that the replacement of a χ^2 statistic by an F statistic to estimate probability levels (Koeppl *et al.*, 1975; 1977) makes ellipse areas dependent on sample size in a way that can be avoided by estimating separate confidence limits for the size of ellipses (Dunn and Brisbin, 1982). The further refinement by Dunn and Gipson (1977), to provide a correction for non-independence of data collected in bursts, is shown to create ellipses very similar to the Jennrich-Turner originals when data are point sampled.

However, animal locations are often not distributed normally about the arithmetic activity centre of an ellipse, which may even lie in an area not visited by the animal (Fig. 9.3). Therefore the resulting circles or ellipses often contain large areas into which an animal never ventured (Adams and Davis, 1967; Macdonald *et al.*, 1980). Moreover, inclusion of an outlier that is not extreme enough to greatly affect the placing of the arithmetic mean range centre can nevertheless greatly affect the shape and area of a Jennrich-Turner ellipse (Fig. 9.3). An asymmetry index of range structure, estimated as the ratio of standard deviations on the major and minor axes of ellipses, would also be vulnerable to this effect. To make ellipses more robust to the effect of outliers, the locations can be weighted by an inverse function of their distance from the activity centre (Samuel and Garton, 1985; Koeppl and Hoffmann, 1985).

Density contouring

Density estimators that better reflect the shape of home ranges can be obtained with models that allow for clumping in the data. Don and Rennolls (1983) developed a multinuclear model in which circular estimates of density distribution at biological attraction points were combined to estimate contours. However, a subjective element in the choice of attraction points gave the advantage to methods that estimate location densities at intersections of a grid with an independent coordinate system superimposed on the location data.

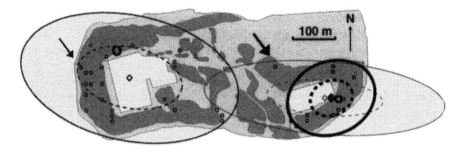

Figure 9.3. Jennrich-Turner ellipses estimated for 50% (- -) and 95% (—) of the location density distribution for two red squirrels (*Sciurus vulgaris*) on an island. The arithmetic mean coordinates that estimate the range centre (diamond) for one squirrel, with circles of locations (o) and nest (**O**) are in the centre of an oil-production area that was never visited. Ellipses for the other squirrel are shown with an outlier as recorded at the time (thick lines, thick arrow) and also with the outlier moved to a more distant location visited by the squirrel at another time (thin lines, thin arrow) to show the sensitivity of ellipse estimation to outliers.

Dixon and Chapman (1980) showed how a function could be used to derive estimates of location density at intersections (nodes) of a grid across the range. Isolines (contours) could then be interpolated between intersections at a value that would enclose areas for, say, 95% of the density distribution (Fig. 9.4). Following Neft (1966), Dixon and Chapman (1980) used as a density generating function the inverse reciprocal (harmonic) mean of the distance (r_{im}) of n locations from m nodes. A difficulty with this harmonic mean estimator

$$H\rho(m) = \frac{1}{\frac{1}{n} \sum\limits_{i=1}^{n} \frac{1}{r_{im}}} \tag{9.1}$$

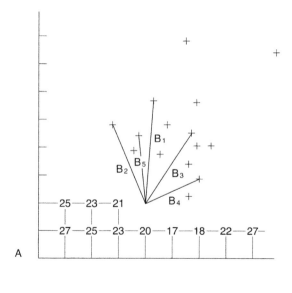

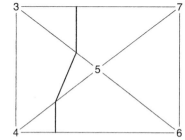

Figure 9.4. The computation technique for harmonic mean contouring. The inverse mean distance from all the fixes is estimated at each grid intersection. In A, computation has reached the fourth intersection in the second row, and summed the inverse distance to the first five fixes from that point. In B, a central value for each grid square has been estimated as the mean value of its corners, and the isoline with a value of 4.5 interpolated on the edges and diagonals.

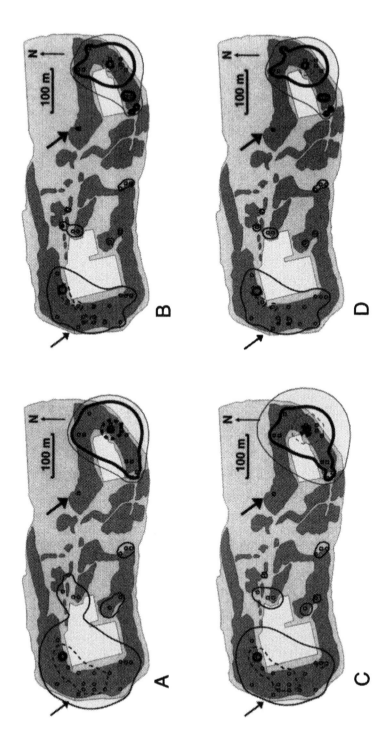

is that $1/r_{im}$ tends to infinity as locations approach nodes (i.e. $r_{im} \to 0$). The solution to this problem in the original implementation was not to allow $1/r_{im}$ to be less than 1; in other words r_{im} could not be less than the tracking resolution in metres. The problem with this approach is that it is sensitive to the scale of the tracking resolution.

The tracking resolution (accuracy) used for harmonic mean estimations from the squirrel data was 10 m (Fig. 9.5A), but if a resolution of 1 m was assumed, the range outlines tended to fragment into rings drawn round individual locations (Fig. 9.5B). This was because the distance between nodes on the 50 × 50 grid changed from maximally one resolution unit (of 10 m) to as much as 10 resolution units. With no more than one resolution unit from a location to the nearest node, even with a distant outlier, the $1/r_{im}$ estimates were frequently one, which resulted in density estimates that were strongly smoothed by truncation and there was little outlier effect. With 10 times as many resolution units (Fig. 9.5B), $1/r_{im}$ was decreasing with distance from grid nodes 10 times as fast, which emphasised the influence of locations near each node and gave inadequate smoothing except when the grid was stretched across the whole island by the outlier.

Spencer and Barrett (1984) pointed out that this dependence on the scale of locations and position of the grid could be reduced if locations were given a fixed distance from the nearest nodes by centering in the cells between. With this modification, results become independent of whether resolution is 10 m (Fig. 9.5C) or 1 m. However, the modified r_{im} estimates become multiples of the distance between nodes, and therefore dependent on grid size (Ackerman *et al.*, 1990). This resulted in area estimates for the two compact home ranges decreasing by 33–82% as the grid was increased from 20 × 20 to 50 × 50, with a further reduction of 36–77% in area on a 100 × 100 grid (Fig. 9.5D). In contrast, areas of uncentred plots changed by only 1–27% from 20 × 20 to 50 × 50 grids. Moreover, size changed by less than 2% when grid size was increased beyond 30 × 30 for the smallest range, which spanned 15 resolution units, and beyond 40 × 40 for the range spanning 26 resolution units (Fig. 9.6). The difference between the smallest range and its distant-outlier version, which spanned 52 resolution units, became minimal on the 100 × 100 grid. Whereas there was too little smoothing for uncentred locations with many resolution units between nodes (Fig. 9.5B), smoothing gave consistent results like Fig. 9.5A once there were about two nodes spanning each resolution unit. Thus, with uncentred locations, grid size across the range should be greater than 2 × *range-span divided/resolution* (e.g. at least 50 × 50 for a range that spans 25 resolution units).

Figure 9.5. Harmonic mean contours containing 50% (- -) and 95% (—) of the location density distribution for the squirrels in Figure 9.3, on a 50 × 50 grid. With estimation based on units of tracking resolution (10 m), the outlier had relatively little effect (A), but a change to 1 m units affected the estimates (B). Estimates became independent of resolution if locations were centred with in the grid (C), but were then affected by changes in node spacing of the grid, which was halved by using a 100 × 100 grid for (D).

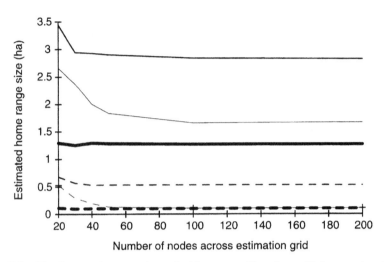

Figure 9.6. The decrease in area estimated with uncentred locations with increase in grid size (i.e. with decrease in node spacing). The effect is greatest for the range with a distant outlier (thin lines, using same conventions as Figure 9.5), but in all cases there is no further size decrease when the grid extends across the range with at least two nodes per resolution unit.

One may conclude that, in contour analyses based on harmonic mean estimators, the need to truncate $1/r_{im}$ when locations are close to nodes makes smoothing dependent on a relationship between tracking resolution and grid size. If locations are centred between nodes to avoid the effects of tracking resolution, the extent of smoothing becomes highly sensitive to grid size and hence to the presence of outliers that expand the grid. Without centering, smoothing becomes reasonably independent of grid size provided that the number of nodes spanning the grid has reached about double the number of resolution units spanning the range. If you claim a 1 m tracking accuracy when 100 m would be more appropriate, you will get 'rings-round locations' undersmoothing unless you use grids that may well be too big for the computer or software, and will at least be slow to compute. You will be safer adjusting the resolution of the data (which is a form of centering independent of the contouring grid) than centring locations and thus making your results highly dependent on the grid.

Spencer and Barrett (1984) also proposed modified indices of range structure. Dixon and Chapman (1980) had defined a harmonic mean range centre Hc at the intersection at which $Hr(m)$ has a minimum value (Hr). Spencer and Barrett (1984) showed that this centre can move considerably if the spacing and origin of the arbitrary grid is changed. However, they pointed out that if the minimum Hp is sought at each location for the $n-1$ distances to all other locations,

$$H\rho(j) = \frac{1}{\dfrac{1}{n-1}\displaystyle\sum_{i=1}^{n-1}\dfrac{1}{r_{ij}}} \tag{9.2}$$

the resulting Hc is unique, as is the standard error (Sr) of the distances r_{ij}. The ratio Hr/Sr is then a measure of fix dispersion about Hc, and Hp/Hr is an index of kurtosis, with values of 2 to 4 indicating a strongly peaked (leptokurtic) distribution of fixes. Moreover, the grand inverse reciprocal mean of all the distances r_{ij} provides a measure of spread among the locations:

$$Hp = \frac{1}{\left[\frac{1}{(n-1)}\right]^2 \sum_{t=1}^{n-1} r_{ij} \sum_{j=1}^{n-1} r_{ij}} \qquad (9.3)$$

The estimation of Hc at a location ensures that the range centre is never in an area unvisited by an animal, in contrast to the arithmetic mean centre (Ac) which can be in an unused area (Fig. 9.3). For central-place foragers like squirrels, Hc often proves to be closer than other estimators to the nest (Lair, 1987), if not at this focal site.

Kernel contours

Worton (1987, 1989) noted that the harmonic mean estimator is only one of a wider family of kernel functions (Silverman, 1986) that can be used to estimate the distribution of locations. He suggested using a function such as the bivariate normal estimator

$$K\rho(m) = \frac{1}{2\pi nh} \sum_{t=1}^{n} \exp\left[-\frac{r^2_{im}}{2h^2}\right] \qquad (9.4)$$

with a smoothing parameter

$$h = \left(\frac{1}{n^{\frac{1}{6}}}\right) \sqrt{\frac{1}{2(n-1)} \left[\sum_{t=1}^{n} (x_t - \bar{x})^2 + (y_t - \bar{y})^2\right]} \qquad (9.5)$$

to derive location density at grid nodes, because values then increased smoothly to 1, for exp[0], when locations were at a node. This estimator therefore removed the need to truncate distances, which creates the dependence on grid size and tracking resolution when using harmonic mean estimators. As a result, the ranges modelled on a 50 × 50 are identical (Fig. 9.7A) to those on a 20 × 20 grid (Fig. 9.7B), apart from the more jagged outlines on a small grid.

Worton (1989) pointed out that the first choice of smoothing parameter (Eq. 9.5), subsequently also called the reference bandwidth or window width (Seaman and Powell, 1996), might overestimate range area when locations have a strongly multimodal distribution (as in Fig. 9.7). Instead he suggested that a choice of a

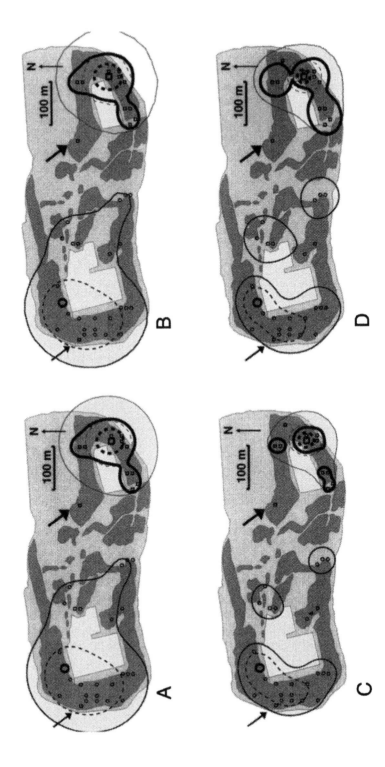

smaller *h* value might be based on least squares cross validation (LSCV) of the mean integrated square error. Although kernel estimates show little variation with grid size, they are greatly affected by applying a multiplier to the first-choice (reference) value of *h* (S. Harris *et al.*, 1990). Smoothing reduces with decrease in *h*, and range size estimates decline down to values found with 'rings-round-locations'. At twice the first-choice *h*, the contours become ellipses, which are larger than bivariate normal ellipses and larger for the range with the distant outlier than for the other two (Fig. 9.8). Moreover, if you compare Fig. 9.8 and Fig. 9.6, you will see that the effect of the outlier on area is greater with the bivariate normal kernel estimator than when using the harmonic mean, especially in the 50% core. Outlier-dependence of kernel estimates probably also explains why kernel outlines perform like ellipses during incremental analyses (Fig. 8.2), whilst harmonic mean outlines resemble MCPs by increasing more steadily but for longer as locations are added.

The LSCV process is not always effective for data with large resolution relative to range span (Seaman and Powell, 1996), but gave a result for 17 of 27 squirrel ranges on the island. Separate LSCV estimates for each range would lead to LSCV-dependent range sizes, with an upward bias for any ranges where LSCV fails and defaults to the reference *h*. If LSCV is to be used, it therefore seems best to use the same multiple of *h* in the whole data set, such as the median *h*-multiple from the ranges where LSCV gives a result. The median *h*-multiple for squirrel ranges where LSCV gave a result was 0.48, which gave range outlines closer to the harmonic mean estimates than with reference *h* (Fig. 9.7C).

Worton (1989) also suggested that permitting local variation in *h* might give better estimates of areas used, and suggested the use of an adaptive kernel estimator that weighted *h* at each node by the inverse of the initial distance function estimate there. This weighting emphasises the tail of the distribution, expanding the 95% contours (Fig. 9.7D). More recent reviews (Worton, 1995a; Seaman and Powell, 1996) have concluded that the adaptive weighting biases area estimates upwards, and that the fixed kernel is preferable. Other variations on the original estimator include replacement of the normal kernel (Eq. 9.4) by the Epanechikov kernel (Worton, 1989) or another bi-weight kernel that computes faster than the bivariate normal kernel (Seaman and Powell, 1996), or truncating the utilisation distribution to reduce overestimation of home range size (Neaf-Daenzer, 1993). These modifications all require comparison with results from bivariate normal estimation to discover whether there are robust advantages.

Figure 9.7.   Increasing node spacing, by reducing grid size from 50 × 50 (A) to 20 × 20 (B) had negligible effect on kernel contour estimates. The estimates are strongly affected by reducing the smoothing parameter to half its reference value (C), and areas with low location density are expanded by adaptive weighting (D). Like the ellipses, these estimates are strongly influenced by outlier distance.

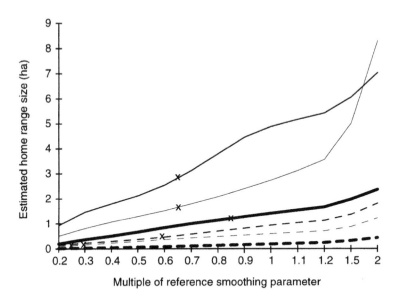

Figure 9.8.   Increasing the value of the smoothing parameter as a multiple of its reference value increases the size of kernel range estimates until they become ellipses. The value of the parameter that estimates areas equivalent to those estimated by harmonic mean contours (X) can differ greatly between the 95% outlines (—) and the 50% outlines (- -).

When you estimate contours by harmonic mean or other kernel methods, you are liable to get different results with different software (Larkin and Halkin, 1994; Lawson and Rogers, 1997). This occurs for several reasons, including variation in: (i) how grids are placed to ensure that outer isolines can be plotted; (ii) whether locations are centred; (iii) whether areas are estimated by counting enclosed grid cells or from the lines plotted; (iv) whether plots are based solely on the density distribution (now more common) or to include required proportions of locations (the original Dixon-Chapman method); (v) if purely density-based, whether the node or location densities are used and how and (vi) what kernel estimator is used. You need to know how defaults are set in the software, so that you can over-ride them to make your results robust and as comparable as possible.

For example, RANGES V: (i) places 70% of the stated grid size across the maximum x or y spread of the locations; (ii) centres locations (default) or leaves them unmodified; (iii) estimates areas within isolines; (iv) plots to locations (default) or to density, which is (v) estimated on log-transformed values at locations and (vi) uses a bivariate normal kernel. Default (ii), of centering locations for harmonic mean estimates, was used for compatibility with RANGES IV, but should be changed. Moreover, estimation of range areas within 95% contours implies plotting to the density distribution rather than to locations, and thus a change of default (iv). The default plots to locations because this should give the most similar results from different software.

Although kernel analyses should not be greatly affected by variation in grid dimensions, the choice of kernel may cause differences between packages and there is also scope for variation in implementing LSCV. The RANGES V implementation re-scales the y coordinates to the same standard error as the x coordinates before estimation of reference h (Worton, 1989); LSCV then decreases the multiple of reference bandwidth h from 1.5 to 0.1 in steps of 0.02, seeking the minimal rate of change in the mean square value; this is either the minimum, or the upper inflection point if there is no true minimum. If there is no minimum or inflection point, reference h is used.

9.3.2 Minimum-linkage estimators

The earliest assessments of range area and position were made by drawing a line round the outermost trap sites or radio fixes (Dalke and Sime, 1938; Mohr, 1947) to create a polygon. A minimum convex polygon (MCP) is a unique estimate if it minimises the sum of link distances between edge locations (Southwood, 1966). A variety of other rules that minimise sums of link distances can be applied, including methods of peeling away locations furthest from a home range centre, or using distances between locations to link them in clusters. Grid cells are the members of this family of minimum-linkage polygons with the smallest link distances.

Convex, concave and peeled polygons

S. Harris *et al.* (1990) noted that MCPs round the outermost locations have probably been the most common way of representing range size and shape. However, outer MCPs give no indication of how intensively the animal uses different parts of its range, and the area is strongly influenced by peripheral fixes. Since many species seem to make occasional excursions well outside their normal foraging range, minimum convex polygons often contain large areas which animals never visited, and are sometimes more representative of areas traversed during excursions than of where an animal spent most of its time. Note the effect of the distant outlier in Fig. 9.9A.

Several techniques have been proposed for reducing the unvisited areas, including the use of concave outlines (Stickel, 1954). This is usually done by measuring the range span, which is the greatest link distance in the range, and then drawing a peripheral line to an internal location wherever the distance between edge locations is more than a half or quarter of the span (Harvey and Barbour, 1965). Alternatively, the locations may be linked in order of distance from the arithmetic centre of activity (Voight and Tinline, 1980). The former approach uses the same linkage rules as an outer MCP, with a maximum-link-distance filter. When the filter is a suitable multiple of the span, such as $0.35 \times$ *span* for the squirrel data (Fig. 9.9B), a range can be shaped to avoid unused areas. If polygons are given a boundary strip, to correct for the possibility of a location

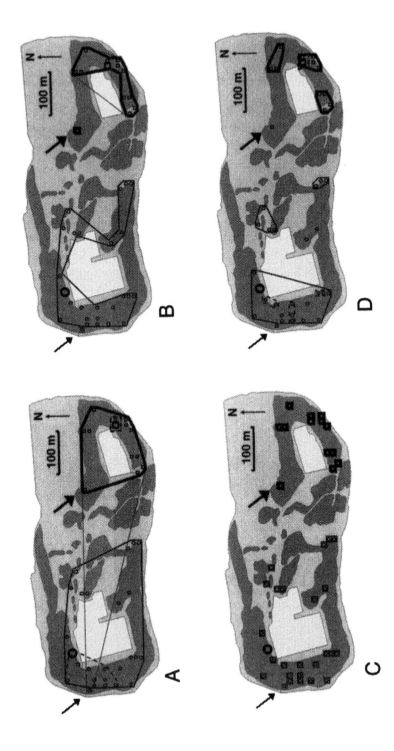

being up to half the resolution distance away from its estimated coordinate (Sanderson, 1966), outlying locations are isolated in separate grid cells. However, there is as yet no objective way to estimate a 'suitable' span multiple. Moreover, the stretching of the span to an outlier affects the outline round the rest of the range.

Different linkage rules can be used to create peeled polygons, by excluding locations ranked in order of linkage distance from a range centre (Kenward, 1987). The peeling of proportions of the outermost locations (Michener, 1979), for example at 5% intervals, gives a utilisation plot (Fig. 9.10) that is a non-parametric analogue (Mohr and Stumpf, 1966) for the probabilistic utilisation distribution provided by location-density estimators (Van Winkle, 1975; Ford and Krumme, 1979). A problem with the peeling-from-centre approach occurs when the range centre lies in unused habitat (e.g. arithmetic mean centres in Fig. 9.5). An outline of the 50% range core may not then be a good representation of the home range (White and Garrott, 1990), a constraint that also affects ellipses centred on the arithmetic mean. However, recalculation of the arithmetic mean and re-ranking locations after each successive exclusion, as used to estimate 50% polygons in Fig. 9.9A, focuses with little computation on the area where locations are densest. The harmonic mean location *fide* Spencer and Barrett (1984) provide centres for range cores that would no more be based on unused areas than with equivalent density estimators. Methods of estimating polygons with maximum location density are considered in section 9.3.3.

Grid cells and box-counting

If linkage rules are set to less than the resolution distance, then locations with boundary strip plot as grid cells (Fig. 9.9C). The plotting of home ranges as an accumulation of grid cells was introduced for the extensive data gathered by the early automated station at Cedar Creek (Siniff and Tester, 1965), but has subsequently been used when manual tracking has collected large quantities of data (Macdonald *et al.*, 1980; Doncaster and Macdonald, 1991). The cells may be hexagons, to equalise distances between their centres, but it is often easier to use grid squares, whose sides should not be smaller than the resolution of the tracking technique.

If there are enough fixes, so that several might occur in any one cell, you can calculate indices of variation in each animal's spatial use of its range. Ecological diversity indices may be used, or the coefficient of variation (standard deviation divided by the mean), but these only show whether the fixes tend to be clumped, and not whether they form one or many clumps. Macdonald *et al.* (1980) pointed

---

Figure 9.9. Outer MCP ranges are strongly influenced by the outlier, but outliers have no effect on cores peeled to include 50% of locations (A). Outlier distance affects range span, and hence does affect restricted-edge polygons (B), but not areas estimated from grid-cells (C) or by cluster analysis (D).

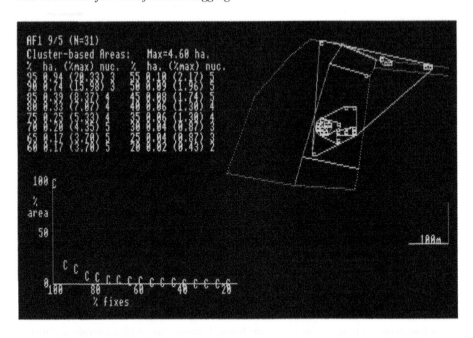

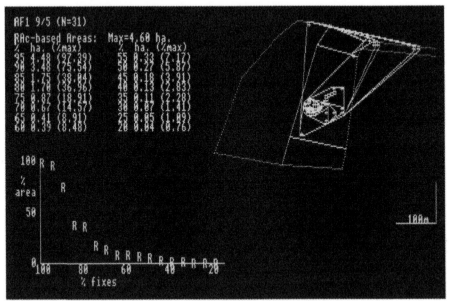

Figure 9.10. Estimating range cores by inspection of utilization plots and by outlier exclusion. Cluster analysis (A) gives a clear indication of a 0.9 ha core at 95% by inspection, and polygons peeled to exclude locations furthest from the harmonic mean (B) provide a

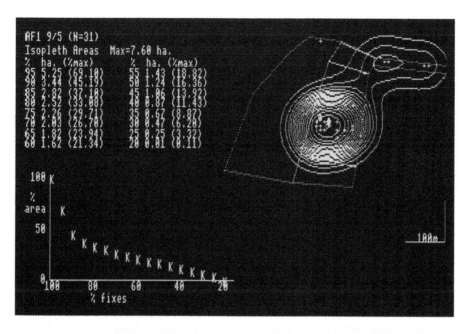

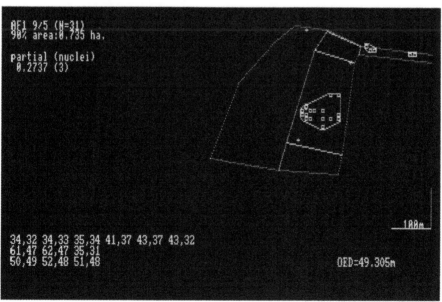

similar sized core less obviously at 75%. These estimates are close to the objective estimate of 0.7 ha for 90% of locations by outlier exclusion (D), but the 2.8 ha indicated by a slight discontinuity from kernel analysis at 85% (C) is much larger.

out that the Rasmussen Index (Rasmussen and Rasmussen, 1979) overcomes this, by taking the distance between cells into account: this index reaches its maximum value when intensively used cells are close together. A value (d) is calculated for each of the P ($=N(N–1)/2$) cell pairs, as the sum of their fixes divided by the distance between them, and used to derive the index.

The grid cell method is at a disadvantage, compared with other range analysis techniques, when comparing range areas. It can take more than 200 locations to reach sampling saturation (Chapter 8) by 'filling-in' all the squares that the animal is using (Voight and Tinline, 1980). This problem can be alleviated, albeit at the expense of accuracy, by relaxing the linkage rules a little to add 'influence cells'. For instance, allowing linkage at the resolution distance provides a 'rook's case' that adds only the four influence cells whose edges join (Fig. 9.11), whereas the 'queen's case' permits linkage at $\sqrt{(2 \times resolution\ distance)}$ to add all eight adjacent cells. The value of influence cells in estimations may be set at a fraction of the visited cells (Doncaster, 1990).

Grid-cell methods have been the basis for some innovative approaches to the estimation of size indices for home ranges. Counting grid cells nested within larger boxes is the basis for treatment of the pattern of locations as a fractal (Loehrl, 1990; Gautestad and Mysterud, 1995), to estimate size and rate of increase of area without also estimating a range shape. Another approach was initiated by Ford and Krumme (1979), who showed that the number of location records in each grid cell could be used to create a smoothed utilisation distribution if cell size was enlarged. Anderson (1982) extended this approach by introducing Fourier analysis of distances between cells for smoothing that could be used to plot range contours. The method can use less than 50 locations to estimate 50% cores, but suffers from the 'large-sample size problem' of grid cell analyses and thus requires more locations for stability at the edges of the distribution. It is therefore less robust than the harmonic mean and other kernel contouring approaches (Worton, 1987). Geissler and Fuller (1985) used cells larger than the resolution unit for casement displays, to show variation in the spatial pattern through time. They also introduced a simple form of cluster analysis to estimate range size from distances between these cells.

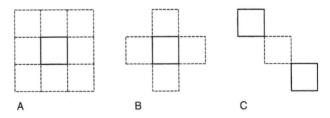

A   B   C

Figure 9.11. Some rules for increasing the area of grid cell ranges. In the 'queen's case' (A), all cells adjacent to a used cell are considered to be used. In the 'rook's case' (B), only the cells which share a side with a used cell are added to the total. 'Linked cells' (C) are added if they lie between cells which were used consecutively.

Cluster polygons

If linkage distances between are gradually increased, locations form into clusters that provide a minimal sum of nearest-neighbour distances. This is a form of hierarchical incremental cluster analysis (Anderberg, 1973; Everett, 1980). To reduce fragmentation as a result of serial spatial correlation, the minimum cluster size in RANGES V is three locations. Thus, the three locations with the minimum sum of nearest-neighbour joining distances form the first cluster. This cluster gains a fourth location if the distance to its nearest outlier is less than the mean nearest-neighbour distance in the next potential cluster (Fig. 9.12). If the location to be added is already assigned to another cluster, the two clusters merge. If two clusters have the same distance to nearest outliers, the outlier with the minimum mean distance to every location in the cluster is the one that joins. Thus, a centroid joining rule (based on local density) resolves ties.

Priority is given to the nearest-neighbour linkage method on the assumption that animals probably route their initial visits to a new location from the nearest familiar location. Nearest-neighbour linkage favours chaining of polygons from their ends along line features (e.g. hedges, rivers). Outlines round the separate clusters are implemented in RANGES V as convex polygons (Fig. 9.9D). In an essentially mononuclear range, there may initially be several clusters, but these merge to a single polygon when 80–90% of the locations are included.

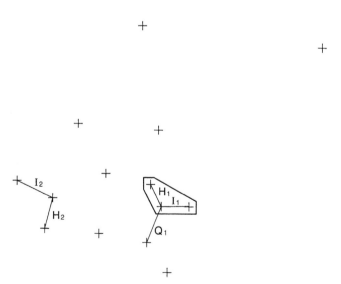

Figure 9.12. The cluster analysis mechanism. The first cluster of three fixes has the smallest distance $H_1 + I_1$. The distance to their nearest neighbour is Q_1. The next smallest cluster of three fixes has a nearest neighbour distance, I_2, from the two closest fixes to the third fix. This cluster will form next if I_2 is less than Q_1. If not, cluster 1 will gain its fourth fix.

Like the harmonic mean and other kernel techniques, cluster analysis provides statistics to describe range shape. As well as the number of clusters (k_x) for the densest $x\%$ of the locations, an index of range patchiness is given by

$$Cp_x = \frac{\sum_{t=1}^{k} a_{xi}}{A_x} \qquad (9.6)$$

where a_{xi} is the area of each cluster polygon and A_x is the area of a single polygon round all the locations assigned to the clusters. Values of Cp_x tend from 1 to 0 as patchiness increases. Johnstone (1992) noted that the single polygon A_x round all the clusters provides a convenient single outline for studying territoriality without the dependence on a range centre that can occur with polygon peeling.

An index of diversity for the Cf_{xi} locations in each cluster polygon is

$$Fd_x = \frac{1}{\sum_{i=1}^{k} \left[\frac{Cf_{xi}}{F_x} \right]^2} \qquad (9.7)$$

where F_x is the number of locations in all the polygons. Values of Fd_x increase from 1 as clusters contain increasingly dissimilar numbers of locations.

With a boundary strip, cluster analysis represents grid cell analysis with an expanding linkage rule. If three or more locations occur at the same coordinates, perhaps from repeated visits to a favourite food tree, cluster analysis attributes one resolution cell to the site no matter how many locations are present. This contrasts with density contouring, which attributes a ring that varies in size with the number of locations. Cluster polygons also define multinuclear cores without the tendency of contours to expand into unused areas. Both effects occur because all the locations contribute to each nodal estimate that creates contours, whereas each cluster is entirely separate unless you force expansion of the linkage distance to include a distant outlier.

The use of a boundary strip makes all polygon analyses an extension of grid-cell analysis. However, if you use a strip it must be a correct representation of the tracking resolution. Too large a strip will seriously inflate your polygon estimates of range size (Hansteen *et al.*, 1997), just as incorrect setting of the resolution will have a similar effect on harmonic mean contours.

9.3.3 Hybrid estimators

There is a continuity between the density-based and linkage-based techniques for estimating home ranges, because both are based on measuring distances. At one extreme are the ellipses (including circles) that use the distance to all the locations to estimate range use as a density pattern around one range centre. The estimators based only on linkage between single locations, to create grid cells and nearest-neighbour clusters, are at the opposite extreme. The use of a grid for contouring adds a further distance element to the density estimation process, and the use of a range-centre for polygon peeling adds a density element to the linkage estimation (the centre being estimated from all the locations). Further hybrid elements are added by using density of locations in resolution-unit grid cells, or larger boxes, to estimate utilisation patterns. Two other methods use an almost equal balance between the density and linkage approaches.

Hartigan (1987) proposed that single polygons be drawn round a chosen proportion of locations to minimise the area, and hence to maximise the density. This is the ultimate rationale behind any peeling of outer locations from polygons. However, for any method in which locations are progressively eliminated, there is a possibility that the densest set for a small core will not have been part of the densest set for a large core. The only robust technique is therefore to seek the minimum density in a polygon containing all combinations of the desired proportion of locations, for which computer time increases exponentially with the number of locations. Another quick way of peeling groups of edge polygons as 'convex hulls' has been proposed by Worton (1995b, see also Glendinning, 1991); possible advantages over methods of peeling to a centre of density remain untested.

The second hybrid approach is the creation of tesselation (tiled) polygons round locations, for example by joining perpendicular bisectors of the lines linking locations to their nearest neighbours (Wray *et al.*, 1992a). Home-range areas and shapes are estimated by including a required proportion of tiles, ranked from smallest to largest. The size of each of these Dirichlet tesselations is an index of local density, because it depends only on nearby locations. Cores are similarly placed to core contours and are expected to relate to those from cluster analysis (Wray *et al.*, 1992b). A problem with this technique is that the area of each tile is strongly dependent on those adjacent locations that are most peripheral in the range. Tesselations therefore expand like contours into unused areas on the edge of the range. This problem also results in peripheral locations having tiles of infinite size. As with Fourier analysis (Anderson, 1982), it is hard to estimate reasonable boundaries unless samples of locations are very large.

9.3.4 Estimating range cores

In all home range models except grid cells, the shape of an outer boundary is affected quite strongly by any outlying locations (Figs 9.5–9.9). Range areas are likely to be most comparable if a 'normal area', in the sense of Burt (1943), can be separated as a range core from an excursive area represented by the outliers. Indeed, defining an outer density contour is problematic, because parametric estimation of the utilisation distribution means that a 100% contour is infinite. From common practice, you might well choose to use a 95% core, although White and Garrott (1990) point out that this analogy with the 5% statistical α-level is arbitrary. There has therefore been a search for more objective ways of defining home range cores.

Utilisation plots provide one way of separating range cores from excursive areas, because home range area tends to drop sharply as outlying fixes are excluded and then to decline less steeply (Clutton-Brock *et al.*, 1982). The discontinuities seen in Fig. 9.10A–C indicate the core (Kenward, 1987; S. Harris *et al.*, 1990). A discontinuity is usually indicated most clearly in plots from cluster analysis (Fig. 9.10A), because peeling of multinuclear ranges progressively excludes the smallest nuclei, which tends to result in a 'stepped' utilisation plot (Fig. 9.10B). Probability plots of density-based utilisation distributions are smooth. RANGES V provides an option for plotting contours that just enclose a required proportion of the locations (and can thus provide a 100% contour). However, even without basing contours on a probability distribution, contours are smoothed by being generated from mean distances to all the locations (Fig. 9.10C), and therefore provide a less distinct core than with the abrupt polygon edges of linkage-based estimators. Separation of isolines into multiple nuclei has been proposed as a core indicator in this case (Wray *et al.*, 1992b).

A more objective way of estimating a core was provided by Samuel *et al.* (1985). In this case the principle is to test for a discontinuity in frequency of locations in boxes. A subsequent modification provides a more robust test (Samuel and Green, 1988). However, as in other grid-cell analyses, you need large numbers of locations for a significant result based on resolution cells; otherwise, you will be making a subjective choice of where to place boxes large enough to contain testable samples of locations.

Two approaches estimate cores by statistical elimination of outliers. Ackerman *et al.* (1990) noted that if the harmonic mean or kernel density value is estimated at each location (Eq. 9.2 or the equivalent from Eq. 9.4), a highly divergent value can be used to exclude a distant outlier. A linkage-based equivalent avoids the smoothing of the density value by using only the nearest link (or two links for equivalence to the minimum three-location cluster) instead of all the distances. The distance distribution is normalised with a kernel function, and highly divergent values excluded (Kenward *et al.*, in press b).

Exclusion of outliers can involve (arbitrary) removal of those in the upper 5% of the distance distribution. However, a more objective approach is based on an assumption that movements in the core and during excursions are different, so that faster movement creates a different distance distributions for outlying and core locations. For example, buzzards were most often flying when first detected at their most outlying locations, so their excursive distances may belong outside a core distance distribution. A core can then be defined by initial exclusion of the largest linkage distances if it is greater than, say, 0.1% of the distribution from the $n - 1$ smaller distances; this process is repeated for the next largest linkage distance until a core with a normal distance distribution remains (Kenward *et al.*, in press b). Iteration with a 0.1% α-level gives loose cores, with minimal tendency to exclude outliers (Fig. 9.10D), whereas 1% exclusion gives tight cores.

Outlier-detection using nearest-neighbour links can be used to eliminate outliers before applying another method to define a boundary. With data from buzzards, it gave very similar results to those estimated subjectively by inspection of utilisation plots (Hodder *et al.*, 1998). The plots and objective analyses indicated that 95% of buzzard ranges had cores that contained at least 85% of the locations; in other words, up to 15% of locations were excursive. The 85% provided a convenient common core for the set of ranges, because although it excluded a small proportion of core area for many birds, there would be less size variation from excluding a small part of the core than from including outliers. Subjective assessment of utilisation plots indicated that cluster polygon cores contained 85% of the locations for a number of species (Kenward, 1992).

9.3.5 Choosing estimators

Choice between the many different estimators is restricted for most people by the current software. Thus, analyses are readily available for the density-based estimators and linkage-based polygons. The former range from simple ellipses to more complex contours. The latter range from simple MCPs to cluster polygons, but also include grid cells.

Two recent studies show that no single choice of estimator, from ellipses to cluster analyses, always gives the best answers to biological questions. In one study, questions were about range size and structure in relation to age, weight and social factors (Kenward *et al.*, in press b). Data were from 14 squirrels on the island in Figs 9.3–9.9, and many more squirrels and raptors elsewhere. The other study used sampling of locations within complete movement trajectories to investigate questions about stability, bias and number of locations required for the different techniques (Robertson *et al.*, 1998). Data were for a central-place forager (a chaffinch) and a transient forager (a goshawk). These studies add to previous comparative work with real data (Jaremovic and Croft, 1987), simulated data (Boulanger and White, 1990; Worton, 1995a) or a mixture (Seaman and Powell, 1996). The use of real data avoids a problem with simulated data, namely that the performance of estimators based on normal distributions is likely to reflect the parametric distribution functions used to generate those data.

The differences between techniques seem to depend mainly on how stability, and detail in relation to the number of locations, is affected by smoothing. Much smoothing brings high stability with few locations, but little detail. You don't get something for nothing!

Density estimators use all the distances, and therefore smooth substantially. Ellipses, with a single estimation centre, smooth maximally. They require minimal numbers of locations for stability (Fig. 8.2, Table 9.1) but also provide minimal detail of shape or structure. As a result of their expansive nature, ellipses estimated for 80–99% of the location density came closest to the precise percentage of goshawk and chaffinch trajectories; other methods gave underestimates. However, the expansion also created a poor fit, so ellipses contained less trajectory per unit area than other estimators (Robertson *et al.*, 1998) and are outlier-sensitive.

Contour estimates are less smoothed than ellipses, but also contain high proportions of trajectory. Smoothing is least parametric when based on harmonic mean distances, and with contours plotted to include locations rather than proportions of a density distribution. However, smoothing still takes the outlines into unused areas, especially where numbers of locations occur close to an abrupt boundary (Figs 9.5, 9.7). Abrupt changes cannot easily be accommodated by a parametric distribution of locations without under-smoothing elsewhere.

Harmonic means have lost popularity to other kernels for contouring. The truncation of inverse distances for locations close to nodes, which makes harmonic mean estimates sensitive to size of small grids, is inelegant. Strong smoothing of kernel estimates can give stable area estimates with small numbers of locations, as few as 15 for buzzard ranges (Kenward *et al.*, in press c). Kernel smoothing can be tightened too – but therein lies scope for subjectivity, with problems of comparability across studies. Harmonic mean contours are less sensitive to a distant outlier than contours from bivariate normal kernels, provided there are at least two grid nodes spanning each resolution unit (Figs 9.6, 9.8), and diverge less than kernel estimates in cores.

Minimal smoothing is provided by rules that link single locations to create polygons. Peeled polygons contained higher trajectory density than contours for central-place foraging of the chaffinch, but not for the multinuclear range of the goshawk (Robertson *et al.*, 1998). However, it was the cluster analysis polygons, based on even tighter nearest-neighbour linkage distances, that contained the highest density of trajectory. Nearest-neighbour distances also provide abrupt changes for objective coring and sharp discontinuities in utilisation plots for subjective coring. Polygons from cluster analyses gave the best correlations with weight of red squirrels and food supply of grey squirrels (Kenward *et al.*, in press b), and the best fit to the coarse-grained habitats of squirrels in plantations, buzzards in fields (Kenward *et al.*, in review a), owls in woodlots (Redpath, 1995) and turtles in bogs (Carter *et al.*, 1999). On the other hand, despite a finding by Hansteen *et al.* (1997) that 95% clusters were more stable than kernel contours, the minimal smoothing and small link distances of the cluster polygons made these the least stable technique with low numbers of locations, with the greatest tendency to under-record trajectories (Robertson *et al.*, 1998). When the biological questions concerned differences based on age of goshawks, or density of squirrels, that probably depended on diffuse social factors rather than abrupt changes, it was the smooth density-based estimators that correlated best (Kenward *et al.*, in press b).

A number of these considerations, summarised in Table 9.1, place us on a road towards choice of outline estimators on purely *a priori* grounds. Already, if you can only standardise on ranges of 10–15 locations, you will probably do best with ellipses. If you have only 15–25 locations, then contours should be better (assuming minimal serial dependence between locations). Standard ranges with at least 30 locations provide scope for polygon methods too.

Table 9.1. Classification of six common types of home range estimator, with their ability to provide measures of size, shape and structure from minimal numbers of locations.

	Normal ellipses	Contour methods	Convex polygons	Concave polygons	Cluster analysis	Grid cells
Type	Density	Density	Linkage	Linkage	Linkage	Linkage
Number of density centres, size of linkage distances	One	Many	Large	Large	Increasing	Small
Sensitivity of size and shape to distant outliers	High	Some	High	Some	Low	Low
Home-range outlines conform well to outer locations	No	No	Yes	Yes	Yes	Yes
Home-range outlines conform well to multi-nuclear cores	No	Yes	No	No	Yes	Yes
Home-range structure statistics (patchiness or dispersion)	Yes	Yes	No	No	Yes	Yes
A single home-range centre is estimated	Yes	Yes	Yes	No	No	No
Number of locations needed for home-range size to stabilise	10–15	10–30	⩾30	⩾30	⩾30	⩾100

Beyond considering the number of locations, the most appropriate outline for answering a particular biological question may depend on how the animals move in that context. Thus, if they are foraging in coarse-grained habitats with abrupt boundaries (e.g. from abrupt natural features or abrupt edges to crops, plantations and other human creations), polygon methods may be best. On the other hand, if locations are sampled without bias along a time axis, location density estimates the expectation of encountering the animal at any time. Density estimators may therefore be best where analyses of sociality depend on encounter probability rather than abrupt habitat boundaries. Quite apart from analyses based on range outlines, you may choose to use harmonic mean estimation of a range centre, nearest-neighbour distances for outlier exclusion, and contours to display your data elegantly.

9.4 CONCLUSIONS

1. Care should be taken to base all analyses on a single value from each tagged individual; the analysis of behaviour or location records for the same animal as

independent events involves psuedoreplication. Sample sizes should be numbers of individuals.

2. Movements may occur within a home range, defined as 'an area repeatedly traversed', or represent transition between home ranges, drifting or nomadic behaviour. Major movements may be defined by travel without return beyond a given distance, or objectively with a vectored dispersal detector. Changes in location with time can be estimated by indices of concordance from box-counting, by MRPP techniques or by expressing the distance between range centres in terms of the distribution of distances around each of them.

3. Home ranges provide convenient indices of size, shape and structure for animal movements. Estimators differ in the extent to which they smooth data. Smoothing stabilises size but conceals details of shape and structure. The most smoothed estimators are those that estimate location density, especially when the estimate is about one centre (for ellipses) rather than on a matrix (for contours). Estimators based on minimal links between locations give less smoothing, even less when links are nearest-neighbour distances, and least in grid cells.

4. Choice between estimators depends on a balance between need to collect few locations and need for detail. With few locations, only maximally smoothed ellipse estimators provide stability; with moderate numbers of locations, use of contour or polygon estimators may depend on whether studies concern social behaviour or use of coarse-grained habitats with abrupt edges; stable size estimates with much detail require large numbers of locations.

5. Smoothed estimators with few locations can be strongly influenced by outlying locations, which may represent different movements to those in a 'normally used' core. Outliers can be excluded subjectively by examination of utilisation plots, or objectively by tests based on box counting, density values or nearest-neighbour distances.

10 Demography and Interactions

This chapter follows on from the analysis of movements and home ranges, because each of its aspects engages in some way with the previous topics. Density estimates can require home-range data. Home-range characteristics may be important covariates in analyses of survival or productivity. Records of movements help define how animals interact with resources, and with each other. This chapter contains even more mathematical expressions than Chapter 9, but again don't worry about trying to understand them. The intention is to provide an overview of a complex field, with enough information to help you get further assistance if you need it.

10.1 DENSITY ESTIMATIONS

Radio tagging can be used for density estimation in two main ways. The tags can be primary markers for Lincoln Index estimates of population size within a given area. Radio-tagged animals can also be used to correct density estimates from grid trapping and visual surveys.

10.1.1 Direct population estimates

An early example of radio tags being used as Lincoln Index markers was an estimation of the number of goshawks that visited study areas in Sweden (Kenward *et al.*, 1981). Whenever a hawk was sighted 'by chance' during routine travel about the study areas (i.e. not during radio tracking), it was checked for presence of a radio tag. This gave a number (m_i) of tagged hawk observations among n_i sightings during a period i in which R_i radio-tagged hawks were present. The number R_i changed during the study seasons as new hawks were marked and others emigrated or died, but for each R_i the estimated number visiting a study area was given by the Lincoln-Petersen Index following Chapman (1951):

$$\hat{N}_i = \frac{(n_i + 1)(R_i + 1)}{(m_i + 1)} - 1 \tag{10.1}$$

with variance as estimated in Seber (1982):

$$\text{Var }(\hat{N}_i) = \frac{(n_i + 1)(R_i + 1)(n_i - m_i)(R_i - m_i)}{(m_i + 1)^2(m_i + 2)} \tag{10.2}$$

To combine estimates from each period, $\hat{N}$ was estimated as the average of the $\hat{N}_i$ values weighted by the inverse of their variances. However, White and Garrott (1990) point out that there are at least 5 other ways of combining the $\hat{N}_i$ estimates. They showed that the joint hypergeometric maximum likelihood estimator (Chapman, 1951; Seber, 1982) gave the least biased estimate of $\hat{N}$ in an extensive simulation exercise. The unweighted arithmetic mean of the $\hat{N}_i$ values was the next least biased estimator: it performed well enough to be recommended in the absence of a computer optimisation routine needed to find the joint hypergeometric MLE.

An assumption behind these estimates was that tagged and untagged hawks were equally likely to be sighted. In one of the study areas there was evidence to justify this assumption, inasmuch as a kill made in the area represented time spent there. Among the many pheasant kills found by game wardens who were not involved in the tracking, the proportion which had been made by tagged hawks was the same as the mean proportion of tagged hawks in sightings (Kenward, 1977).

A particular advantage of using radio tags for Lincoln-Petersen estimation is that the number of tagged animals in an area can be known more accurately than for passive tags, which may be shed or on dead or emigrated animals (Mahoney *et al.*, 1998). Tagging with radios is also especially suitable for elusive species, which are often sighted too briefly for a visual marker to be spotted. The individuals must of course be widely dispersed, or the signal coming from a sighting position might be coming from a tagged animal near to the untagged one that was seen. Using a visual marker as well as the radio can reduce this risk.

10.1.2 Edge correction for density estimates

Within a site of area α, the function $\hat{N}/\alpha$ was an overestimate of the hawk density, because some birds had home-ranges that overlapped the edge of the area and spent some of their time outside it. The density of hawks in area α was therefore estimated by $\kappa^{f\alpha}\hat{N}/\alpha$, where

$$\kappa^{f\alpha} = \frac{1}{R}\sum_{j=1}^{R}\frac{f_j^{\alpha}}{f_j} \tag{10.3}$$

and f_j^{α}/f_j is the proportion of all the locations f_j for the r hawks that were within area α at any time. Assumptions in the application of correction factor $\kappa^{f\alpha}$ are: (i) that the locations are a representative sample of the time spent in different parts of each animal's home range; (ii) that the proportion of locations of radio-tagged animals within area α is also representative for untagged animals and (iii) that $\kappa^{f\alpha}$ does not vary between the i observation periods. In principle, assumption (iii) would be avoided if R in Eq. 10.3 is replaced by R_i and a separate estimate of κ_i^{fa} applied to R_i in Eqs 10.1 and 10.2 for each period i. However, it is not clear how this substitution may affect the use of a hypergeometric MLE to combine $\hat{N}_i$ values.

The correction factor κ^{fa} can be applied widely, and not merely for populations estimated by using radio tags as markers. Edge corrections are also needed, for example, when populations are estimated by grid trapping. The grid may cover only part of a wood or a grassland, so that some of the animals caught on the grid normally spend little of their time on it, and may even have been drawn onto it because of the baited traps.

Provided that the grid spacing is regular and small enough for several traps to lie in each animal's range, the most sophisticated approach is to relate the proportion of a tagged animal's range or time within the grid to the distribution of traps in which it was captured (Tonkin, 1983). The best-fit function may be the result of a simple or a multiple regression, perhaps involving the number of different traps in which each individual was captured, the number or proportion of these traps at the edge of the grid, or the number of captures per trap in each category. This function is then used to estimate the proportion of each untagged animal's time (or range area) spent on the grid, assuming that tagged and untagged animals used the grid area in the same way. This is a reasonable assumption if tagging was unbiased: for instance, every nth individual might have been tagged at first capture.

A less sophisticated edge correction can be applied if ranges are relatively compact (not elongated), and where continuity of habitat justifies the assumption that ranges are distributed evenly across the edge of the grid. In this case, the population trapped can be estimated to come from an area that extends, at grid edge in continuous habitat, by the radius (r) of a circle with the mean of the A_i range areas of R radio-tagged animals:

$$r = \sqrt{e\left(\frac{1}{R\pi} \sum_{i=1}^{R} ln(A_i) \right)} \tag{10.4}$$

Although equation 10.4 derives the radius for a circle with the geometric mean area, and was used initially with outer MCP ranges (Kenward, 1985), there is scope for work to establish whether other range estimators or area transformations would give less biased density estimates. Care must be taken if range areas differ between animal age or sex categories. In squirrels, for example, adult males have much larger ranges than females. However, there was an equal sex ratio in woods that were covered completely by trap grids, so total densities were estimated by doubling the female densities, which were corrected with the mean range radius for females where woodland extended beyond trap grids (Kenward and Holm, 1993).

10.1.3 Enhancement of line-transect techniques

During line transects, the distance from the line of each sighted animal should be recorded carefully. The distribution of sighting distances from the line, appropriately transformed, is then examined to find the distance μ beyond which sightings decline sharply (Buckland *et al.*, 1993; Laake *et al.*, 1994). The density of animals (D_t) along a line of length L is estimated from the n sightings within the truncation distance μ on either side of the line by the equation

$$D_t = n / 2.L.\mu \tag{10.5}$$

The transect approach provides a rigorous way of collecting data for population estimation by Lincoln-Petersen Index techniques (section 10.1.1), by removing risk of marking bias (*i.e.* the marking of animals that live mainly in areas that are also most visited for chance observations). This bias, and the risk of overestimating tagged individuals among sightings (by misallocation of tag signals to untagged birds), both increase m_i in Eq. 10.1 and hence underestimate the population. Misallocation bias is reduced if sightings of tagged and untagged animals are accepted only within the truncation distance μ, to give a 'truncated-mark resighting' estimate (Kenward *et al.*, in press d). Within this distance, visual detection of the radio tag can provide an independent check of signal allocation (and other visual markers can be used to improve tag visibility, provided these will not bias the results by affecting animal behaviour or survival). An analogous approach is to base the mark resighting estimates on visual markers, while using radios to estimate persistence of marked animals in the survey area (Mahoney *et al.*, 1998).

A second use of radio tags in line-transect surveys is to calibrate the detection efficiency of observers, such that a correction factor can be applied to the line-transect estimates of density. In this case, standard range data are used to estimate a detection probability P_i^d for each bird, from the total area A_j of each contour or ellipse with inclusion probability p_j and the part o_j that was on the transect (Fig. 10.1), using the equation

$$P_i^d = \sum_{j=0.05}^{0+j.1} (p_j - p_{j-1}) - \frac{(o_j - o_{j-1})}{(A_j - A_{j-1})} \tag{10.6}$$

From t transect repeats that included m_i sightings of radio-marked bird i, the radio-corrected transect density $D_{rct} = D_t/\rho$ where $\rho = \Sigma m_i/t.\Sigma P_i^d$ (10.7) and, with substitution of $m = \Sigma m_i$ and $E = \Sigma P_i^d$, $\text{Var}(D_{rct}) = [\text{Var}(D_t).E^2 + \text{Var}(E).D_t^2]/m^2$ where $\text{Var}(E) = t.\Sigma P_i^d (1-P_i^d)$ (10.8, from Kenward *et al.* (in press d)).

The disadvantage of this approach, compared with the truncated-mark resighting estimate, is the need to collect range data independent of the transect surveyor. Moreover, the inclusion levels in contours may underestimate location probabilities, and they need correction, for example by jack-knifing (Robertson *et al.*, 1998). Nevertheless, the detection rate ρ for a given observer could then in principle be used to estimate densities in areas without radio-tagged animals, provided that transects are in similar habitats, weather, seasons and times of day.

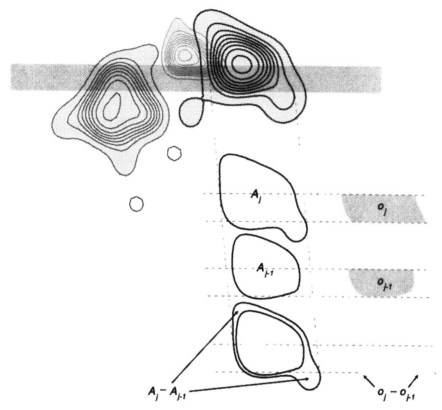

Figure 10.1. The probability of encountering an animal along a transect is the sum of the probabilities of encountering it on the transect between each pair of adjacent contour lines j and $j-1$. The probability of encounter between pairs of contours that enclose areas A_j and A_{j-1} is the probability of encounter in the whole area between those contours (p_j-p_{j-1}) multiplied by the proportion of the area that is covered by transect $(o_j-o_{j-1})/(A_j-A_{j-1})$.

A more direct estimate of the detection rate ρ was used for transect surveys of grouse. The proportion of radio-tagged birds missed by a census team was estimated by an independent radio-tracker following close behind them (Brittas and Karlbom, 1990).

10.2 SURVIVAL ANALYSES

When radio tagging is used to estimate survival and dispersal rates, the calculations are often complicated by animals entering a tagged population over a number of months, instead of fledging or weaning in a relatively short period. Moreover, animals leave the tagged population through tag extinction (including component failures and cell exhaustion) as well as through death or emigration. To detect these events, you will probably have chosen to gather data on a daily, weekly or monthly basis. Your schedule for survival checks will depend on expectations of tag life, mortality rate, emigration rate, cost of checking and any need for fresh corpses to investigate cause of death (see Chapter 8). Checks might involve a twice-daily walk, e.g. for songbirds just out of the nest with 6-week tags, or a 4-monthly survey by air for large raptors with 4-year tags. Figure 10.2 plots the sort of data that could be collected. Sections 10.2.1 to 10.2.5 then describe different survival estimators appropriate to the data, using $\hat{s}$ for an estimator that applies through the span of all five intervals, $\hat{s}_i$ for the estimator within each interval i and $\hat{s}_i'$ for an estimator of the daily survival rate through interval i.

10.2.1 Interval and tag-day estimation

If d deaths are recorded among r animals that are known from working tags to be at risk, the survival rate estimate

$$\hat{s} = (r - d)/r \qquad (10.9)$$

Assuming that each death is an independent event (*i.e.* not due to a predator that is concentrating on a brood), d may be treated as a binomial variable (White and Garrott, 1990) so that

$$\text{Var}(\hat{s}) = \hat{s}\,(1 - \hat{s})/r \qquad (10.10)$$

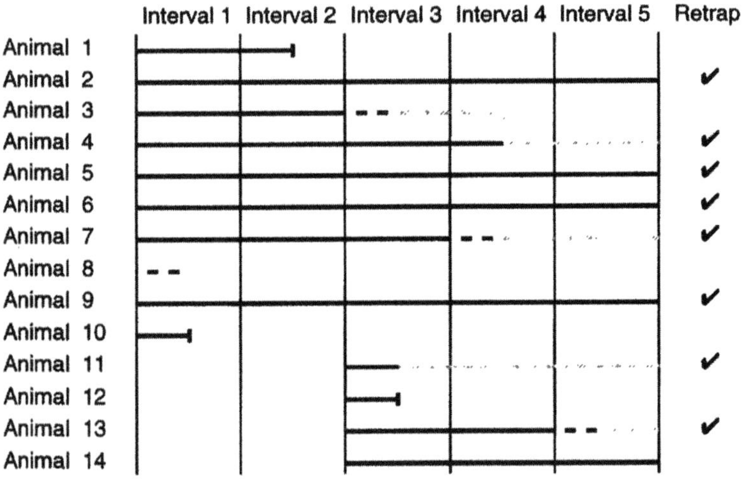

Figure 10.2. A sub-sample of survival data recorded for 10 animals through five intervals, with another four animals added in interval 2. Animals 1, 10 and 12 were found dead, whereas animals 2, 5, 6 and 9 survived throughout. Animals 4 and 11 were known to be *live* when their tags failed, but signals were *lost* without explanation from animals 3, 7, 8 and 13. In fact, animal 3 died some time after tag failure and animal 8 was destroyed with its tag; animals 7 and 13 remained alive, to be reported again in a very efficient trapping session after the fifth interval.

In Fig. 10.2, the true survival rate of ten animals that were tagged throughout the study was 0.60: four died and six survived.

If one uses the most simple estimate that depends only on radio tracking, only two animals were known to die and four to survive. The survival rate is then 0.67, which is an over estimate because two animals with failed tags also died. To set against this maximum estimate of the survival rate ($\hat{s}_{max}$), a minimum survival rate is estimated by assuming the death of all u animals that were inexplicably lost, such that

$$\hat{\underline{s}}_{min} = (r - d - u)/(r + u) \tag{10.11}$$

With three lost tags as well as two known deaths and four known survivors, the estimated survival rate was only 0.44.

This single-interval estimation was based only on animals that survived, died or were lost inexplicably during the study. The estimate took no account of animal 4, whose tag was heard to fail when its cell exhausted just before the end of the study, or of animals 11–14 that were added during the study. Such animals can be included by modifying the nest-survival method of Mayfield (1961, 1975). Accordingly, Trent and Rongstad (1974) used the number of tag days (x) for which animals were tracked during an interval i to estimate the daily survival rate as

$$\hat{s}_i' = (x_i - d_i)/x_i \tag{10.12}$$

The equivalent minimum survival rate comes from

$$(x_i - d_i - u_i)/x_i \tag{10.13}$$

For the whole L days of interval i, the survival rate

$$\hat{s}_i = \hat{s}_i'^L \tag{10.14}$$

If the five intervals in Fig. 10.2 were days, then the original ten animals were tracked through 36 days (including those of death or loss). The maximum (Eq. 10.12) and minimum (Eq. 10.13) survival rate estimates are $([36 - 2]/36)^5 = 0.75$ and $([36 - 2 - 3]/36)^5 = 0.47$ respectively; these rates are higher than the binomial estimates, thanks to the inclusion of animal 4. However, unlike the binomial method, there is no easy way to estimate a variance for these estimates, and hence attach confidence limits, because tag days from the same animal lack statistical independence.

10.2.2 Multi-interval estimation

Heisey and Fuller (1985) pointed out that if estimates are available for a number of consecutive intervals, then the survival rate throughout the span is $\Pi\hat{s}_i'^L$. Moreover, the variance of the $\hat{s}_i'$ values can be used to attach confidence limits, provided that survival rates are constant through the span. The distribution of $\hat{s}_i'$ may well be skewed, but can probably be normalised with a natural log transformation (Bart and Robson, 1982), in which case logarithms of the standard error must be used to attach asymmetric confidence limits with the appropriate value of Student's t for the sample size. If the natural logarithm of a constant daily survival rate is treated as a hazard function λ, the survival rate at time t is modelled by

$$S(t) = e^{-t\lambda} \tag{10.16}$$

Equation 10.16 is the simplest in a series of exponential survival models. Regression analysis can be used to investigate whether λ varies with t by defining shape and scale parameters as in the Weibull model (Lee, 1980; Cox and Oates, 1984) where

$$S(t) = e^{-(t/\alpha)\gamma} \tag{10.17}$$

The shape parameter γ determines departure from the simple exponential model. When $\gamma = 1$, this model becomes the exponential model with $\lambda = 1/\alpha$, whereas if $\gamma < 1$ the survival rate increases with time, which is likely when young animals are being marked.

Data on deaths in each interval may also be expressed as a mortality (i.e. $1 -$ survival rate). Partial mortalities can be used to examine the role of different causes of death. If the number of deaths due to cause j is d_{ij} out of a total d_i during an interval i in which survival is $\hat{s}_i$, its partial mortality rate

$$\hat{m}_{ij} = (d_{ij}/d_i)(1 - \hat{s}_i) \tag{10.18}$$

Heisey and Fuller (1985) show that the partial mortality through a series of intervals can be estimated by the sum of the partial mortalities, for each interval, multiplied by the product of the survivals for all preceding intervals as

$$\hat{m}_j = \sum_{i=1}^{t} m_{ij} \prod_{k=1}^{t-1} \hat{s}_k \tag{10.19}$$

The main problem with estimates based on tag days is the assumption that rates of survival or mortality are constant through a span. Variation in weather, animal experience when young and behaviour tend to invalidate this assumption and can greatly influence the estimates. As was made clear by White and Garrott (1990), if a storm kills half of a 50-animal cohort in the first day of a span of 100 days, Eqs 10.12 and 10.14 use 2525 tag days to estimate a survival rate of 0.37 for the 100 days, whereas exactly the same number of deaths in the last day are based on 4975 tag days and estimate a survival rate of 0.60.

In an alternative approach from Kaplan and Meier (1958), survival for the whole span is estimated as the product of survival in each interval i using the binomial estimators for animals at risk in each interval:

$$\hat{s}_{max} = \Pi(r_i - d_i)/r_i \tag{10.20}$$

and

$$\hat{\underline{s}}_{min} = \Pi(r_i - d_i - u_i)/(r_i + u_i) \tag{10.21}$$

To estimate the variance of the survival estimate at time t, Cox and Oakes (1984) suggested the use of Greenwood's formula:

$$V(\hat{\underline{s}}(t)) = [\hat{\underline{s}}(t)]^2 \sum_{i=1}^{t} \frac{d_i}{r_i(r_i - d_i)} \tag{10.22}$$

However, Cox and Oakes (1984) also proposed an alternative:

$$V(\hat{\underline{s}}(t)) = \frac{[\hat{\underline{s}}(t)]^2[1 - \hat{\underline{s}}(t)]}{r_i(t)} \tag{10.23}$$

Equation 10.23 may be more appropriate at the end of a span because it uses $\hat{\underline{s}}_i(t)$ and $r_i(t)$, the survival and number of animals at risk in the last time interval.

The method of applying Kaplan-Meier estimation to the data in Fig. 10.2 is shown in Table 10.1, together with a correction procedure described in the next section. For example, interval 1 includes one death and one unexplained loss. The lost animal is eliminated from the analysis at the start of the interval (right-censored) for the estimate that maximum survival (from Eq. 10.20) was eight of nine animals (0.91), whereas this animal was included as if dead for the estimate that minimum survival (from Eq. 10.20) was eight of ten animals (0.80). The animals that were known to be live when tags failed, in intervals 3 and 5, are right-censored from both maximum and minimum estimates for that and subsequent intervals. As well as permitting exit of censored animals at different times, this approach also permits staggered entry of animals. The full 'staggered-access' approach is applied in Table 10.1 to estimate survival rates that include the four animals added during the second interval.

Table 10.1. Data from Fig. 10.2 to show Kaplan-Meier survival rate estimates as true, minimum, maximum and report-corrected values, in five intervals and as their product for the full period. The top four rows are 10 animals marked from the start of the study, with right-censoring, and then also using the additional four animals in a full staggered-access analysis.

	Interval 1	Interval 2	Interval 3	Interval 4	Interval 5	*Product*
Right-censored only (n = 10)						
'True survival'	8/10	7/8	7/7	6/7	6/6	*0.60*
Recorded deaths only ($\hat{s}_{min}$)	8/9	7/8	6/6	4/4	4/4	*0.77*
Assumed death of lost tags ($\hat{s}_{max}$)	8/10	7/8	6/7	4/5	4/4	*0.48*
Report-corrected survival ($\hat{s}_c$)	0.843	0.875	0.926	0.894	1.000	*0.61*
Full staggered-access (n = 14)						
'True survival'	8/10	7/8	10/11	9/10	9/9	*0.57*
Recorded deaths only ($\hat{s}_{min}$)	8/9	7/8	8/9	6/6	5/5	*0.69*
Assumed death of lost tags ($\hat{s}_{max}$)	8/10	7/8	8/10	6/7	5/6	*0.40*
Report-corrected survival ($\hat{s}_c$)	0.843	0.875	0.843	0.926	0.913	*0.54*

It is worth noting that intervals need not be of equal length for either the tag-days or staggered-access approaches. The results remain robust even if up to 10% of the animals are not checked in each interval (Bunck *et al.*, 1995). In all cases, however, modification is needed if individual deaths, for example of brood mates, are not independent (Flint *et al.*, 1995).

One drawback to the staggered-access design occurs if tag failures, losses and additions of new animals are frequent relative to a small number of deaths. Chance occurrence of a death in an interval with few animals can substantially lower the survival estimate for the whole period. In this case, although an estimate based on tag days for the whole span will lack confidence intervals, it is likely to be more reliable provided that the survival rate is relatively constant.

Note that maximum and minimum estimates of survival for the first 10 animals were 0.67 and 0.44, respectively, if estimated for the whole period by censoring from the very start, compared with 0.77 and 0.48 in Table 10.1. The equivalent tag-day estimates were 0.75 and 0.47. This shows how a multi-interval approach, using Kaplan-Meier with frequent intervals or tag days, can strongly overestimate maximum survival if tags are destroyed by traumatic events which also kill the animal (e.g. predation, road accidents, shooting). However, the gap between true and minimum survival estimates is also least with these approaches, which is important for producing the most conservative survival estimates if there is no scope for correction with independent data from recoveries or tag failure rates. Lack of correction data puts a premium on using reliable tags, and on routine recording of signal characteristics that indicate impending failure of tags. Given the presence of reliable tags, it is also wise to truncate data collection at the point when the first tag is heard to run down close to the expected date. Otherwise, the burst of signal losses that follow make survival estimates particularly unreliable.

10.2.3 Correction from recovery rates

The trapping after the end of the study depicted in Fig. 10.2 enabled recovery rates to be estimated, of f_e for the two animals with explained tag losses and f_u for the four animals with unexplained tag losses. Although f_e was based on only two animals, confidence in the estimate of 1.0 was enhanced by the recapture of all five animals known to be alive at the end of the study. The recovery rate f_u of the four animals with unexplained tag losses was only 0.5.

Kenward (1993) proposed that if the recovery rate for u animals with unexplained tag losses (f_u) is less than the recovery rate (f_e) for e animals with explained tag losses, due to destruction of some tags with their wearers, then the function $\zeta = 1 - f_u/f_e$ (10.24) can be used to interpolate a corrected estimate of the survivorship ($\hat{s}_c$) as:

$$\hat{s}_c = \exp\,(\ln \hat{s}_{max} - \zeta.(\ln \hat{s}_{max} - \ln \hat{s}_{min})) = (1 - d/r)(r/(r + u))^\zeta \qquad (10.25)$$

If $\hat{s}_c$ is derived as a product of survival in intervals, so that $\hat{s}_c = \hat{s}\alpha^\zeta$ where $\hat{s} = \Pi\,\hat{s}_i = \Pi(r_i - d_i)/r_i$ and $\alpha = \Pi\,\alpha_i = \Pi\,r_i/(r_i + u_i)$, Aebischer (1993) showed that an estimate of its variance is:

$$\text{Var}(\hat{\underline{s}}_c) = (\hat{\underline{s}}_c)^2[\text{Var}(\hat{s})/\hat{s}^2 + \zeta^2(\Sigma\text{Var}(\alpha_i)/\alpha_i^2) + (1 - \zeta)^2(\ln \alpha)^2(\text{Var}(f_u)/f_u^2 + \text{Var}(f_e)/f_e^2) - 2\zeta(1 - \zeta)(\ln \alpha)(\Sigma\text{Cov}(\alpha_i f_u)/(\alpha_i f_u))] \qquad (10.26)$$

Note that if no animals are recaptured after unexplained tag loss, which implies that all were destroyed with their tags, then $f_u = 0$ and, from equations 10.14 and 10.15, $\zeta = 1$ and $\hat{\underline{s}}_c = \hat{\underline{s}}_{min}$. Only sampling error should cause f_u to exceed f_e, in which case ζ can be set to 0 and $\hat{\underline{s}}_c = \hat{\underline{s}}_{max}$.

With the recoveries in Fig.10.2, $\zeta = 1 - 0.5/1 = 0.5$. With this correction, the single-interval estimate of survival rate is $\exp(\ln (0.67) - 0.5 (\ln (0.67) - \ln (0.44)))$ = 0.54. The estimate is still low compared with the true value of 0.60, mainly because the survival of animal 4 is excluded. However, if survival rates are estimated for each of the five intervals (Table 10.1), then the corrected estimate is 0.61, very close to the true value.

If the multi-interval estimation of survival is used to include data from animals 11–14, then the true survival estimate is 0.57. The maximum and minimum estimates are 0.69 and 0.40 respectively, with $\hat{\underline{s}}_c = 0.54$, in this case slightly lower than the true value. The downward bias is the result of 'half a death' ($\zeta = 0.5$) being assigned to unexplained signal losses in each of the last three periods, totalling 1.5 deaths, when only one was a death actually occurred. Note that this correction may be applied to any estimator that provides maximum and minimum survival rates, and that another way of estimating ζ could be to use the expected failure rate of tags.

10.2.4 Comparing categories

The most simple biological questions about survival may be of the type 'how well do these animals survive?'. However, many studies will want also to make comparative tests to determine whether category A survives better than category B, or to investigate the effects on survival of continuous covariates such as bodymass or range size.

The simplest comparisons can be made by dividing animals into those that die and those that remain alive at the end of a study period, and apply a 2×2 χ^2 test to compare two categories, such as experimental treatments. Extension as a $2 \times N$ test is appropriate when there are N categories, such as age or year groups. With multiple categories, simulation procedures can be used to generate expected results that are then compared with observed results using a G test (White and Garrott, 1990). Animals with unexplained signal loss are a problem. They would usually be censored, but results can only be considered robust if they remain significant with such animals included as dead. The exclusion of animals with explained signal loss reduces the power of the test without biasing it. However, these animals can be included if survival has been recorded systematically at intervals.

Multi-interval comparisons can be based on daily hazard rates $\lambda = \ln(\hat{s}_i)$ from tag-day estimates, using a t-test if you are bold enough to believe the rates to be constant. Alternatively, the robust approach of not assuming an underlying shape for the survival function is to use tests based on Kaplan-Meier method. In that case, a fairly conservative option is to use t-tests based on the Cox-Oakes variance (Eq. 10.23). Log-rank tests were suggested by Pollock *et al.* (1989) for comparing different groups throughout a set of equivalent time periods if the survival curves are similar in shape. In this case:

$$\chi^2_{(1)} = \frac{[\Sigma d_{1i} - \Sigma E(d_{1i})]^2}{\Sigma V(d_{1i})}$$

(10.27)

where d_{1i} are the deaths in population 1 in interval i, and d_{2i} the deaths in population 2, and

$$E(d_{1i}) = \frac{(d_{1i} + d_{2i})r_{1i}}{(r_{1i} + r_{2i})}$$

(10.28)

and

$$V(d_{1i}) = \frac{(d_{1i} + d_{2i})[r_{1i}][r_{2i}]([r_{1i} + r_{2i}] - [d_{1i} + d_{2i}])}{[r_{1i} + r_{2i}]^2[r_{1i} + r_{2i} - 1]}$$

(10.29)

or, more conservatively, where $d_i = (d_{1i} + d_{2i})$ and $r_i = (r_{1i} + r_{2i})$,

$$\Sigma V(d_{1i}) = 1 / \left(\frac{1}{\Sigma \frac{d_i r_{1i}}{r_i}} + \frac{1}{\Sigma \frac{d_i r_{2i}}{r_i}} \right)$$

(10.30)

However, when data have been recorded systematically in the same intervals for two animal categories, the non-parametric Wilcoxon signed-rank test (e.g. as described in Siegel, 1957) is suitable for detecting a persistent difference in survival between them without making any assumptions about the shape of the survival curves. This test can be used on tag-day estimates from Eqs 10.12 and 10.13, or on staggered-access estimates from Eqs 10.9 and 10.10, and is also appropriate for partial mortality rates estimated with Eq. 10.18. The Wilcoxon test was more sensitive that the other tests for showing that survival of juvenile goshawks differed between males and females during their first 9 months of life (Kenward *et al.*, 1999).

10.2.5 Continuous covariates

Tests become more demanding when there are continuous covariates, such as habitat characteristics or body mass. The most straightforward approach is to consider the dichotomy of live or dead through a single period using logistic regression (Cox, 1970; White and Garrott, 1990). This can also be extended to test data in multiple intervals (Efron, 1988), and can be used to include intermediate categories such as unexplained signal loss. The influence of covariates on different groups could also be investigated with discriminant function analysis. However, not only is multiple logistic regression the more robust approach mathematically (White and Garrott, 1990) but a systematic step-wise regression is more rigorous than the 'trawling' for variable combinations that is implicit in discriminant function analysis.

The Cox proportional-hazards model (Cox, 1972; Kalbfleisch and Prentice, 1980) is an alternative approach for analysing effects of continuous covariates. It uses the covariates as independent variables for multiple regression, in which the dependent variable is a hazard function derived from the length of survival of each animal. An early example showed the effect of weight and other factors on the survival of snowshoe hares (Sievert and Keith, 1985). This model can be more powerful than logistic regression, but is also more restrictive because it assumes that the mortality rate associated with each covariate remains proportional to that from other covariates. This condition would not be met where one factor acts mainly after another, as would be the case of animals suffering predation when young and then from weather in the winter. The proportion-hazards model is also primarily suitable when most individuals die before the end of a period. When they do not, and mortality from different factors may act at different times, it is probably wise to use logistic regression.

More detail on some of the various approaches and their origins is available in other reviews (White and Garrott, 1990; Samuel and Fuller, 1994). A variety of estimation methods and tests based on tag-day data are available in the MICROMORT software (Heisey and Fuller, 1985), with survival by proportional hazards addressed specifically in SURPH (Smith *et al.*, 1994). Other methods are included in the SURVIV suite (White and Garrott, 1990); these authors also provide SAS routines for Kaplan-Meier estimates and tests.

Most approaches for analysing survival are also appropriate for studying dispersal and aspects of productivity. Thus, logistic regression can be used to investigate whether individual mass, tag type, sociality or environmental factors influence whether or not animals disperse by a particular date. Although unexplained signal losses can in theory create as many problems as for survival estimates, dispersal will often occur within the mid-life period of maximum tag reliability. This is the period between the few days after start-up and the likely exhaustion of sub-standard cells. Finally, radios can be used to find breeding attempts of tagged animals at an early stage, which can provide brood survival functions with less bias than those found at a later date by cold-searching.

10.3 INTERACTION WITH RESOURCES

Information about which habitats, feeding sites or refuge areas are important for a species is often crucial for wildlife conservation. Radio tagging can contribute this information with less opportunity for bias than in visual studies. The information can be contributed in at least four ways.

Some of the most common resource investigations concern the habitat that animals use more or less than expected if movements are random. However, above-average association with a habitat does not necessarily mean that it is critical for an animal (White and Garrott, 1990). Further evidence that a habitat is important would be a significant correlation between habitat content and home-range size, indicating that movements depended on access to the habitat (Kenward, 1982b). Better still would be evidence that survival, productivity or other aspects of demography relate to spatial resource variables. Nevertheless, the only proof that habitat is critical would come from recording impacts of removal or addition of the resource (White and Garrott, 1990), which can become an experimental aspect of land management if animals are radio-tagged in control and treatment areas. This section briefly examines analysis for association or avoidance, for habitat dependence, and for resource-related performance with its associated experimental opportunities.

10.3.1 Habitat associations

The term 'preference' is often used when an animal occurs in a habitat more than expected by chance. However, disproportionate use need not mean that a habitat is preferred. Think of an animal that has 100 locations in three equally abundant habitats, one of which is totally unused (0 locations), and another used somewhat less often (40 locations) than the third (60 locations). If you use a simple index, (*proportion Used*)/(*proportion aVailable*), use is greater than expected if $U/V > 1$. The three habitats score 0/0.33, 0.4/0.33 and 0.6/0.33, with the implication that the second and third habitat are 'preferred'. However, what if the first habitat was water? You might then consider only the second and third habitats really available, and only the first habitat preferred. Relative avoidance of one habitat always creates relative preference for another because proportional usage scores are not independent: they sum to unity (i.e. $0 + 0.4 + 0.6 = 1$). This is the 'unit-sum constraint' (Aebischer *et al.*, 1993b).

A way round this problem is to use ranking, which in the above case would provide an order of least preferred (1) to most preferred (3). Tests of whether overall habitat use diverges from random, and of habitats for which use differs significantly, can be based on multiple pairwise comparisons between habitats (see below). None the less, you may find simple comparisons of use and availability within each habitat preferable for display. If so, U/V is an unsatisfactory index because it is asymmetrical. Avoidance values can only lie between 0 and 1, whereas association values can be anything greater than 1. An alternative is to use a log ratio $\ln(U/V)$, which scores 0 when proportions used and available are equal, and is symmetrical for (negative) avoidance and (positive) association scores, but becomes infinite when $U = 0$. Another possible index, $(U-V)/(U+V)$ from Ivlev (1961), is more straightforward to estimate but is biased where U but not V can become 0. However, displays can be based on the unbiased, symmetrical index $(U-V/[U+V-2.U.V])$ suggested by Jacobs (1974). This gives values between -1 and $+1$.

Early analyses of habitat use employed χ^2 tests, of differences between observed and expected numbers of locations, to examine whether individuals differed from each other or from random use of habitats. Bonferroni confidence intervals were fitted to show whether use of each habitat by individuals, or by groups with pooled data, was greater or less than the proportion of that habitat in an area (Neu *et al.*, 1974; Marcum and Loftsgaarden, 1980; Byers *et al.*, 1984; Alldredge and Ratti, 1986, 1992). The main problem with these tests, beyond the results being vulnerable to disproportionate effects of individuals if data are pooled (Thomas and Taylor, 1990; White and Garrott, 1990), is that they not only treat locations as independent data (pseudoreplication) but also ignore the unit-sum constraint. The same problems affect more complex analyses (Heisey, 1985) that use Manly's measure of preference with resource replacement (Manly, 1974; Manly *et al.*, 1993).

One solution to the pseudoreplication problem and unit-sum constraints is provided by comparing rankings of usage and availability values for each individual (Johnson, 1980). There is a difference value, obtained by subtracting *rank-of-available* from *rank-of-use* for each habitat across all individuals, with a null-hypothesis (for random use) that the sum of differences is zero across habitats. The approach meets two main aims of resource-use analyses, namely testing (i) whether overall use of habitats is non-random and (ii) whether use of a particular habitat is non-random. The test for overall non-random use was originally based on the parametric F statistic, but could use randomisation. The important issue is that it avoids the unit sum constraint, whereas Friedman (1937) or Quade (1969) tests on usage indices do not.

Another solution to the pseudoreplication problem and unit-sum constraints is compositional analysis (Aebischer *et al.*, 1993b), which is more sensitive than ranking and can be extended to analysis of covariates. Use of each habitat U_i is expressed relative to each of the other habitats U_j, as a log ratio $\ln(U_i/U_j)$, with availability the equivalent $\ln(V_i/V_j)$. Aitchison (1986) showed that such compositions tend to a normal distribution, as does the difference $\ln(U_i/U_j) - \ln(V_i/V_j)$. This difference indicates preference between pairs of habitats: it scores 0 if animals have a similar association with each member of the habitat pair i and j. Using the assumption of normality, Wilk's Λ can be estimated for the resulting matrix of pair-by-pair values for an overall test of non-random use. This process can be extended to more sophisticated multivariate tests, for example to investigate habitat use as a covariate in an investigation of survival (Aebischer *et al.*, 1993a). Student's t can be used to test whether the preference differs significantly from zero for each habitat pair and results used to rank the habitats by preference. Alternatively, assumptions of normality can be avoided by using randomisation tests (Aebischer *et al.*, 1993b). As ln (0) cannot be calculated, null scores for U_i/U_j are assigned a value one order of magnitude lower than the lowest real score (e.g. 0.001 if the lowest real score was 0.01). Cases where $V_i = 0$ are omitted.

This compositional analysis can be illustrated by an analysis of locations that were recorded for goshawks during winter. Habitats were categorised as open country, woodland within 200 m of an edge, or woodland more remote from an edge (Table 10.2). Thus, the log ratio for use of deep woodland relative to open country of hawk JM4 was calculated from locations as $\ln([4/265]/[58/265]) = \ln(4/58) = -2.67$. With habitat availability sampled at grid intersections of a map, the corresponding $\ln(V_i/V_j)$ was –0.50 (for all hawks), so the difference ($\ln(U_i/U_j)$ – $\ln(V_i/V_j)$) was –2.17.

Table 10.2. Compositional analysis on data from seven goshawks tracked in one area in winter.

	JM4	AM6	AM8	AM9	JF10	AF11	AF14	
Locations n *in:*	265	126	203	166	93	75	98	
Deep woodland	4	2	3	13	1	1	12	
Woodland edge	203	92	162	139	70	70	81	
Open country	58	32	38	14	22	4	5	
Use								*Availability*
Deep woodland	0.02	0.02	0.01	0.08	0.01	0.01	0.12	0.28
Edge of woodland	0.76	0.73	0.80	0.84	0.75	0.93	0.83	0.26
Open country	0.22	0.25	0.19	0.08	0.24	0.06	0.05	0.46
Log ratios								
Deep/open	−2.67	−2.77	−2.54	−0.07	−3.09	−1.37	0.88	−0.50
Edge/open	1.25	1.06	1.45	2.30	1.16	2.86	2.79	−0.57
Deep/edge	−3.93	−3.83	−3.99	−2.37	−4.25	−4.25	−1.91	0.07
Log ratio difference								*Mean (s.e.)*
Deep/open	−2.17	−2.27	−2.04	0.43	−2.59	−0.87	1.38	−1.16 (0.58)
Edge/open	1.82	1.63	2.02	2.87	1.73	3.43	3.36	2.41 (0.30)
Deep/edge	−4.00	−3.90	−4.06	−2.44	−4.32	−4.32	−1.98	−3.58 (0.36)

From the resulting matrix, Wilks $\Lambda = 0.016$, for which $P < 0.001$ indicates strongly significant non-random use of habitats. Among the data for pairs, the differences from 0 of the means for edge/open and deep/edge were also significant at $P < 0.001$, whereas deep/open was $P = 0.10$. Results could be expressed as woodland edge >>> open country > deep woodland.

It was noticeable that two adult birds, AM9 and AF14, were less averse to deep woodland than the others. If the data had been collected in the breeding season, some hawks might well have preferred deep woodland, as nesting habitat, while others avoided it. Where there is evidence of strong divergence in habitat use (e.g. Hulbert et al., 1996), statistical separation of distributions (e.g. Macdonald and Pitcher, 1979) on one habitat could be followed by generation of separate matrices to determine the habitat requirements of each specialist group. It may also be wise to do separate analyses for different seasons, rather than pooling data. There may even be differences in use between years (Schooley, 1994).

Johnson (1980) noted that preference can be studied at a number of descending levels of scale. These include: (i) geographic range on the globe; (ii) home-range placement on the map; (iii) where locations occur within the range, and perhaps even (iv) in a third dimension at the location (e.g. height in a tree). The analysis for hawks made a comparison of habitat at locations with habitat available in the general area, across categories (ii) and (iii). However, if habitat use can be analysed at more than one level, comparison of results between levels can provide more insights than within-level data. This is because placement of outer ranges may be influenced by factors that differ from those that affect placement of cores or foraging locations. Outer ranges may seem to avoid a favoured habitat if a few individuals displace many others from the best areas, but then be disproportionately represented in cores of all. In squirrels, home-range placement showed little preference for areas with high levels of tree seeds, but locations within ranges favoured high food density (Kenward and Holm, 1993).

If you are relying on a map to count the number of locations in different habitats, rather than observations made while tracking (Chapter 8), results will be most convincing if you provide data showing that patch sizes were large relative to tracking resolution. Otherwise, you would be wise to check that results are the same if based on habitat in location-centred cells or circles that span the tracking resolution, or if sub-sampled in accuracy ellipses (Samuel and Kenow, 1992). If you are using a map of remote-sensed data with raster size similar to the tracking resolution, inaccuracy of the map may spoil results unless you use this shape-based approach.

One of the problems of analysing associations with habitat is measurement of availability (Samuel and Fuller, 1994). For location-within-range associations (level iii), habitat availability can be defined in an outer-MCP, or more generously in contours or even ellipses plotted to include 95% or 99% of the density distribution. However, for range placement associations (level ii) there may be problems putting a boundary on the area of habitat that is available, unless habitat is homogeneous across a map (which it was for goshawks) or on an island. One approach is to measure habitat in a boundary applied as for level (iii) analyses, for example by estimating an MCP round all the locations used in all the ranges. This is a conservative approach liable to produce Type II errors, because it already takes into account the placement of edge ranges; it is therefore most appropriate for large, well-spaced samples of ranges. A second way round this problem may be to measure available habitat in large 'opportunity circles' round locations (Arthur *et al.*, 1996).

10.3.2 Resource-dependence analysis

Two recent methods avoid the problem of defining available habitat. Instead, they investigate how the use of habitat varies with other factors in a home range outline. Studies have often asked 'is there disproportionate use of a resource?' However, a less frequent question is 'does use vary with availability; is there a functional response?' This question is common in studies of diet (which are the origin of most preference analyses). It was addressed by Mysterud and Ims (1998) for data from radio tracking in two types of habitat. In this case, random use with independent observations is expected to show a binomial distribution, so a regression of $\ln(U_1/[1-U_1])$ as a function of $\ln(V_1/[1-V_1])$ should have a slope $=1$ and intercept at $U_1 = 0$ (Fig. 10.3A). A plot for animals that all spent disproportionate time in

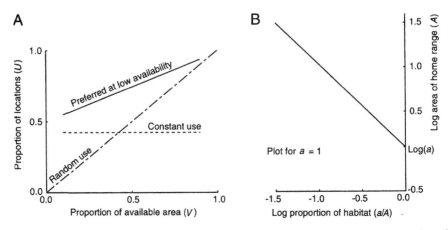

Figure 10.3. Two ways of examining habitat dependence. Regression of the proportion of habitat at locations against proportion of habitat available in range outlines for a number of individuals (A) has slopes of 1 for random use or 0 for constant use, and positive y-intercept if preferred at least at low availability. If animals adjust their foraging to access a particular habitat, a log-log plot of range area on proportion of habitat in the range will have a negative slope, of -1 if the sample of animals maintain a constant area of habitat.

habitat 1 would have slope > 0 and intercept-on-$U_1 > 0$. For those spending a constant proportion of time in habitat 1, slope would be 0 and intercept-on-$U_1 > 0$, in which case there would only be a preference for habitat 1 over habitat 2 at low availability of habitat 1.

Change in availability of habitat or other resources can also be reflected in range size, indicating that animals are covering a large enough area to meet their requirements (Kenward, 1982b; Wauters and Dhondt, 1992; Kenward and Holm, 1993; Thirgood, 1995; Hulbert *et al.*, 1996; Tufto *et al.*, 1996; Marzluff *et al.*, 1997b). This can be formalised by plotting area a of particular habitats within ranges of total area A. If animals seek a constant minimum of a_j for the jth habitat in their range, a plot of a_{ij} against A_i for the $I = 1 \ldots i$ animals will tend to a slope of 0, except that a_j will be forced downwards if A decreases below the point at which $a_j = A$. However, the emphasis of small-range effects is avoided in log-plots of habitat proportion against range area. Thus, a plot of $\log(A)$ against $\log(a_{ij}/A_i)$ truncates to $\log(a_{ij}/A_i) = 0$ when $a_j = A$, at which point the intercept on A indicates the required area of habitat a_j, with a slope of -1 for a constant proportion in the range (Fig. 10.3B). Presence of a significant negative correlation then indicates dependence of range size on proportion of habitat, with one major proviso. Since a_i is a part of A_i, there is always a degree of dependence in plots of a_i or a_{ij}/A_i against A_i. This means that the null hypothesis, of no dependence of range size on habitat content, is not necessarily that slope $= 0$ and $r = 0$, but must be estimated by randomisation for each set of ranges. The null values by randomisation can be either positive or negative (Kenward *et al.*, in review a).

This habitat dependence analysis (HDA) is less sensitive to location accuracy than location-based preference analysis, and can also be applied for decreasing core sizes to determine what proportion of an animal's effort (as represented by the proportion of locations in core m) is maximal for habitat j, from the plot of $\log(A_{im})$ against $\log(a_{ijm}/A_{im})$ that gives the most negative slope. Significant divergence of slopes from -1, as well as from the null value estimated by randomisation, indicates variation of habitat quality, for example such that the smallest ranges contain the habitat in better quality. HDA is not constrained to two habitats, and stepwise removal from A of extensive size-dependent habitats can reveal dependence on minor habitats which had their relationships 'swamped' by the extensive habitats in preliminary regressions. Thus, data from grey squirrels that foraged in various types of woodland but also in wheat fields first revealed weak dependence of range size on mature deciduous woodland. With deciduous woodland omitted, HDA then showed a much more significant dependence on wheat (Kenward *et al.*, in review a) than was found for the same data as a preference (Aebischer *et al.*, 1993b) or as a functional response (Mysterud and Ims, 1998).

10.3.3 Resource-dependent performance

The size of an animal's range is one measure of performance. Range size represents use of energy for movement, so small range size may reflect less need to move by the more efficient individuals at times of constrained food intake. However, small range size is not an entirely reliable performance indicator. Squirrels with the smallest ranges had highest body mass, survival and breeding success if they were female, but male range size was associated with reproductive status (Kenward, 1985; Wauters and Dhondt, 1992; Kenward *et al.*, in press b). Survival, breeding success and absence of emigration are direct measures of performance, and the best evidence for importance of habitat or other resources is a correlation with these demographic variables.

Linear regressions can be used to investigate relationships between productivity and habitat in range outlines, at locations or in combination. Aebischer *et al.* (1993a) used compositional analysis to show that pheasants with particular crops in their home ranges suffered least predation. However, it may also be robust to use raw habitat variables in regressions provided that the unit-sum constraint is avoided by analysing only variables with positive slopes, or with negative slopes, at the same time. Linear regressions may be used for analyses with continuous response variables, such as productivity or time to die if death is recorded for most animals. Logistic regression is more appropriate when the response is death versus survival, or emigration versus persistence, within a period (North and Reynolds, 1996).

Remember, however, that even if performance is linked to a resource, it cannot be assumed that the habitat is critical. Experiments are necessary to prove that a habitat is critical.

10.4 SOCIAL INTERACTIONS

Location data can be used to examine interactions between animals in a number of ways. These can be divided broadly into static interaction of overlapping home ranges recorded during a similar time period, and dynamic interaction based on locations recorded at the same time for different individuals (Dunn, 1979; Macdonald *et al.*, 1980). Additional measures of sociality, such as the spacing of home-range centres and the association of individuals with fixed sites (such as nests or feeding stations) can be estimated in ways similar to dynamic interactions.

10.4.1 Static interactions

If a home range of area A_1 has an area a_1 overlapped by another animal, a measure of the overlap for animal 1 is a_1/A_1. There is an equivalent overlap a_2/A_2 ($=a_1/A_2$) for animal 2. If animals 1 and 2 differ greatly in size, then a_1/A_1 could be 0.1 while a_2/A_2 is 0.8, or even $a_2/A_2 = 1$ if there is complete overlap and A_1 is ten times the size of A_2. It may be important to separate the indices a_{j1}/A_{j1} from a_{j2}/A_{j2} because, for example, j represents sex or age categories and you want to know about female overlap on males.

You need to be careful about sample sizes when analysing interactions of these types. If you have five similar animals with extensively overlapping home ranges, they have up to 24 ($=[5-1]!$) overlaps, and it is tempting to assume that this is the sample size. In fact, the sample size is 5, because you should use for each animal the mean of the k overlaps in which it takes part, which is

$$\frac{1}{k}\sum_{i=1}^{k} f\left(\frac{a_{ij}}{A_j}\right) \tag{10.31}$$

where $f(a_{ik})$ is an appropriate transformation, for which logarithms to give a geometric mean, or arcsines, may be most appropriate. If the animals overlap in separate dyads (avoiding the word 'pairs' which can imply opposite sexes), the situation is ambiguous. By analogy with the above, you might use a_j/A_j for each animal and a sample size of $2k$ where k is the number of dyads. However, those neighbours have some familiarity with each other and are therefore not truly independent. An alternative approach would therefore be to use an index such as the $2.a_1/(A_1 + A_2)$ or $2.a_2/(A_1 + A_2)$ of Coles (1949), with a sample size of k, but that may be too conservative. You may therefore choose to give both results.

Another question concerns the home range outline to use in the analyses. You might want to examine, for example, whether particular categories of animal, such as sibs, overlap more than others as range cores reduce in size (Poulle *et al.*, 1994), in which case questions of shape and smoothing would be relevant.

You might analyse not only the extent of overlap for animals whose ranges touched, but also the proportion of neighbouring animals that touched in the most expansive outlines. The situation becomes more complex if you wish to examine whether ranges overlap more or less than expected by chance as range cores reduce in size. One approach would be to compare the observed overlaps with a distribution generated by random overlap, using an envelope provided by the most expansive outlines of that technique to confine the possible positions of the cores.

The grid cell approach provides similar tools to the outline methods for investigating interactions between different individuals (Adams and Davis, 1967). For example, the proportion of each animal's grid cells that are visited by another provides a measure of static interaction, possibly including intensity of use for each cell as a coefficient of concordance (Doncaster, 1990).

10.4.2 Dynamic interactions

Static interactions can conceal the fact that, although ranges overlapped, the animals avoided being in the same place at the same time. More subtle measures of animal relationships are possible if locations have been recorded at the same time. For that situation, Dunn (1979) used an Orstein-Uhlenbeck model of animal movements to test whether neighbours interacted. This test avoids any assumption that locations are serially independent; instead it assumes that animals follow Markovian movement patterns. Macdonald *et al.* (1980) proposed a simpler test based on an assumption of a bivariate normal (elliptical) distribution of locations.

Both these tests assume that animal locations or movements follow particular distributions, and seek to show a significant result for each test dyad. Non-parametric tests based on randomisation provide more robust alternatives (Mielke and Berry, 1982; Kenward *et al.*, 1993b). If there are n pairs of locations x_{1j}, y_{1j} and x_{2j}, y_{2j} for each dyad, the observed mean distance between them is

$$D_o = \frac{1}{n} \sum_{j=1}^{n} \sqrt{(x_{1j} - x_{2j})^2 + (y_{1j} - y_{2j})^2} \tag{10.32}$$

which can be compared with the expected mean distance

$$D_E = \frac{1}{n^2} \sum_{j=1}^{n} \sum_{k=1}^{n} \sqrt{(x_{1j} - x_{2k})^2 + (y_{1j} - y_{2k})^2} \tag{10.33}$$

obtained by randomising all possible pairs of locations at which the animals were detected (Kenward *et al.*, 1993b).

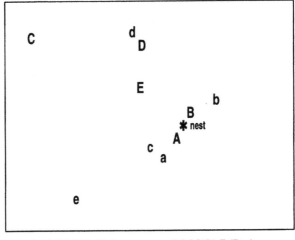

OBSERVED (D_o)   POSSIBLE (D_e)

OBSERVED (D_o)	POSSIBLE (D_e)
A-a	A-a A-b A-c A-d A-e
B-b	B-a B-b B-c B-d B-e
C-c	C-a C-b C-c C-d C-e
D-d	D-a D-b D-c D-d D-e
E-e	E-a E-b E-c E-d E-e

Figure 10.4. Pairs of locations recorded at the same time, denoted by A, B, C ... for one animal and a, b, c ... for another, have mean distances apart (A–a, B–b, C–c ...) that are smaller than a mean for all possible distance pairs if the animals tend to cohere and larger if they avoid each other. The means (or medians) can be used in an index of dynamic inter-action for brood-mates or other animals whose home ranges have extensive static overlap.

As an example, consider an individual with locations *A, B* and *C* at the same time as a second has locations *a, b* and *c* (Fig. 10.4). A statistic can then be produced for that dyad to show how the observed distances apart (A-a, B-b, C-c) at each time compare with the possible distances apart using all the locations (A-a, A-b, A-c, B-a, B-b, B-c, C-a, C-b, C-c).

The D_O and D_E scores for each dyad can be combined with Jacob's index to give an unbiased, symmetrical interaction coefficient for which values between -1 and $+1$ indicate avoidance when negative and cohesion when positive. A Wilcoxon matched-pair test can then be used to test for a significant difference from the null hypothesis that the overall interaction score is 0. This approach is robust, because the one test statistic for each dyad makes no assumption that observations within a dyad are statistically independent: the sample size is number of dyads, not of locations.

RANGES V estimates the Jacob's index for arithmetic mean distances (as shown in Eqs 10.32 and 10.33), but also as medians and geometric means. The geometric means generally estimate least variable coefficients. The interaction coefficients are useful for a detailed examination of relationships where there are strong static interactions, for example between sibs at the same nest (Fig. 10.5). However, the scores tend to 0 when home ranges have little overlap, and have not yet been as useful as static interactions for examining social relations of neighbouring animals.

The D_O and D_E estimates can also be used to examine whether individuals tend to avoid or approach particular sites, to answer questions such as 'do non-breeders avoid active nests?' In this case D_O is estimated from distances between locations and the sites. Estimation of D_E is by randomisation, but this is problematic in the same way as definition of available habitats. In RANGES V, options include (i) estimation of D_O and D_E only of locations and random values within circles round nests, which is conservative for tests of cohesion and (ii) estimation of D_O and D_E within a polygon envelope that includes such circles round all the nests, which is

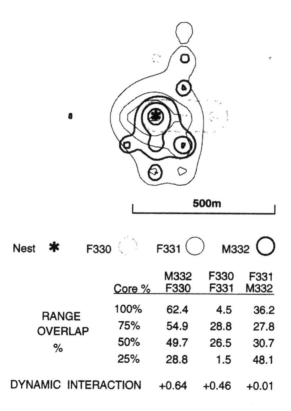

	Core %	M332 F330	F330 F331	F331 M332
RANGE	100%	62.4	4.5	36.2
OVERLAP	75%	54.9	28.8	27.8
%	50%	49.7	26.5	30.7
	25%	28.8	1.5	48.1
DYNAMIC INTERACTION		+0.64	+0.46	+0.01

Figure 10.5. Sociality indices for three goshawks (*Accipiter gentilis*) in the post-fledging period. Extensive overlap of areas within kernel contours between male 332 and female 330 is reflected in a large positive Jacobs index for dynamic interaction, but there is no cohesion between male 332 and female 331 despite extensive overlap of their densest range cores.

conservative for tests of avoidance. Distances between range centres can also be examined using nearest-neighbour analysis (Clarke and Evans, 1954; Brown and Rothery, 1978), using estimated distributions and randomisation within envelopes to assess whether there is significant departure from random towards regular spacing.

10.5 CONCLUSIONS

1. Lincoln-Petersen estimates of animal density can use radio tags as markers in sighting surveys. Multiple estimates can be combined as unweighted arithmetic means. Independent radio-location data can be used to provide density corrections at edges of study areas and during transect surveys.
2. Survival of individuals through periods can be analysed with logistic regression, in which the dependent variable is fate, and individual and environmental factors are covariates. Alternatively, for periods in which most individuals die, analyses can be of hazard rates based on tag days of survival for each individual.
3. Survival rates for groups of animals can be estimated from deaths within total numbers of tag days, but this assumes constant survival rates. It is easier to estimate confidence limits and to accommodate variable rates if there are enough animals to estimate rates in consecutive weekly or monthly periods. This also permits staggered entry of tagged animals and exit through loss or tag failure. Independent records after the estimation period permit correction for possible lost-signal bias.
4. Resource-use analyses should not use locations as independent data. Analyses can be based (i) on rank differences between use and availability or (ii) on compositional analysis. Dependence on resources may be examined in terms of preference variation with availability, or in terms of variation in range size with habitat content. Analyses may first need to separate individuals with divergent strategies. Dependence on habitats or other resources can also be indicated by relationships with productivity, dispersal and survival, but only experiments can show if habitats are critical for animals.
5. Analyses of social interactions should be based on values for pairs of individuals, and not use locations as independent data. Indices of static overlap of range outlines or grid cells can examine change in extent of possible interaction with intensity of range use and time. Indices of dynamic (same-time) interaction, with other animals and with sites, can be obtained from randomisation procedures.

References

Ables, E.D. (1969). Home-range studies of red foxes (*Vulpes vulpes*). *Journal of Mammalogy* **50**, 108–120.

Ackerman, B.B., Leban, F.A., Samuel, M.D. and Garton, E.O. (1990). User's manual for program HOME RANGE. Forestry, Wildlife and Range Experimental Station, University of Idaho, Moscow. Technical Report 15.

Adams, L. and Davis, S.D. (1967). The internal anatomy of home range. *Journal of Mammalogy* **48**, 529–536.

Aebischer, N.J. (1993). Using recoveries of censored animals to correct survival rates estimated from radio-tagging (J.-D. Lebreton and P.M. North, eds), 167. Birkhauser Verlag, Basel, Switzerland.

Aebischer, A.E., Robertson, P.A. and Kenward, R.E. (1993b). Compositional analysis of habitat use from animal radio-tracking data. *Ecology* **74**, 1313–1325.

Aebischer, A.E., Marcström, V., Kenward, R.E. and Karlbom, M. (1993a). Survival and habitat utilisation: a case for compositional analysis. In *Marked Individuals in the Study of Bird Populations* (J.-D. Lebreton and P.M. North, eds), 343–353. Birkhauser Verlag, Basel, Switzerland.

Aitchison, J. (1986). *The Statistical Analysis of Compositional Data*. Chapman and Hall, London, England.

Aldridge, H.D.J.N. and Brigham, R.M. (1988). Load carrying and manoeuvrability in an insectivorous bat: a test of the 5% 'rule' of radio-telemetry. *Journal of Mammalogy* **69**, 379–382.

Alexander, I. and Cresswell, B. (1990). Foraging by Nightjars *Caprimulgus europaeus* away from their nesting areas. *Ibis* **132**, 568–574.

Alkon, P.U. and Cohen, Y (1986). Acoustical biotelemetry for wildlife research: a preliminary test and prospects. *Wildlife Society Bulletin* **14**, 193–196.

Alkon, P.U., Cohen, Y. and Jordan, P.A. (1989). Towards an acoustic biotelemetry system for animal behaviour studies. *Journal of Wildlife Management* **53**, 658–662.

Alldredge, J. R., and Ratti, J. T. (1986). Comparisons of some statistical techniques for analysis of resource selection. *Journal of Wildlife Management* **50**, 157–165.

Alldredge, J.R. and Ratti, J.T. (1992). Further comparison of some statistical techniques for analysis of resource selection. *Journal of Wildlife Management* **56**, 1–9.

Althoff, D.P., Storm, G.L., Collins, T.W. and Kuechle, V.B. (1989). Remote sensing for monitoring animal activity, temperature and light. In *Biotelemetry X* (C.J. Amlaner, ed.), 116–124. University of Arkansas Press, Fayetteville.

Amlaner, C.J. (1978). Biotelemetry from free ranging animals. In *Animal Marking: Recognition Marking of Animals in Research* (B. Stonehouse, ed.), 205–228. Macmillan, London.

Amlaner, C.J. (1980). The design of antennas for use in radio telemetry. In *A Handbook on Biotelemetry and Radio Tracking* (C.J. Amlaner and D.W. Macdonald, eds), 251–261. Pergamon, Oxford.

Amlaner, C.J. (ed.) (1989). *Biotelemetry X*. University of Arkansas Press, Fayetteville.

Amlaner, C.J. and Macdonald, D.W. (1980). A practical guide to radio tracking. In *A Handbook on Biotelemetry and Radio Tracking* (C.J. Amlaner and D.W. Macdonald, eds), 143–159. Pergamon, Oxford.

Amlaner, C.J., Sibly, R. and McCleery, R. (1978). Effects of transmitter weight on breeding success in herring gulls. *Biotelemetry and Patient Monitoring* **5**, 154–163.

Amstrup, S.C. (1980). A radio collar for game birds. *Journal of Wildlife Management* **44**, 214–217.

Ancel, A., Kooyman, G.L., Ponganis, P.J., Gendner, J.P. and LeMaho, Y. (1992). Foraging behaviour of emperor penguins as a resource detector in winter and summer. *Nature* **360**, 336–339.

Anderka, F.W. (1980). Modulators for miniature tracking transmitters. In *A Handbook on Biotelemetry and Radio Tracking* (C.J. Amlaner and D.W. Macdonald, eds), 181–184. Pergamon, Oxford.

Andersen, D.E. (1994). Longevity of solar-powered radio transmitters on buteonine hawks in eastern colorado. *Journal of Field Ornithology* **65**, 122–132.

Anderson, D.J. (1982). The home range: a new non-parametric estimation technique. *Ecology* **63**, 103–112.

Anderson, F. and Hitchins, P.M. (1971). A radiotracking study of the black rhinoceros. *Journal of the South African Wildlife Management Association* **1**, 26–35.

Anderson, D.G. and Rongstad, O.J. (1989). Home-range estimates of red-tailed hawks based on random and systematic relocations. *Journal of Wildlife Management* **53**, 802–807.

Andrews, D.F., Bickel, P.J., Hampel, F.R., Huber, P.J., Rogers, W.H. and Tukey, J.W. (1972). *Robust Estimates of Location, Survey and Advances*. Princeton University Press, Princeton, New Jersey.

Angerbjorn, A. and Becker, D. (1992). An automatic location system. In *Wildlife Telemetry – Remote Monitoring and Tracking of Animals* (I.G. Priede and S.M. Swift, eds), 68–75. Ellis Horwood, Chichester.

Arnould, J.P.Y., Briggs, D.R., Croxall, J.P., Prince, P.A. and Wood, A.G. (1996). The foraging behaviour and energentics of wandering albatrosses brooding chicks. *Antarctic Science* **6**, 229–236.

ARRL (1984). *The ARRL Antenna Book*, 14th edition. American Radio Relay League, Newington, Connecticut.

Arthur, S.M., Manly, B.F.J., McDonald, L.L. and Garner, G.W. (1996). Assessing habitat selection when availability changes. *Ecology* **77**, 215–227.

Bailey, G.N.A., Linn, I.J. and Walker, P.J. (1973). Radioactive marking of small mammals. *Mammal Review* **3**, 11–23.

Baker, R. R. (1978). *The Evolutionary Ecology of Animal Migration*. Hodder & Stoughton, London.

Banks, E.M., Brooks, R.J. and Schnell, J. (1975). A radiotracking study of home range and activity of the brown lemming (*Lemmus trimucronatus*). *Journal of Mammalogy* **56**, 888–901.

Barbour, R.W., Harvey, M.J. and Hardin, J.W. (1969). Home range, movements, and activity of the eastern worm snake, *Carphophis amoenus*. *Ecology* **50**, 470–476.

Bart, J. and Robson, D.S. (1982). Estimating survivorship when the subjects are visited periodically. *Ecology* **63**, 1078–1090.

Batschelet, E. (1981). *Circular Statistics in Biology*. Academic Press, New York.

Batchelor, T.A. and McMillan, J.R. (1980). A visual marking system for nocturnal animals. *Journal of Wildlife Management* **44**, 497–499.

Beale, D.M.and Smith, A.D. (1973). Mortality of pronghorn antelope fawns in western Utah. *Journal of Wildlife Management* **37**, 343–352.

Beaumont, W.R.C., Clough, S., Ladle, M. and Welton, J.S. (1996). A method for the attachment of miniature radio tags to small fish. *Fisheries Management and Ecology* **3**, 201–207.

Berdeen, J.B. and Krementz, D.G. (1998). The use of fields at night by wintering American woodcock. *Journal of Wildlife Management* **62**, 939–947.

Berteaux, D., Masseboeuf, F., Bonzom, J.M., Bergeron, J.M., Thomas, D.W. and Lapierre, H. (1996). Effect of carrying a radiocollar on expenditure of energy by meadow voles. *Journal of Mammalogy* **77**, 359–363.

Berthold, P., Nowak, E. and Querner, U. (1992). Satelliten-Telemetrie beim Weißstorch (*Ciconia ciconia*) auf dem Wegzug: eine Pilotstudie. *Journal für Ornithologie* **133**, 155–163.

Bertram, B. (1980). The Serengeti radio-tracking program, 1971–1973. In *A Handbook on Biotelemetry and Radio Tracking* (C.J. Amlaner and D.W. Macdonald, eds), 625–631. Pergamon, Oxford.

Beyer, D.E. and Haufler, J.B. (1994). Diurnal versus 24–hour sampling of habitat use. *Journal of Wildlife Management* **58**, 178–180.

Birks, J.D.S. and Linn, I.J. (1982). Studies of home range of the feral mink, *Mustela vison*. In *Telemetric Studies of Vertebrates* (C.L. Cheeseman and R.B. Mitson, eds), 231–257. Academic Press, London.

Boone, R.B. and Mesecar, R.S. III (1989). Telemetric egg for use in egg-turning studies. *Journal of Field Ornithology* **60**, 315–322.

Boag, D.A. (1972). Effect of radio packages on behaviour of captive red grouse. *Journal of Wildlife Management* **36**, 511–518.

Bögel, R. and Burchard, D. (1992). An air-pressure transducer for telemetering flight altitude of birds. In *Wildlife Telemetry – Remote Monitoring and Tracking of Animals* (I.G. Priede, and S.M. Swift, eds), 100–106. Ellis Horwood, Chichester, UK.

Bohus, B. (1974). Telemetered heart rate response of the rat during free and learned behaviour. *Biotelemetry* **I**, 193–201.

Bosak, A. (1992). Some new design concepts for simple lightweight radio-tracking equipment. In *Wildlife Telemetry – Remote Monitoring and Tracking of Animals* (I.G. Priede, and S.M. Swift, eds), 92–97. Ellis Horwood, Chichester, U.K.

Boulanger, J.G. and White, G.C. (1990). A comparison of home-range estimators using Monte Carlo simulation. *Journal of Wildlife Management* **54**, 310–315.

Boyd, J.C.and Sladen, W.J.L. (1971). Telemetry studies of the internal body temperatures of Adelie and emperor penguins at Cape Crozier, Ross Island, Antarctica. *Auk* **88**, 366–380.

Brand, C.J., Vowles, R.H. and Keith, L.B. (1975). Snowshoe hare mortality monitored by telemetry. *Journal of Wildlife Management* **39**, 741–747.

Brander, R.B. (1968). A radio-package harness for game birds. *Journal of Wildlife Management* **32**, 630–632.

Brander, R.B. and Cochran, W.W. (1971). Radio-location telemetry. In *Wildlife Management Techniques* (R.H. Giles, ed.), 95–103. The Wildlife Society, Washington.

Bray, O.E. and Corner, G.W. (1972). A tail clip for attaching transmitters to birds. *Journal of Wildlife Management* **36**, 640–642.

Brittas, R. and Karlbom, M. (1990). A field evaluation of the Finnish 3-man chain: a method for estimating forest grouse numbers and habitat use. *Ornis Fennica* **67**, 18–23.

Britten, M.W., Kennedy, P.L. and Ambrose, S. (in press) An evaluation of the performance and accuracy of small satellite transmitters. *Journal of Wildlife Management*.

Bro, E., Clobert, J. and Reitz, F. (1999). Effects of radiotransmitters on survival and reproductive success of gray partridge. *Journal of Wildlife Management* **63**, 1044–1051.

Brodeur, S., Décarie, R., Bird, D.M. and Fuller, M. (1996). Complete migration cycle of golden eagles breeding in northern Quebec. *Condor* **98**, 293–299.

Broekhuizen, S., Van't Hoff, C.A., Jansen, M.B. and Niewold, F.J.J. (1980). Application of radio tracking in wildlife research in the Netherlands. In *A Handbook on Biotelemetry and Radio Tracking* (C.J. Amlaner and D.W. Macdonald, eds), 65–84. Pergamon, Oxford.

Brown, D. and Rothery, P. (1978). Randomness and local regularity of points in a plane. *Biometrica* **65**, 115–122.

Brown, W.S. and Parker, W.S. (1976). Movement ecology of *Coluber constrictor* near communal hibernacula. *Copeia* (1976), 225–242.

Bub, H. and Oelke, H. (1980). *Markierungsmethoden für Vögel*. Die Neue Brehm-Bücherei, Wittenberg-Lutherstadt.

Buchler, E.R. (1976). Chemiluminescent tag for tracking bats and other small nocturnal animals. *Journal of Mammalogy* **57**, 173–176.

Buckland, S.T., Anderson, D.R., Burnham, K.P. and Laake, J.L. (1993). *Distance Sampling: Estimating Abundance of Biological Populations.* London: Chapman and Hall.

Buechner, H.K., Craighead, F.C., Craighead, J.J. and Cote, C.E. (1971). Satellites for research on free roaming animals. *BioScience* **2**, 1201–1205.

Buehler, D.A., Fraser, J.D., Fuller, M.R., McAllister, L.S. and Seegar, J.K.D. (1995). Captive and field-tested radio transmitter attachments for bald eagles. *Journal of Field Ornithology* **66**, 173–180.

Bunck, C.M., Chen, C.-L. and Pollock, K.H. (1995). Robustness of survival estimates from radio-telemetry studies with uncertain relocation of animals. *Journal of Wildlife Management* **59**, 790–793.

Burchard, D. (1989). Direction finding in wildlife research by Doppler effect. In *Biotelemetry X* (C.J. Amlaner, ed.), 169–177. University of Arkansas Press, Fayetteville.

Burger, L.W., Ryan, M.R., Jones, D.P. and Wywialowsky, A.P. (1991). Radio transmitters bias estimation of movements and survival. *Journal of Wildlife Management* **55**, 693–697.

Burt, W.H. (1943). Territoriality and home range concepts as applied to mammals. *Journal of Mammalogy* **24**, 346–352.

Butler, R.W. and Jennings, J.G. (1980). Radio tracking of dolphins in the eastern tropical Pacific using VHF and HF equipment. In *A Handbook on Biotelemetry and Radio Tracking* (C.J. Amlaner and D.W. Macdonald, eds), 757–759. Pergamon, Oxford.

Butler, P.J., Woakes, A.J. and Bishop, C.M. (1998). Behaviour and physiology of Svalbard Barnacle Geese *Branta leucopsis* during their autumn migration. *Journal of Avian Biology* **29**, 536–545.

Byers, C.R., Steinhorst, R.K. and Krasuman, P.R. (1984). Clarification of a technique for analysis of utilization-availability. *Journal of Wildlife Management* **48**, 1050–1053.

Caccamise, D.F. and Hedin, R.F, (1985). An aerodynamic basis for selecting transmitter loads in birds. *Wilson Bulletin* **97**, 306–318.

Cairns, D.K., Bredin, K.A., Birt, V.L. and Montevecchi, W.A. (1987). Electronic activity recorders for aquatic wildlife. *Journal of Wildlife Management* **51**, 395–399.

Calhoun, J.B. and Casby, J.U. (1958). *Calculation of Home Range and Density of Small Mammals.* United States Public Health Service, Public Health Monograph 55.

Calvo, B. and Furness, R.W. (1992). A review of the use and the effects of marks and devices on birds. *Ringing and Migration* **13**, 129–151.

Camponotus AB and Radio Location Systems AB. (1994). *TRACKER: Wildlife Tracking and Analysis Software User Manual, version 1.1.* Solna and Huddinge, Sweden.

Carrel, W.K., Ockenfels, R.R., Wennerlund, J.A. and Devos, J.C. (1997). Topographic mapping, LORAN-C, and GPS accuracy for aerial telemetry locations. *Journal of Wildlife Management* **61**, 1406–1412.

Carroll, J.P. (1990). Winter and spring survival of radio-tagged gray partridge in North Dakota. *Journal of Wildlife Management* **54**, 657–662.

Carter, S.L., Haas, C.A, and Mitchell, J.C. (1999). Home range and habitat selection of bog turtles in southwestern Virginia. *Journal of Wildlife Management* **63**, 853–860.

Castell, P.M. and Trost, R.E. (1996). Neck bands reduce survival of Canada geese in New Jersey. *Journal of Wildlife Management* **60**, 891–898.

Cederlund, G., Dreyfert, T. and Lemnell, P.A. (1979). *Radio-Tracking Techniques and the Reliability of Systems Used for Larger Birds and Mammals.* Swedish Environmental Protection Board, Solna.

Cederlund, G. and Lemnell, P.A. (1980). A simplified technique for mobile radio tracking. In *A Handbook on Biotelemetry and Radio Tracking* (C.J. Amlaner and D.W. Macdonald, eds), 319–322. Pergamon, Oxford.

Chapman, D.G. (1951). Some properties of the hypergeometric distribution with applications to zoological sample censuses. University of California at Berkeley, *Publications in Statistics* 1, 131–160.

Charles-Dominique, P. (1977). Urine marking and territoriality in *Gallago alleni* (Waterhouse 1837 – Lorisidae Primates) – a field study by radio telemetry. *Zeitschrift für Tierpsychologie* 43, 113–138.

Cheeseman, C.L. and Mallinson, P.J. (1980). Radio tracking in the study of bovine tuberculosis in badgers. In *A Handbook on Biotelemetry and Radio Tracking* (C.J. Amlaner and D.W. Macdonald, eds), 649–656. Pergamon, Oxford.

Cheeseman, C.L. and Mitson, R.B. (eds) (1982). *Telemetric Studies of Vertebrates.* Symposia of the Zoological Society of London 49. Academic Press, London.

Chu, D.S., Hoover, B.A., Fuller, M.R. and Geissler, P.H. (1989). Telemetry location error in forested habitat. In *Biotelemetry X* (C.J. Amlaner, ed.), 188–194. University of Arkansas Press, Fayetteville.

Church, K.E. (1980). Expanded radio tracking potential in wildlife investigations with the use of solar transmitters. In *A Handbook on Biotelemetry and Radio Tracking* (C.J. Amlaner and D.W. Macdonald, eds), 247–250. Pergamon, Oxford.

Chute, F.S., Fuller, W.A., Harding, P.J.R. and Herman, T.B. (1974). Radio-tracking of small mammals using a grid of overhead wires. *Canadian Journal of Zoology* 52, 1481–1488.

Ciofi, C., Chelazzi, G. and Santina, P.D. (1992). A new technique of radio-tracking snakes: external implantation of transmitters. In *Wildlife Telemetry – Remote Monitoring and Tracking of Animals* (I.G. Priede and S.M. Swift, eds), 490–494. Ellis Horwood, Chichester, UK.

Clarke, P.J. and Evans, F.C. (1954). Distance to nearest neighbour as a measure of spatial relationships in populations. *Ecology* 35, 445–453.

Clute, R.K. and Ozoga, J.J. (1983). Icing of transmitter collars on white-tailed deer fawns. *Wildlife Society Bulletin* 11, 70–71.

Clutton-Brock, T.H. and Guiness, F. (1975). Behaviour of red deer (Cervas elephas L.) at calving time. *Behaviour* 55, 287–300.

Clutton-Brock, T.H., Guiness, F.E. and Albon, S.D. (1982). *Red Deer: Behaviour and Ecology of Two Sexes.* Edinburgh University Press, Edinburgh, Scotland.

Coah, R.S., White, M., Trainer, D.O. and Glazener, W.C. (1971). Mortality of young white-tailed deer fawns in south Texas. *Journal of Wildlife Management* 35, 47–56.

Cochran, W.W. (1980). Wildlife telemetry. In *Wildlife Management Techniques Manual*, 4th edition (S.D. Schemnitz, ed.), 507–520. Wildlife Society, Washington.

Cochran, W.W. and Lord, R.D. (1963). A radio-tracking system for wild animals. *Journal of Wildlife Management* 27, 9–24.

Cochran, W.W., Warner, D.W., Tester, J.R. and Keuchle, V.B. (1965). Automatic radio-tracking system for monitoring animal movements. *BioScience* 15, 98–100.

Cole, L.C. (1949). The measurement of interspecific association. *Ecology* 30, 411–424.

Coleman, J.S. and Jones, A.B.I. (1988). *User's Guide to TELEM88: Computer Analysis System for Radio-Telemetry Data.* Department of Fisheries and Wildlife, Virginian Polytechnic Institute and State University, Blacksburg, Virginia.

Cooper, H.M. and Charles-Dominique, P. (1985). A microcomputer data acquisition-telemetry system: a study of activity in the bat. *Journal of Wildlife Management* 49, 850–854.

Cooper, W.E. (1978). Home range size and population dynamics. *Journal of Theoretical Biology* 75, 327–337.

Côté, S.D., Festa-Bianchet, M. and Fournier, F. (1998). Life-history effects of chemical immobilization and radiocollars on mountain goats. *Journal of Wildlife Management* 62, 745–752.

Cotter, R.C. and Gratto, C.J. (1995). Effects of nest and brood visits and radio transmitters on rock ptarmigan. *Journal of Wildlife Management* 59, 93–98.

Covich, A.P. (1976). Analyzing shapes of foraging areas: some ecological and economic theories. *Annual Review of Ecology and Systematics* **7**, 235–257.

Cox, D.R. (1970). *The Analysis of Binary Data*. Chapman and Hall, New York.

Cox, D.R. (1972). Regression models and life tables. *Journal of the Royal Statistical Society* **B 34**, 187–220.

Cox, D.R. and Oakes, D. (1984). *Analysis of Survival Data*. Chapman and Hall, New York.

Crabtree, R.L., Burton, F.G., Garland, T.R., Cataldo, D.A. and Rickard, W.H. (1989). Slow-release radioisotope implants as individual markers for carnivores. *Journal of Wildlife Management* **53**, 949–954.

Creel, S., Creel, N.M. and Monfort, S.L. (1997). Radiocollaring and stress hormones in African wild dogs. *Conservation Biology* **11**, 544–548.

Cresswell, B. (1992). Use of thermistors in radio tags. In *Wildlife Telemetry – Remote Monitoring and Tracking of Animals* (I.G. Priede and S.M. Swift, eds), 98–99. Ellis Horwood, Chichester, UK.

Cresswell, B. and Alexander, I. (1992). Activity patterns of foraging nightjars (*Caprimulgus europaeus*). In *Wildlife Telemetry – Remote Monitoring and Tracking of Animals* (I.G. Priede and S.M. Swift, eds), 642–647. Ellis Horwood, Chichester, UK.

Cresswell, W.J. and Smith, G.C. (1992). The effects of temporally autocorrelated data on methods of home range analysis. In *Wildlife Telemetry – Remote Monitoring and Tracking of Animals* (I.G. Priede and S.M. Swift, eds), 272–284. Ellis Horwood, Chichester, U.K.

Cristalli, C., Amlaner, C.J. and Neuman, M.R. (eds) (1996). *Biotelemetry XIII*. Williamsburg.

Crook D.A. and White R.W.G. (1995). Evaluation of subcutaneously implanted visual implant tags and coded wire tags for marking and benign recovery in a small scaleless fish, *Galaxias truttaceus* (*Pisces: Galaxiidae*). *Marine and Freshwater Research* **46**, 943–946.

Croll, D.A., Osmek, S.D. and Bengtson, J.L. (1991). Effect of instrument attachment on foraging trip duration in chinstrap penguins. *Condor* **93**, 777–779.

Culik, B. and Wilson, R.P. (1991). Swimming energetics and performance of instrumented Adelie penguins *Pygoscelis adeliae*. *Journal of Experimental Biology* **158**, 355–368.

Cypher, B.L. (1997). Effects of radiocollars on San Joachin kit foxes. *Journal of Wildlife Management* **61**, 1412–1423.

Dalke, P.D. and Sime, P.R. (1938). Home and seasonal ranges of the eastern cottontail in Connecticut. *Transcripts of the North American Wildlife Conference* **3**, 659–669.

Davis, J.R., Von Relum, A.F., Smith, D.D. and Guynn, D.C. (1984). Implantable telemetry in beaver. *Wildlife Society Bulletin* **12**, 322–324.

Deat, A., Mauget, C., Mauget, R., Maurel, D. and Sempere, A. (1980). The automatic, continuous and fixed radio tracking system of the Chizé Forest: theoretical and practical analysis. In *A Handbook on Biotelemetry and Radio Tracking* (C.J. Amlaner and D.W. Macdonald, eds), 439–451. Pergamon, Oxford.

Delguidice, G.D., Kunkel, K.E., Mech, D.L. and Seal, U.S. (1990). Minimizing capture-related stress on white-tailed deer with a capture collar. *Journal of Wildlife Management* **54**, 299–303.

De Solla, S.R., Bonduranski, R. and Brooks, R.J. (1999). Eliminating autocorrelation reduces biological relevance of home range estimations. *Journal of Animal Ecology* **68**, 221–234.

Dice, L.R. and Clark, P.J. (1953). *The Statistical Concept of Home Range as Applied to the Recapture of the Deermouse* (Peromyscus). University of Michigan Laboratory of Vertebrate Biology Contributions 62.

Dixon, K.R. and Chapman, J.A. (1980). Harmonic mean measure of animal activity areas. *Ecology* **61**, 1040–1044.

Dodge, W.E. and Steiner, A.J. (1986). XYLOG: A computer program for field-processing locations of radio-tagged wildlife. US Fish and Wildlife Service, Fisheries and Wildlife Technical Report 4.

Don, B.A.C. and Rennolls, K. (1983). A home range model incorporating biological attraction points. *Journal of Animal Ecology* **52**, 69–81.

Doncaster, C.P. (1990). Non-parametric estimates of interaction from radio-tracking data. *Journal of Theoretical Biology* **143**, 431–443.

Doncaster, C.P. and Macdonald, D.W. (1991). Drifting territoriality in the red fox *Vulpes vulpes*. *Journal of Animal Ecology* **60**, 423–439.

Douglas, M.E. and Pickard, C.J. (1992). Telemetry of body tilt for automatic data logging of blue duck diel behaviour. In *Wildlife Telemetry – Remote Monitoring and Tracking of Animals* (I.G. Priede and S.M. Swift, eds), 599–611. Ellis Horwood, Chichester, UK.

Downhower, J.F. and Pauley, J.D. (1970). Automatic recordings of body temperature from free-ranging yellow-bellied marmots. *Journal of Wildlife Management* **34**, 639–641.

Dunn, J.E. (1979). A complete test for dynamic territorial interaction. *Proceedings of the Second International Conference on Wildlife Biotelemetry*, 159–169.

Dunn J.E. and Brisbin, I.L. (1982). Characterizations of the multivariate Ornstein-Uhlenbeck diffusion process in the context of home range analysis. Statistical Laboratory Technical Report 16. University of Arkansas, Fauetteville.

Dunn, J.E. and Gipson, P.S. (1977). Analysis of radio telemetry data in studies of home range. *Biometrics* **33**, 85–101.

Dunstan, T.C. (1972). A harness for radio-tagging raptorial birds. *Inland Bird Banding News* **44**, 4–8.

Dunstan, T.C. (1973). A tail feather package for radio-tagging raptorial birds. *Inland Bird Banding News* **45**, 3–6.

Dwyer, T.J. (1972). An adjustable radio-package for ducks. *Bird-Banding* **43**, 282–284.

Eagle, T.C., Choromanski-Norris, J. and Keuchle, V.B. (1984). Implanting radio transmitters in mink and Franklin's ground squirrels. *Wildlife Society Bulletin* **12**, 180–184.

Edge, W.D. and Marcum, C.L. (1989). Determining elk distribution with pellet-group and telemetry techniques. *Journal of Wildlife Management* **53**, 621–624.

Efron, B. (1988). Logistic regression survival analysis, and the Kaplan-Meier curve. *Journal of the American Statistical Association* **83**, 414–425.

Erikstad, K.E. (1979). Effects of radio packages on reproductive success of willow grouse. *Journal of Wildlife Management* **43**, 170–175.

Estes, J.A., Tinker, M.T., Williams, T.M. and Doak, D.F. (1998) Killer whale predation on sea otters linking oceanic and nearshore ecosystems. *Science* **282**, 473–475.

Everett, B. (1980). *Cluster Analysis*. Heinemann, London, England.

Exo, K.M., Eggers, U., Laschefski-Sievers, R. and Scheiffarth, G. (1992). Monitoring activity patterns using a microcomputer-controlled radio-telemetry system, tested for waders (*Charadrii*) as an example. In *Wildlife Telemetry – Remote Monitoring and Tracking of Animals* (I.G. Priede and S.M. Swift, eds), 79–87. Ellis Horwood, Chichester, UK.

Fagerstone, K.A. and Johns, B.E. (1987). Transponders as permanent identification markers for domestic ferrets, black-footed ferrets and other wildlife. *Journal of Wildlife Management* **51**, 294–297.

Fedak, M.A., Anderson, S.S. and Curry, M.G. (1983). Attachment of a radio tag to the fur of seals. *Journal of Zoology, London* **200**, 298–300.

Fitch, H.S. and Shirer, H.W. (1971). A radiotelemetric study of spatial relationships in some common snakes. *Copeia* (1971), 118–128.

Fitzner, R.E. and Fitzner, J.N. (1977). A hot melt glue technique for attaching radiotransmitter tail packages to raptorial birds. *North American Bird Bander* **2**, 56–57.

Flint, P.L., Pollock, K.H., Thomas, D. and Sedinger, J.S. (1995). Estimating pre-fledging survival: allowing for brood mixing and dependence among brood mates. *Journal of Wildlife Management* **59**, 448–455.

Folk, G.E. and Folk, M.A. (1980). Physiology of large mammals by implanted radio capsules. In *A Handbook on Biotelemetry and Radio Tracking* (C.J. Amlaner and D.W. Macdonald, eds), 33–43. Pergamon, Oxford.

Follman, E.H. and Buitt, J.M. (1978). An adjustable radio collar for foxes. *Journal of Wildlife Management* **42**, 949–951.

Ford, G. and Krumme, D.W. (1979). The analysis of space use patterns. *Journal of Theoretical Biology* **76**, 125–155.

Forys, E.A. and Humphrey, S.R. (1997). Comparison of 2 methods to estimate density of an endangered lagomorph. *Journal of Wildlife Management* **61**, 86–92.

Foster, C.C., Forsman, E.D., Meslow, E.C., Miller, G.S., Reid, J.A., Wagner, F.F., Carey, A.B. and Lint, J.B. (1992). Survival and reproduction of radio-marked adult spotted owls. *Journal of Wildlife Management* **56**, 91–95.

French, J. and Goriup, P. (1992). Design of a small Argos PTT for the houbara bustard. In *Wildlife Telemetry – Remote Monitoring and Tracking of Animals* (I.G. Priede and S.M. Swift, eds), 160–176. Ellis Horwood, Chichester, UK.

French, J., Latham, D.M., Oldham, R.S. and Bullock, D.J. (1992). An automated radio-tracking system for use with amphibians. In *Wildlife Telemetry – Remote Monitoring and Tracking of Animals* (I.G. Priede and S.M. Swift, eds), 477–483. Ellis Horwood, Chichester, UK.

French, J. and Priede, I.G. (1992). A microwave radar transponder for tracking studies. In *Wildlife Telemetry – Remote Monitoring and Tracking of Animals* (I.G. Priede and S.M. Swift, eds), 41–54. Ellis Horwood, Chichester, UK.

Friedman, M. (1937). The use of ranks to avoid the assumption of normality implicit in the analysis of variance. *Journal of the American Statistical Association* **32**, 671–701.

Fullagar, P.J. (1967). The use of radio-telemetry in Australian biological research. *Proceedings of the Ecological Society of Australia* **2**, 16–26.

Fuller, M.R. (1975). A technique for holding and handling raptors. *Journal of Wildlife Management* **39**, 824–825.

Fuller, M.R., Levanon, N., Strikwerda, T.E., Seegar, W.S., Wall, J., Black, H.D., Ward, F.P., Howey, P.W. and Partelow, J. (1984). Feasibility of a bird-borne transmitter for tracking via satellite. In *Biotelemetry VIII* (H.P. Kimmich and H.J. Klewe, eds), 375–378. Nijmegan, Netherlands.

Fuller, M.R., Seegar, W.S. and Howey, P.W. (1995). The use of satellite systems for the study of bird migration. *Israel Journal of Zoology* **41**, 243–252.

Fuller, M.R. and Tester, J.R. (1973). An automated radio tracking system for biotelemetry. *Raptor Research* **7**, 105–106.

Fuller, R.M., Groom, G.B. and Jones, A.R. (1994). The land cover map of Great Britain: an automated classification of Landsat Thematic Mapper data. *Photogrammetric Engineering and Remote Sensing* **60**, 553–562.

Gammonley, J.H. and Kelly, J.R. (1994). Effects of back-mounted radio-packages on breeding wood ducks. *Journal of Field Ornithology* **65**, 530–533.

Garcelon, D.K. (1977). An expandable drop-off transmitter collar for young mountain lions. *California Fish and Game* **63**, 185–189.

Garner, G.W., Amstrup, S.C., Douglas, D.C. and Gardner, C.L. (1989). Performance and utility of satellite telemetry during field studies of free-ranging polar bears in Alaska. In *Biotelemetry X* (C.J. Amlaner, ed.), 66–76. University of Arkansas Press, Fayetteville Arkansas, USA.

Garrott, R.A., Bartmann, R.M. and White, G.C. (1985). Comparison of radio-transmitter packages relative to deer fawn mortality. *Journal of Wildlife Management* **49**, 758–759.

Garrott, R.A., White, G.C., Bartmann, R.M. and Weybright, D.L. (1986). Reflected signal bias in biotelemetry triangulation systems. *Journal of Wildlife Management* **50**, 747–752.

Garshelis, D.L. and Siniff, D.B. (1983). Evaluation of radio-transmitter attachments for sea otters. *Wildlife Society Bulletin* **11**, 378–383.

Gatti, R.C., Dumke, R.T. and Pils, C.M. (1989). Habitat use and movements of female ring-necked pheasants during fall and winter. *Journal of Wildlife Management* **53**, 462–475.

Gautestad, A.O. and Mysterud, I. (1995). The home range ghost. *Oikos* **74**, 195–204.

Gautier, J.P. (1980). Biotelemetry of the vocalizations of a group of monkeys. In *A Handbook on Biotelemetry and Radio Tracking* (C.J. Amlaner and D.W. Macdonald, eds), 535–544. Pergamon, Oxford.

Geissler, P.H. and Fuller, M.R. (1985). Detecting and displaying the structure of an animal's home range. In *Proceedings of the American Statistical Association 1985*, 378–383.

Gessamen, J.A. (1974). Telemetry of electrocardiograms from free-living birds: a method of electrode placement. *Condor* **76**, 479–481.

Gessamen, J.A. (1980). An evaluation of heart rate as an indirect measure of daily energy metabolism of the American kestrel. *Comparative Biochemistry and Physiology* **65**, 273–289.

Gessamen, J.A. and Nagy. K.A. (1988). Transmitter loads affect the flight speed and metabolism of homing pigeons. *Condor* **90**, 662–668.

Gessamen, J.A., Workman, G.W. and Fuller, M.R. (1991). Flight performance, energetics and water turnover of tippler pigeons with a harness and dorsal load. *Condor* **93**, 546–554.

Gillingham, M.P. and Bunnell, F.L. (1985). Reliability of motion-sensitive radio collars for estimating activity of black-tailed deer. *Journal of Wildlife Management* **49**, 951–958.

Gillingham, M.P. and Parker, K.L. (1992). Simple timing device increases reliability of recording telemetric activity data. *Journal of Wildlife Management* **56**, 191–196.

Gilmer, D.S., Keuchle, V.B. and Ball, I.J. (1971). A device for monitoring radio-marked animals. *Journal of Wildlife Management* **35**, 829–832.

Gilmer, D.S., Bell, J.J., Cowardin, L.M. and Reichmann, J.H. (1974). Effects of radio packages on wild ducks. *Journal of Wildlife Management* **38**, 243–252.

Gilmer, D.S., Cowardin, L.M., Duval, R.L., Mechlin, L.M., Shaiffer, C.W. and Keuchle, V.B. (1981). *Procedures for the Use of Aircraft in Wildlife Biotelemetry Studies*. United States Department of the Interior, Fish and Wildlife Service Resource Publication 140.

Giroux, J.-F., Bell, D.V., Percival, S. and Summers, R.W. (1990). Tail-mounted radio transmitters for waterfowl. *Journal of Field Ornithology* **61**, 303–309.

Glendinning, R.H. (1991). The convex hull of a dependent vector-valued process. *Journal of Statistical Computation and Simulation* **38**, 219–237.

Godfrey, G.A. (1970). A transmitter harness for small birds. *Inland Bird Banding News* **2**, 3–5.

Goldberg, J.S. and Haas, W. (1978). Interactions between mule deer dams and their radio-collared and unmarked fawns. *Journal of Wildlife Management* **42**, 422–425.

Goodyear, J.D. (1993). A sonic/radio tag for monitoring dive depths and underwater movements of whales. *Journal of Wildlife Management* **57**, 503–513.

Goodyear, N.C. (1989). Studying fine-scale habitat use in small mammals. *Journal of Wildlife Management* **53**, 941–946.

Göransson, G. (1980). Animal activity recorded by radio tracking and an audio time lapse recorder. In *A Handbook on Biotelemetry and Radio Tracking* (C.J. Amlaner and D.W. Macdonald, eds), 457–460. Pergamon, Oxford.

Gorman, M.L., Frears, S. and Racey, P.A. (1992). Radio-tracking and the function of the fortress of the mole (*Talpa europea*). In *Wildlife Telemetry – Remote Monitoring and Tracking of Animals* (I.G. Priede, and S.M. Swift, eds), 510–520. Ellis Horwood, Chichester, UK.

Goss-Custard, J.D.(ed) (1996). *The Oystercatcher: From Individuals to Populations*. Oxford University Press, UK.

Graber, R.R. and Wunderle, S.L. (1966). Telemetric observations of a robin (*Turdus migratorius*). *Auk* **83**, 674–677.

Greager, D.C., Jenness, C.A. and Ward, G.D. (1979). An acoustically sensitive transmitter for telemetering the activities of wild animals. *Journal of Wildlife Management* **43**, 1001–1007.

Green, R. (1988). Effects of environmental factors on the timing and success of breeding of Common Snipe *Gallinago gallinago* (*Aves. Scolopacidae*). *Journal of Applied Ecology* **25**, 79–93.

Greenwood, R.J. and Sargeant, A.B. (1973). Influence of radio packs on captive mallards and blue-winged teal. *Journal of Wildlife Management* **37**, 3–9.

Guynn, D.C., Davis, J.R. and Von Recum, A.F. (1987). Pathological potential of intraperitoneal transmitter implants in beavers. *Journal of Wildlife Management* **51**, 605–606.

Hallberg, D.L., Janza, F.J. and Trapp, G.R. (1974). A vehicle-mounted directional antenna system for biotelemetry monitoring. *California Fish and Game* **60**, 172–177.

Hammond, P.S., McConnell, B.J., Fedak, M.A. and Nicholas, K.S. (1992). Grey seal activity patterns around the Farne Islands. In *Wildlife Telemetry – Remote Monitoring and Tracking of Animals* (I.G. Priede and S.M. Swift, eds), 677–686. Ellis Horwood, Chichester, UK.

Hanski, I.K. and Haila, Y. (1988). Singing territories and home ranges of breeding Chaffinches: visual observation vs. radio-tracking. *Ornis Fennica* **65**, 97–103.

Hansteen, T.L., Andreassen, H.P. and Ims, R.A. (1997). Effects of spatiotemporal scale on autocorrelation and home range estimators. *Journal of Wildlife Management* **61**, 280–290.

Harden Jones, F.R. and Arnold, G.P. (1982). Acoustic telemetry and the marine fisheries. In *Telemetric Studies of Vertebrates* (C.L. Cheeseman and R.B. Mitson, eds), 75–93. Academic Press, London.

Hardy, A.R. and Taylor, K.D. (1980). Radio tracking of *Rattus norvegicus* on farms. In *A Handbook on Biotelemetry and Radio Tracking* (C.J. Amlaner and D.W. Macdonald, eds), 657–665. Pergamon, Oxford.

Harestad, A.S. and Bunnell, F.L. (1979). Home range and body weight – a reevaluation. *Ecology* **60**, 389–402.

Harris, R.B., Fancy, S.G., Douglas, D.C, Garner, G.W., Amstrup, S.C., McCabe, T.R. and Pank, L.F. (1990). Tracking wildlife by satellite: current systems and performance. US Fish and Wildlife Service, Washington DC. Technical Report 30.

Harris, S. (1980). Home ranges and patterns of distribution of foxes (*Vulpes vulpes*) in an urban area, as revealed by radio tracking. In *A Handbook on Biotelemetry and Radio Tracking* (C.J. Amlaner and D.W. Macdonald, eds), 685–690. Pergamon, Oxford.

Harris, S., Cresswell, W.J., Forde, P.G., Trewella, W.J., Woollard T. and Wray S. (1990). Home-range analysis using radio-tracking data – a review of problems and techniques particularly as applied to the study of mammals. *Mammal Review* **20**, 97–123.

Harrison, J.L. (1958). Range and movements of some Malayan rats. *Journal of Mammalogy* **39**, 190–206.

Hartigan, J.A. (1987). Estimation of a convex density contour in two dimensions. *Journal of the American Statistical Association* **82**, 267–270.

Harvey, M.J. and Barbour, R.W. (1965). Home range of *Microtus ochrogaster* as determined by a minimum area method. *Journal of Mammalogy* **46**, 398–402.

Hayes, R.W. (1982). A telemetry device to monitor big game traps. *Journal of Wildlife Management* **46**, 551–553.

Hayne, D.W. (1949). Calculation of size of home range. *Journal of Mammalogy* **30**, 1–18.

Haynes, J.M. (1978). Movement and Habitat Studies of Chinook Salmon and White Sturgeon. PhD Thesis, University of Minnesota.

Heath, R.G.M. (1987). A method for attaching transmitters to penguins. *Journal of Wildlife Management* **51**, 399–401.

Heezen, K.L. and Tester, J.R. (1967). Evaluation of radio-tracking by triangulation with special reference to deer movements. *Journal of Wildlife Management* **31**, 124–141.

Heisey, D.M. (1985). Analyzing selection experiments with log-linear models. *Ecology* **66**, 1744–1748.

Heisey, D.M. and Fuller, T.K. (1985). Evaluation of survival and cause-specific mortality rates using telemetry data. *Journal of Wildlife Management* **49**, 668–674.

Hellgren, E.C., Carney, D.W., Garner, N.P. and Vaughan, M.R. (1988). Use of breakaway cotton spacers on radio collars. *Wildlife Society Bulletin* **116**, 216–218.

Herbst, L. (1991). Pathological and reproductive effects of intraperitoneal telemetry devices on female armadillos. *Journal of Wildlife Management* **55**, 628–631.

Higuchi, H., Ozaki, K., Fujita, G., Minton, J., Veta, M., Soma, M. and Mita, N. (1996). Satellite tracking of white-naped cranes migration and the importance of the Korean demilitarised zone. *Conservation Biology* **10**, 806–812.

Higuchi, H, Ozaki, K., Fujita, G., Soma, M., Kanmuri, N. and Ueta, M. (1992). Satellite tracking of the migration routes of cranes from southern Japan. *Strix* **2**, 1–20.

Hill, I., Cresswell, B.H. and Kenward, R.E. (in press) The problems and rewards of radio tagging nestling passerines – testing harnesses to accommodate growth. In *Proceedings of the Fifth International Conference on Wildlife Telemetry*, Strasbourg, France.

Hill, I., Cresswell, B.H. and Kenward, R.E. (1999). Field testing the suitability of a new back pack harness for radio-tagging passerines. *Journal of Avian Biology* **30**, 135–142.

Hill, R.D., Schneider, R.C., Liggins, G.C., Hochachka, P.W., Schuette, A.H. and Zapol, W.M. (1983). Microprocessor controlled recording of bradycardia during free diving of the Antarctic Wedell seal. *Federal Proceedings* **42**, 470.

Hines, J.E. and Zwickel, F.C. (1985). Influence of radio packages on young blue grouse. *Journal of Wildlife Management* **49**, 1050–1054.

Hirons, G.J.M. and Owen, R.B. (1982). Radio tagging as an aid to the study of woodcock. In *Telemetric Studies of Vertebrates* (C.L. Cheeseman and R.B. Mitson, eds), 139–152. Academic Press, London.

Hodder, K.H., Kenward, R.E., Walls, S.S. and Clarke, R.T. (1998). Estimating core ranges: a comparison of techniques using the common buzzard (*Buteo buteo*). *Journal of Raptor Research* **32**, 82–89.

Holm, J.L. (1990). The ecology of red squirrels (*Sciurus vulgaris*) in deciduous woodland. PhD thesis, University of London, unpublished.

Hölzenbein, S. (1992). Expandable PVC collar for marking and transmitter support. *Journal of Wildlife Management* **56**, 473–476.

Hooge, P.N. (1991). The effects of radio weight and harnesses on time budgets and movements of acorn woodpeckers. *Journal of Field Ornithology* **62**, 230–238.

Hooge, P.N. and Eichenlaub, B. (1997). Animal movements extension to ArcView version 1.1. Alaska Biological Science Center, US Geological Survey, Anchorage.

Horton, G.I. and Causey, M.K. (1984). Brood abandonment by radio-tagged American woodcock hens. *Journal of Wildlife Management* **48**, 606–607.

Houston, R.A. and Greenwood, R.J. (1993). Effects of radio transmitters on nesting captive mallards. *Journal of Wildlife Management* **57**, 703–709.

Howey, P.W. (1992). Tracking of birds by satellite. In *Wildlife Telemetry – Remote Monitoring and Tracking of Animals* (I.G. Priede and S.M. Swift, eds), 177–184. Ellis Horwood, Chichester, UK.

Howey, P.W., Board, R.G. and Kear, J. (1977). A pulse-position-modulated multichannel radio telemetry system for the study of avian nest microclimate. In *Biotelemetry IV*, 169–180.

Howey, P.W., Seegar, W.S., Fuller, M.R. and Titus, K. (1988). A coded tracking telemetry system. In *Biotelemetry X* (C.J. Amlaner, ed.), 103–107. University of Arkansas Press, Fayetteville.

Howey, P.W., Witlock, D.R., Fuller, M.R., Seegar, W.S. and Ward, F.P. (1984). A computerised biotelemetry receiving and datalogging system. In *Biotelemetry VIII* (H.P. Kimmich and H.J. Klewe, eds), 442–446. Nijmegan, Netherlands.

Hubbard, M.W., Tsao, L.-L., Klaas, E.E., Kaiser, M. and Jackson, D.H. (1998). Evaluation of transmitter attachment techniques on growth of wild turkey poults. *Journal of Wildlife Management* **62**, 1574–1578.

Huempfner, R.A., Maxson, S.J., Erickson, J. and Schuster, R.J. (1975). Recapturing radio-tagged ruffed grouse by nightlighting and snow burrow netting. *Journal of Wildlife Management* **39**, 821–823.

Hulbert, I.A.R., Iason, G.R., Elston, D.A. and Racey, P.A. (1996). Home-range sizes in a stratified upland landscape of two lagomorphs with different feeding strategies. *Journal of Animal Ecology* **33**, 1479–1488.

Hünerbein, K., Hamann, H.J., Rüter, E. and Wiltschko, W. (2000). A GPS based System for Recording the Flight Paths of Birds. Naturwissenschafton **87**, 278–279.

Hupp, J.W. and Ratti, J.T. (1983). A test of radiotelemetry triangulation accuracy in heterogeneous environments. In *Biotelemetry IV*, 31–46.

Hurlbert, S.H. (1984). Pseudoreplication and the design of ecological field experiments. *Ecological Monographs* **54**, 187–211.

Ims, R.A. (1988) Spatial clumping of sexually receptive females induces space sharing among male voles. *Nature* **335**, 541–543.

Ireland, L.C. (1980). Homing behaviour of juvenile green turtles *Chelonia mydas*. In *A Handbook on Biotelemetry and Radio Tracking* (C.J. Amlaner and D.W. Macdonald, eds), 761–764. Pergamon, Oxford.

Ireland, L.C. and Kanwisher, J.W. (1978). Underwater acoustic biotelemetry: procedures for obtaining information on the behaviour and physiology of free-swimming aquatic animals in their natural environments. In *The Behaviour of Fish and Other Aquatic Animals* (D.I. Mostofsky, ed.), 341–379. Academic Press, New York.

Ivlev, (1961). *Experimental Ecology of the Feeding of Fishes.* Yale University Press, Connecticut, USA.

Jackson, D.H., Jackson, L.S. and Seitz, W.K. (1985). An expandable drop-off transmitter harness for young bobcats. *Journal of Wildlife Management* **49**, 46–49.

Jacobs, J. (1974). Quantitative measurement of food selection. A modification of the forage ratio and Ivlev's electivity index. *Oecologia* **14**, 413–417.

Jaremovic, R.V. and Croft, D.B. (1987). Comparison of techniques to determine eastern grey kangaroo home range. *Journal of Wildlife Management* **51**, 921–930.

Jenkins, D. (1980). Ecology of otters in northern Scotland. I. Otter (*Lutra lutra*) breeding and dispersion in mid-Deeside, Aberdeenshire in 1974–79. *Journal of Animal Ecology* **49**, 713–735.

Jennrich, R.J. and Turner, F.B. (1969). Measurement of non-circular home range. *Journal of Theoretical Biology* **22**, 227–237.

Johnson, B.K., Ager, A.A., Findholt, S.L, Wisdom, M.J., Marx, D.B., Kern, J.W. and Bryant, L.D. (1998). Mitigating spatial differences in observation rates of automated telemetry systems. *Journal of Wildlife Management* **62**, 958–967.

Johnson, J.D., Pebworth, J.L. and Krueger, H.O. (1991). Retention of transmitters attached to passerines using a glue-on technique. *Journal of Field Ornithology* **62**, 486–491.

Johnson, P.B. (1980). The use of longterm ultrasonic implants for the location and harvest of schooling fish. In *A Handbook on Biotelemetry and Radio Tracking* (C.J. Amlaner and D.W. Macdonald, eds), 777–780. Pergamon, Oxford.

Johnson, R.N. and Berner, A.H. (1980). Effects of radio transmitters on released cock pheasants. *Journal of Wildlife Management* **44**, 686–689.

Johnstone, A.D.F., Lucas, M.C., Boyland, P. and Carter, T.J. (1992). Telemetry of tail-beat frequency of Atlantic salmon (*Salmo salar* L.) during spawning. In *Wildlife Telemetry – Remote Monitoring and Tracking of Animals* (I.G. Priede and S.M. Swift, eds), 456–465. Ellis Horwood, Chichester, UK.

Johnstone, I.G. (1992). Home range utilization and roost selection by non-breeding territorial European robins (*Erithacus rubecula*). In *Wildlife Telemetry – Remote Monitoring and Tracking of Animals* (I.G. Priede and S.M. Swift, eds), 495–509. Ellis Horwood, Chichester, UK.

Jones, G. and Moreton, M. (1992). Radio-tracking studies on habitat use by greater horseshoe bats (*Rhinolophus ferrumequinum*). In *Wildlife Telemetry – Remote Monitoring and Tracking of Animals* (I.G. Priede and S.M. Swift, eds), 521–537. Ellis Horwood, Chichester, UK.

Jouventin, P. and Weimerskirch, H. (1990) Satellite tracking of wandering albatrosses. *Nature* **343**, 746–748.

Jullien, J.M., Vassant, J. and Brandt, S. (1990). An extensible transmitter collar designed for wild boar (*Sus scrofa scrofa*): Study of neck size development in the species. *Gibier Faune Sauvage* **7**, 377–387.

Kalås, J., Lofaldli, L. and Fiske, P. (1989). Effects of radio packages on great snipe during breeding. *Journal of Wildlife Management* **53**, 115–1158.

Kalbfleisch, J.D. and Prentice, R.L. (1980). *The Statistical Analysis of Failure Time Data.* John Wiley & Sons, New York.

Kaneko, Y., Suzuki, T., Maruyama, N., Atoda, O., Kanzaki, N. and Tomisawa, M. (1998). The 'trace recorder', a new device for surveying mammal home ranges, and its application to raccoon dog research. *Mammal Study* **23**, 109–118.

Kaplan, E.L. and Meier, P. (1958). Nonparametric estimation from incomplete observations. *Journal of the American Statistical Association* **53**, 457–481.

Karl, B.J. and Clout, M.N. (1987). An improved radio transmitter harness with a weak link to prevent snagging. *Journal of Field Ornithology* **58**, 73–77.

Keating, K.A. (1995). Mitigating elevation-induced errors in satellite telemetry locations. *Journal of Wildlife Management* **59**, 801–808.

Keating, K.A., Brewster, W.G. and Key, C.H. (1991). Satellite telemetry: performance of animal-tracking systems. *Journal of Wildlife Management* **55**, 160–171.

Keith, L.B., Meslow, E.C. and Rongstad, O.J. (1968). Techniques for snowshoe hare population studies. *Journal of Wildlife Management* **32**, 801–812.

Kenward, R.E. (1976). The effect of predation by goshawks, *Accipiter gentilis*, on woodpigeon, *Columba Palumbus*, populations. D.Phil. Thesis, University of Oxford.

Kenward, R.E. (1977). Predation on released pheasants (*Fasianus colchicus*) by goshawks (Accipiter gentilis) in central Sweden. *Viltrevy* **10**, 79–112.

Kenward, R.E. (1978). Radio transmitters tail-mounted on hawks. *Ornis Scandinavica* **9**, 220–223.

Kenward, R.E. (1980). Radio monitoring birds of prey. In *A Handbook on Biotelemetry and Radio Tracking* (C.J. Amlaner and D.W. Macdonald, eds), 97–104. Pergamon, Oxford.

Kenward, R.E. (1982a) Techniques for monitoring the behaviour of grey squirrels by radio. In *Telemetric Studies of Vertebrates* (C.L. Cheeseman and R.B. Mitson, eds), 175–196. Academic Press, London.

Kenward, R.E. (1982b). Goshawk hunting behaviour, and range size as a function of food and habitat availability. *Journal of Animal Ecology* **51**, 69–80.

Kenward, R.E. (1985). Ranging behaviour and population dynamics in grey squirrels. In *Behavioural Ecology. Ecological Consequences of Adaptive Behaviour* (R.M. Sibly and R.H. Smith, eds), 319–330. Blackwell Scientific Publications, Oxford.

Kenward, R.E. (1987). *Wildlife Radio Tagging: Equipment, Field Techniques and Data Analysis.* Academic Press, London.

Kenward, R.E. (1990). *Software for Analysing Animal Location Data (Ranges IV).* Institute of Terrestrial Ecology, Wareham, UK.

Kenward, R.E. (1992). Quantity versus quality: programming for collection and analysis of radio tag data. In *Wildlife Telemetry – Remote Monitoring and Tracking of Animals* (I.G. Priede, and S.M. Swift, eds), 231–246. Ellis Horwood, Chichester, U.K.

Kenward, R.E. (1993). Modelling raptor populations: to ring or to radio tag? In *The Use of Marked Individuals in the Study of Bird Population Dynamics: Models, Methods and Software.* (J.D. LeBreton, and P.M. North, eds), 157–167. Birkhauser, Basle, Switzerland.

Kenward, R.E., Aebischer, N.J., Robertson, P.A., Fuller, R.M., Rose, R.J. and Walls, S.S. (in review a) Distance and density estimators of home-range: assessing habitat dependence from area and composition. In review with *Ecology.*

Kenward, R.E., Clarke, R.T., Hodder, K.H. and Walls, S.S. (in press, b). Distance and density estimators of home range in raptors and squirrels: defining multi-nuclear cores by nearest-neighbor clustering. *Ecology*.

Kenward, R.E., Hirons, G.J.M. and Ziesemer, F. (1982). Devices for telemetering the behaviour of free-living birds. In *Telemetric Studies of Vertebrates* (C.L. Cheeseman and R.B. Mitson, eds), 129–136. Academic Press, London.

Kenward, R. E. and Hodder, K. H. (1996). *Ranges V. An Analysis System for Biological Location Data*. Institute of Terrestrial Ecology, Wareham, UK.

Kenward, R.E. and Holm, J.L. (1993). On the replacement of the red squirrel in Britain: a phytotoxic explanation. *Proceedings of the Royal Society of London* B **251**, 187–194.

Kenward, R.E., Marcström, M. and Karlbom, M. (1981). Goshawk winter ecology in Swedish pheasant habitats. *Journal of Wildlife Management* **45**, 397–408.

Kenward, R.E., Marcström, M. and Karlbom, M. (1993a). Post-nestling behaviour in goshawks, *Accipiter gentilis*: I. The causes of dispersal. *Animal Behaviour* **46**, 365–370.

Kenward, R.E., Marcström, M. and Karlbom, M. (1993b). Post-nestling behaviour in goshawks, *Accipiter gentilis*: II. Sex differences in sociality and nest switching. *Animal Behaviour* **46**, 371–378.

Kenward, R.E., Marcström, V. and Karlbom, M. (1999). Demographic estimates from radio-tagging: models of age-specific survival and breeding in the goshawk. *Journal of Animal Ecology* **68**, 1020–1033.

Kenward, R.E., Pfeffer, R.H., Al-Bowardi, M.A., Fox, N.C., Riddle, K.E., Bragin, Y.A., Levin, A.S., Walls, S.S. and Hodder, K.H. (in press, a). New techniques for demographic studies of falcons. *Journal of Field Ornithology*.

Kenward, R.E., Robertson, P.A., Coates, A.R., Marcström, V. and M. Karlbom. (1993c). Techniques for radio-tagging pheasant chicks. *Bird Study* **40**, 51–54.

Kenward, R.E. and Walls, S.S. (1994). The systematic study of radio-tagged raptors: I, survival, home range and habitat use. In *Raptor Conservation Today* (B.U. Meyburg and R.D. Chancelor, eds), 303–315. World Working Group on Birds of Prey, Berlin.

Kenward, R.E., Walls, S.S. and Hodder, K.H. (in press, c). Life path analysis: scaling indicates priming effects of social and habitat factors on dispersal distances. *Journal of Animal Ecology*.

Kenward, R.E., Walls, S.S., Hodder, K.H., Pahkala, M., Freeman, S.N. and Simpson, V. R. (in press, d) The prevalence of non-breeders in raptor populations: evidence from radio-tagging and survey data. *Oikos*.

Keuchle, V.B. (1982). State of the art of biotelemetry in North America. In *Telemetric Studies of Vertebrates* (C.L. Cheeseman and R.B. Mitson, eds), 1–18. Academic Press, London.

Keuchle, V.B., Fuller, M.R., Reichle, R.A., Schuster, R.J. and Duke, G.E. (1987). Telemetry of gastric motility data from owls. In *Biotelemetry IX* (H.P. Kimmich and M.R. Neuman, eds), 363–366.

Keuchle, V.B., Haynes, J.M. and Reichle, R.A. (1989). Use of small computers as telemetry data collectors. In *Biotelemetry X* (C.J. Amlaner, ed.), 695–699. University of Arkansas Press, Fayetteville.

Kie, J.G., Baldwin, J.A. and Evans, C.J. (1996). CALHOME: a program for estimating animal home ranges. *Wildlife Society Bulletin* **24**, 342–344.

Klugman, S.S. and Fuller, M.R. (1990). Effects of implanted transmitters on captive Florida sandhill cranes. *Wildlife Society Bulletin* **18**, 394–399.

Ko, W.H. (1980). Power sources for implant telemetry and stimulation systems. In *A Handbook on Biotelemetry and Radio Tracking* (C.J. Amlaner and D.W. Macdonald, eds), 225–245. Pergamon, Oxford.

Koehler, D.K., Reynolds, T.D. and Anderson, S.H. (1987). Radio-transmitter implants in 4 species of small mammals. *Journal of Wildlife Management* **51**, 105–108.

Koeppl, J.W., Slade, N.A., Harris, K.S. and Hoffmann, R.S. (1977). A three-dimensional home range model. *Journal of Mammalogy* **58**, 213–220.

Koeppl, J.W., Slade, N.A. and Hoffmann, R.S. (1975). A bivariate home range model with possible application to ethological data analysis. *Journal of Mammalogy* **56**, 81–90.

Koeppl, J.W. and Hoffmann, R.S. (1985). Robust statistics for spatial analysis: the bivariate normal home range model applied to syntopic populations of two species of ground squirrels. University of Kansas Museum Natural History Occasional Pap. 116.

Kohn, M.H. and Wayne, R.K. (1997). Facts from faeces revisited. *Trends in Ecology and Evolution* **12**, 223–227.

Kolz, A.L. (1975). Mortality sensing wildlife transmitter. *Instrument Society of America Biomedical Sciences Institution Symposium* **12**, 57–60.

Kolz, A.L. and Johnson, R.E. (1975). An elevating mechanism for mobile receiving antennas. *Journal of Wildlife Management* **39**, 819–820.

Kolz, A.L. and Johnson, R.E. (1981). The human hearing response to pulsed-audio tones: implications for wildlife telemetry design. *Proceedings of the Third International Conference on Wildlife Biotelemetry,* 27–34.

Kolz, A.L., Lentfer, J.W. and Fallek, H.G. (1980). In *A Handbook on Biotelemetry and Radio Tracking* (C.J. Amlaner and D.W. Macdonald, eds), 743–752. Pergamon, Oxford.

Kooyman, G.L., Billups, J.O. and Farwell, W.D. (1983). Two recently developed monitors for recording diving activity of birds and mammals. In *Experimental Biology at Sea.* (A.G. Macdonald and I.G. Priede, eds), 197–214. Academic Press, London and New York.

Korschgen, C.E., Kenow, K.P., Gendron-Fitzpatrick, A., Green, W.L. and Dein, F.J. (1996a). Implanting intra-abdominal radiotransmitters with external whip antennas in ducks. *Journal of Wildlife Management* **60**, 132–137.

Korschgen, C.E., Kenow, K.P., Green, W.L., Johnson, D.H., Samuel, M.D. and Sileo, L. (1996b). Survival of radio-marked canvasback ducklings in Northwest Minnesota. *Journal of Wildlife Management* **60**, 120–132.

Korschgen, C.E., Kenow, K.P., A., Green, W.L., Samuel, M.D. and Sileo, L. (1996c). Technique for implanting radiotransmitters subcutaneously in day-old ducklings. *Journal of Field Ornithology* **67**, 392–397.

Korschgen, C.E., Maxson, S.J. and Keuchle, V.B. (1984). Evaluation of implanted radio transmitter in ducks. *Journal of Wildlife Management* **46**, 982–987.

Kreeger, T.J., Vargas, A., Plumb, G.E. and Thorne, E.T. (1998). Ketamine-Medetomidine or isoflurane immobilization of black-footed ferrets. *Journal of Wildlife Management* **62**, 654–662.

Kreiberg, H. and Powell, J. (1991). Metomidate sedation reduces handling stress in Chinook salmon. *World Aquaculture* **22**, 58–59.

Kruuk, H. (1978). Spatial organisation and territorial behaviour of the European badger *Meles meles. Journal of Zoology, London* **184**, 1–19.

Kruuk, H., Gorman, M. and Parish, T. (1980). The use of ^{65}Zn for estimating populations of carnivores. *Oikos* **34**, 206–208.

Kruuk, H., Parish, T., Brown, C.A.J. and Carrera, J. (1979). The use of pasture by the European badger (*Meles meles*). *Journal of Applied Ecology* **16**, 453–459.

Kufeld, R.C., Bowden, D.C. and Siperek, J.M. (1987). Evaluation of a telemetry system for measuring habitat usage in mountainous terrain. *Northwest Science* **61**, 249–256.

Kunkel, K.E., Chapman, R.C., Mech, L.C. and Gese, E.M. (1992). Testing the Wildlink activity system on wolves and white-tailed deer. *Canadian Journal of Zoology* **69**, 2466–2469.

Laake, J.L., Buckland, S.T., Anderson, D.R. and Burnham, K.P. (1994). *DISTANCE User's Guide.* Fort Collins: Colorado Cooperative Fish & Wildlife Research Unit.

Lacki, M.J., Smith, P.G., Peneston, W.T. and Vogt, F.D. (1989). Use of methoxyfluorane to surgically implant transmitters in muskrats. *Journal of Wildlife Management* **53**, 331–333.

Lair, H. (1987). Estimating the location of the focal center in red squirrel home ranges. *Ecology* **68**, 1092–1101.

Laird, L.M. and Oswald, R.L. (1975). A note on the use of benzocaine (Ethyl p-amino benzoate) as a fish anaesthetic. *Fishery Management* **6**, 92–94.

Lance, A.N. and Watson, A. (1977). Further tests of radio-marking on red grouse. *Journal of Wildlife Management* **41**, 579–592.

Larkin, R.P. and D. Halkin. (1994). A review of software packages for estimating animal home ranges. *Wildlife Society Bulletin* **22**, 274–287.

Larkin, R.P., Raim, A. and Diehl, R.H. (1996). Performance of a non-rotating direction finder for automated radio tracking. *Journal of Field Ornithology* **67**, 59–71.

Laundré, J.W. and Keller, B.L. (1981). Home range use by coyotes in Idaho. *Animal Behaviour* **29**, 449–461.

Laundré, J.W., Reynolds, T.D., Knick, S.T. and Ball, I.J. (1987). Accuracy of daily point relocations in assessing real movement of radio-marked animals. *Journal of Wildlife Management* **51**, 937–940.

Lawson, E.J.G. and Rogers, A.R. (1997). Differences in home-range size computed in commonly used software programs. *Wildlife Society Bulletin* **25**, 721–729.

Lawson, K., Kanwisher, J. and Williams, T.C. (1976). A UHF radio-telemetry system for wild animals. *Journal of Wildlife Management* **40**, 360–362.

Lee, E.T. (1980). *Statistical Methods for Survival Data Analysis.* Lifetime Learning Publications, Belmont, California, USA.

Lee, J.E., White, G.C., Garrott, R.A., Bartmann, R.M. and Alldredge, A.W. (1985). Accessing accuracy of a radiotelemetry system for estimating animal locations. *Journal of Wildlife Management* **49**, 658–663.

Legendre, P. (1993). Spatial autocorrelation: trouble or new paradigm? *Ecology* **74**, 1659–1673.

Lemnell, P.A., Johnsson, G., Helmersson, H., Holmstrand, O. and Norling, L. (1983). An automatic radio-telemetry system for position determination and data acquisition. *Proceedings of the Fourth International Conference on Wildlife Biotelemetry*, 76–93.

Lenth, R.V. (1981). On finding the source of a signal. *Technometrics* **23**, 149–154.

Leuze, C.C.K. (1980). The application of radio tracking and its effect on the behavioural ecology of the water vole, *Arvicola terrestris* (Lacepede). In *A Handbook on Biotelemetry and Radio Tracking* (C.J. Amlaner and D.W. Macdonald, eds), 361–366. Pergamon Press, Oxford.

Lewis, J.C. and Haithcoat, T.L. (1986). *TELEMPC: Personal Computer Package for Analyzing Radio-Telemetry Data.* Geographic Resources Center, University of Missouri, Columbia.

Linn, I.J. and Wilcox, P. (1982). A semi-automated system for collecting data on the movements of radio tagged voles. In *Telemetric Studies of Vertebrates* (C.L. Cheeseman and R.B. Mitson, eds), 197–205. Academic Press, London.

Livezey, K.B. (1988). Protective frame for a 2-element hand-held Yagi antenna. *Journal of Wildlife Management* **52**, 565–567.

Livezey, K.B. (1990). Toward the reduction of marking-induced abandonment of newborn ungulates. *Wildlife Society Bulletin* **18**, 193–203.

Loehle, C. (1990). Home range: a fractal approach. *Landscape Ecology* **5**, 39–52.

Loft, E.R. and Kie, J.G. (1988). Comparison of pellet-group and radio triangulation methods for assessing deer habitat use. *Journal of Wildlife Management* **52**, 524–527.

Lotimer, J.S. (1980). A versatile coded wildlife transmitter. In *A Handbook on Biotelemetry and Radio Tracking* (C.J. Amlaner and D.W. Macdonald, eds), 185–191. Pergamon Press, Oxford.

Lotimer, J. (1998). Automatic positioning and data collection systems for desktop tracking of wildlife. In *Radiotelemetry Applications for Wildlife Toxicology Field Studies* (L.W. Brewer, and K.A. Fagerstone), 181–190. Society of Environmental Toxicology and Chemistry, Pensacola, Florida.

Loughlin, T.R. (1980). Radio telemetric determination of the 24-hour feeding activities of sea otters *Enhydra lutris*. In *A Handbook on Biotelemetry and Radio Tracking* (C.J. Amlaner and D.W. Macdonald, eds), 717–724. Pergamon Press, Oxford.

Lovett, J.W. and Hill, E.P. (1977). A transmitter syringe for recovery of immobilised deer. *Journal of Wildlife Management* **41**, 313–315.

Lutterschmidt, W.I. and Reinert, H.K. (1990). The effect of ingested transmitters upon the temperature preference of the northern water snake, *Nerodia s. sipedon*. *Herpetologica* **46**, 39–42.

Macdonald, D.W. (1978). Radio-tracking: some applications and limitations. In *Animal Marking: Recognition Marking of Animals in Research* (B. Stonehouse, ed.), 192–204. Macmillan, London.

Macdonald, D.W. and Amlaner, C.J. (1980). A practical guide to radio tracking. In *A Handbook on Biotelemetry and Radio Tracking* (C.J. Amlaner and D.W. Macdonald, eds), 143–159. Pergamon Press, Oxford.

Macdonald, P.D.M. and Pitcher, T.J. (1979). Age groups from size frequency data: a versatile and efficient method of analysing distribution mixtures. *Journal of the Fishery Research Board of Canada* **36**, 987–1001.

Mackay, R.S. (1964). Galapagos tortoise and marine iguana deep core body temperature measured by radio-telemetry. *Nature* **204**, 4956.

Madsen, T. (1984). Movements, home range size and habitat use of radio-tracked grass snakes (*Natrix natrix*) in southern Sweden. *Copeia* (1984), 707–713.

Madison, D.M. (1998). Habitat-contingent reproductive behaviour in radio-implanted salamanders: a model and a test. *Animal Behaviour* **55**, 1203–1210.

Maechtle, T.L. (1998). The Aba: A device for restraining raptors and other large birds. *Journal of Field Ornithology* **69**, 66–70.

Mahoney, S.P., Virgl, J.A., Fong, D.W., MacCharles, A.M. and McGrath, M. (1998). Evaluation of a mark-resighting technique for woodland caribou in Newfoundland. *Journal of Wildlife Management* **62**, 1227–1235.

Maier, J.A.K., Maier, H.A. and White, R.G. (1996). Effects of ambient temperature on activity monitors of radiocollars. *Journal of Wildlife Management* **60**, 393–398.

Mancini, P., Fioretti, S., Cristalli, C. and Bedini, R. (eds.) (1993). *Biotelemetry XII*. Editrice Universitaria Litografia Felici, Pisa.

Manly, B.F.J. (1974). A model for certain types of selection experiment. *Biometrics* **30**, 281–294.

Manly, B.F.J, McDonald, L.L. and Thomas, D.L. (1993). *Resource Selection by Animals: Statistical Design and Analysis for Field Studies*. Chapman and Hall, London, England.

Marcström, V., Kenward, R.E. and Karlbom, M. (1989). Survival of ring-necked pheasants with backpacks, necklaces, and leg bands. *Journal of Wildlife Management* **53**, 808–810.

Marcum, C.L. and Loftsgaarden, D.O. (1980). A nonmapping technique for studying habitat preferences. *Journal of Wildlife Management* **44**, 963–968.

Marks, J.S. and Marks, V.S. (1987). Influence of radio collars on survival of sharp-tailed grouse. *Journal of Wildlife Management* **51**, 468–471.

Marques, M. (1972). Ensemble de reception 72 MHz a deux ariens permettant la localisation d'emetteurs destines au radiotracking animaux. *Mammalia* **36**, 299–304.

Marquiss, M. and Newton, I. (1982). Habitat preference in male and female sparrowhawks (*Accipiter nisus*). *Ibis* **124**, 324–328.

Marshall, W.H. and Kupa, J.J. (1963). Development of radio-telemetry techniques for ruffed grouse studies. *Transactions of the North American Wildlife and Natural Resources Conference* **28**, 443–456.

Martin, M.L. and Bider, J.R. (1978). A transmitter attachment for blackbirds. *Journal of Wildlife Management* **42**, 683–685.

Marzluff, J.M., Vekasy, M.S. and Coody, C. (1994). Comparative accuracy of aerial and ground telemetry locations of foraging raptors. *Condor* **96**, 447–454.

Marzluff, J.M., Knick, S.T., Vekasy, M.S., Schueck, L.S. and Zarriello, T.J. (1997b). Spatial use and habitat selection of golden eagles in south-western Idaho. *Auk* **114**, 673–687.

Marzluff, J.M., Vekasy, M.S., Kochert, M.N. and Steenhof, K. (1997a). Productivity of golden eagles wearing backpack radiotransmitters. *Journal of Raptor Research* **31**, 223–227.

Massey, B.W., Keene, K. and Boardman, C. (1988). Adverse effects of radio transmitters on the behavior of nesting least terns. *Condor* **90**, 945–947.

Mate, B.R., Harvey, J.T., Hobbs, L. and Maiefski, R. (1983). A new attachment device for radio-tagging large whales. *Journal of Wildlife Management* **47**, 868–872.

Mate, B.R., Nieukirk, S.L. and Kraus, S.D. (1997). Satellite-monitored movements of the northern right whale. *Journal of Wildlife Management* **61**, 1393–1405.

Mattson, N.S. and Ripple, T.H. (1989). Metomidate, a better anaesthetic for cod (*Gadus morhua*) in comparison with benzocaine, MS-222, chlorbutanol and phenoxyethanol. *Aquaculture* **83**, 89–94.

Mauser, D.M. and Jarvis, R.L. (1991). Attaching radio transmitters to 1–day-old mallard ducklings. *Journal of Wildlife Management* **55**, 488–491.

Mauser,D.M., Jarvis, R.L. and Gilmer, D.S. (1994). Survival of radio-marked mallard ducklings in north-eastern California. *Journal of Wildlife Management* **58**, 82–87.

Mayfield, H. (1961). Nesting success calculated from exposure. *Wilson Bull.* **73**, 255–261.

Mayfield, H. (1975). Suggestions for calculating nest success. *Wilson Bull.* **87**, 456–466.

McGinnis, S.M. (1967). The adaptation of biotelemetry to small reptiles. *Copeia* (1967), 472–473.

McShane, R., Gedling, K., Kenward, B., Kenward, R., Hope, T. and Jacoby, R. (1998). The feasibility of electronic tracking devices in dementia: a telephone survey and case series. *International Journal of Geriatric Psychiatry* **13**, 556–563.

Mech, L.D. (1967). Telemetry as a technique in the study of predation. *Journal of Wildlife Management* **31**, 492–496.

Mech, L.D. (1974). Current techniques in the study of elusive wilderness carnivores. *International Congress on Game Biology* **11**, 315–322.

Mech, L.D. (1980). Making the most of radio-tracking: a summary of wolf studies in Northeastern Minnesota. In *A Handbook on Biotelemetry and Radio Tracking* (C.J. Amlaner and D.W. Macdonald, eds), 85–95. Pergamon Press, Oxford.

Mech, L.D. (1983). *Handbook of Animal Radio-Tracking*. University of Minnesota.

Mech, L.D., Chapman, R.C., Cochran, W.W., Simmons, L. and Seal, U.S. (1984). Radio-triggered anaesthetic dart collar for recapturing large mammals. *Wildlife Society Bulletin* **12**, 69–74.

Mech, L.D., Kunkel, K.E., Chapman, R.C. and Kreeger, T.J. (1990). Field testing of commercially manufactured capture collars on white-tailed deer. *Journal of Wildlife Management* **54**, 297–299.

Mee, D. and Clarke, D. (1992). Tag-failure controls: their importance in interpretation of tracking statistics. In *Wildlife Telemetry – Remote Monitoring and Tracking of Animals* (I.G. Priede and S.M. Swift, eds), 441–444. Ellis Horwood, Chichester, UK.

Mellas, E.J. and Haynes, J.M. (1985). Swimming performance and behavior of rainbow trout (*Salmo gairdneri*) and white perch (*Morone americana*): effects of attaching telemetry transmitters. *Canadian Journal of Fisheries and Aquatic Science* **42**, 488–493.

Melquist, W.E. and Hornocker, M.G. (1979). Development and use of a telemetry technique for studying river otter. *Proceedings of the Second International Conference on Wildlife Telemetry,* 104–114.

Melvin, S.M., Drewien, R.C., Temple, S.A. and Bizeau, E.G. (1983). Leg-band attachment of transmitters for large birds. *Wildlife Society Bulletin* **11**, 282–285.

Meyburg, B.-U. and Lobkov, E.G. (1994). Satellite tracking of a juvenile Steller's sea eagle *Haliaeetus pelagicus*. *Ibis* **136**, 105–106.

Meyburg, B.-U., Scheller, W. and Meyburg, C. (1995). Zug und Überwinterung des Schreiadlers *Aquila pomarina*: Satellitentelemetrische Untersuchungen. *Journal für Ornithologie* **136**, 401–422.

Meyburg, B.-U. and C. Meyburg (1998). The study of raptor migration using satellite telemetry: some goals, achievements and limitations. In *Biotelemetry XIV* (T. Penzel, S. Salmons and M. Neuman, eds), 415–420. Tectum Verlag, Marburg.

Michener, G.R. (1979). Spatial relationships of adult Richardson's ground squirrels. *Canadian Zoology* **57**, 125–139.

Michener, M.C. and Walcott, C. (1966). Navigation of single homing pigeons: airplane observations by radio tracking. *Science* **154**, 410–413.

Mielke, P.W. and Berry, K.J. (1982). An extended class of permutations tests for matched pairs. *Community Statistics* **11**, 1197–1207.

Mikesic, D.G. and Drickamer, L.C. (1992). Effects of radiotransmitters and fluorescent powders on activity of wild house mice (*Mus musculus*). *Journal of Mammalogy* **73**, 663–667.

Miles, M.A., De Souza, A.A. and Povoa, M.M. (1981). Mammal tracking an nest location in Brazilian forest with an improved spool-and-line device. *Journal of Zoology, London* **195**, 331–347.

Miller, R.G. (1983). What price Kaplan-Meier? *Biometrics* **39**, 1077–1081.

Mills, L.S. and Knowlton, F.F. (1989). Observer performance in known and blind radio-telemetry accuracy tests. *Journal of Wildlife Management* **53**, 340–342.

Millspaugh, J.J. and Marzluff, J.M. (in press). *Wildlife Radiotelemetry: Design and Analysis*. Academic Press, San Diego, USA.

Minta, S.C. (1992). Tests of spatial and temporal interaction among animals. *Ecological Applications* **2**, 178–188.

Mitchell-Jones, A.J., Jefferies, D.J., Twelves, J., Green, J. and Green, R. (1984). A practical system of tracking otters *Lutra lutra* using radiotelemetry and 65–Zn. *Lutra* **27**, 71–84.

Mitson, R.B., Storeton-West, T.J. and Pearson, N.D. (1982). Trials of an acoustic transponding fish tag compass. *Biotelemetry Patient Monitoring* **9**, 69–79.

Moen, R., Pastor, J., Cohen, Y. and Scgwartz, G.C. (1996). Effects of moose movement and habitat use on GPS collar performance. *Journal of Wildlife Management* **60**, 659–668.

Moen, R., Pastor, J. and Cohen, Y. (1997). Accuracy of GPS telemetry collar locations with differential correction. *Journal of Wildlife Management* **61**, 530–539.

Mohr, C.O. (1947). Table of equivalent populations of North American small mammals. *American Midland Naturalist* **37**, 223–249.

Mohr, C.O. and Stumpf, W.A. (1966). Comparison of methods for calculating areas of animal activity. *Journal of Wildlife Management* **30**, 293–304.

Moore, A., Potter, E.C.E. and Buckley, A.A. (1992). Estuarine behaviour of migrating Atlantic salmon (*Salmo salar* L.) smolts. In *Wildlife Telemetry – Remote Monitoring and Tracking of Animals* (I.G. Priede and S.M. Swift, eds), 389–399. Ellis Horwood, Chichester, UK.

Morin, P.A. and Woodruff, D.S. (1996). Noninvasive genotyping for vertebrate conservation. In *Molecular Genetic Approaches in Conservation* (R.K. Wayne and T.B. Smith, eds) 298–313. University Press, Oxford, UK.

Morris, J.A. (1992). Methods of waterproofing transmitter packages for marine environments. In *Wildlife Telemetry – Remote Monitoring and Tracking of Animals* (I.G. Priede and S.M. Swift, eds), 88–89. Ellis Horwood, Chichester, UK.

Morris, P. (1980). An elementary guide to practical aspects of radio tracking mammals. In *A Handbook on Biotelemetry and Radio Tracking* (C.J. Amlaner and D.W. Macdonald, eds), 161–168. Pergamon Press, Oxford.

Mullican, T.R. (1988). Radio telemetry and fluorescent pigments: a comparison of techniques. *Journal of Wildlife Management* **52**, 627–631.

Murray, D.L. and Fuller, M.R. (in press). A critical review of the effects of marking on the biology of vertebrates.

Mysterud, A. and Ims, R.A. (1998). Functional responses in habitat use: availability influences relative use in trade-off situations. *Ecology* **79**, 1435–1441.

Nams, V.O. (1989a). A technique to determine the behaviour of a radio-tagged animal. *Canadian Journal of Zoology* **67**, 254–258.

Nams, V.O. (1989b). Effects of sample size and bias when testing for habitat selection. *Canadian Journal of Zoology* **67**, 1631–1636.

Nams, V.O. and Boutin, S. (1991). What is wrong with error polygons? *Journal of Wildlife Management* **55**, 172–176.

Neaf-Daenzer, B. (1993). A new transmitter for small animals and enhanced methods of home range analysis. *Journal of Wildlife Management* **57**, 680–689.

Neft, D.S. (1966). *Statistical Analysis for Areal Distributions*. Regional Scientific Research Institute Monograph Series 2.

Nenno, E.S. and Healey, W.M. (1979). Effects of radio packages on behaviour of wild turkey hens. *Journal of Wildlife Management* **43**, 760–765.

Neu, C.W., Byers, C.R. and Peek, J.M. (1974). A technique for analysis of utilisation-availability data. *Journal of Wildlife Management* **38**, 541–545.

Nicholas, K.S., Fedak, M.A. and Hammond, P.S. (1992). An automatic recording station for detecting and storing radio signals from free-ranging animals. In *Wildlife Telemetry – Remote Monitoring and Tracking of Animals* (I.G. Priede and S.M. Swift, eds), 76–78. Ellis Horwood, Chichester, UK.

Nicholls, T.H. and Warner, D.W. (1968). A harness for attaching radio transmitters to large owls. *Bird Banding* **39**, 209–214.

Nicholls, T.H., Ostry, M.E. and Fuller, M.R. (1981). Marking ground targets with radio transmitters dropped from aircraft. US Forestry Services Research Note NC-274.

Nietfeld, M.T., Barrett, M.W. and Silvy, N. (1994). Wildlife marking techniques. In *Research and Management Techniques for Wildlife Habitats*, 5th edn. (T.A. Bookhout, ed.), 140–168. The Wildlife Society, Bethesda, Maryland.

Nolan, J.W., Russell, R.H. and Anderka, F. (1984). Transmitters for monitoring Aldrich snares set for grizzly bears. *Journal of Wildlife Management* **48**, 942–945.

North, M.P. and Reynolds, J.H. (1996). Microhabitat analysis using radiotelemetry locations and polytomous logistic regression. *Journal of Wildlife Management* **60**, 639–653.

Obrecht, H.H., Pennycuick, C.J. and Fuller, M.R. (1988). Wind tunnel experiments to assess the effect of back-mounted radio transmitters on bird body drag. *Journal of Experimental Biology* **135**, 265–273.

O'Conner, P.J., Pyke, G.H. and Spencer, H. (1987). Radio-tracking honeyeater movements. *Emu* **87**, 249–252.

Odum, E.P. and Kuenzler, E.J. (1955). Measurement of territory and home range size in birds. *Auk* **72**, 128–137.

Ogden, J. (1985) The California condor – capture and radio telemetry. *International Council for Bird Preservation Technical Publication* **5**, 475–476.

Oldham, R.S. and Swan, M.J.S. (1992). The effects of ingested radio transmitters on *Bufo bufo* and *Rana temporaria*. *Journal of Herpetology* **2**, 82–85.

Olsen, G.H., Dein, F.J., Haramis, G.M. and Jorde, D.G. (1992). Implanting radio transmitters to wintering canvasbacks. *Journal of Wildlife Management* **56**, 325–328.

Osgood, D.W. (1970). Thermoregulation in water snakes studied by telemetry. *Copeia* (1970),568–571.

Osgood, D.W. (1980). Temperature sensitive telemetry applied to studies of small mammal activity patterns. In *A Handbook on Biotelemetry and Radio Tracking* (C.J. Amlaner and D.W. Macdonald, eds), 525–528. Pergamon Press, Oxford.

Ostfeld, R.S. (1986) Territoriality and mating system of California voles. *Journal of Animal Ecology* **55**, 691–706.

Otis, D.L. and White, G.C. (1999). Autocorrelation of location estimates and the analysis of radiotracking data. *Journal of Wildlife Management* **63**, 1039–1044.

Pages, E. (1975). Etude eco-ethologique du Pangolin *Manis tricuspis* a l'aide de la technique du radio-tracking. *Mammalia* **39**, 613–641.

Palomares, F. and Delibes, M. (1991). Assessing three methods to estimate daily activity patterns in radio-tracked mongooses. *Journal of Wildlife Management* **55**, 698–700.

Palomares, F. and Delibes, M. (1993). Determining activity types and budgets from movement speed of radio-marked mongooses. *Journal of Wildlife Management* **57**, 164–167.

Paquette, G.A., Devries, J.H., Emery, R.B., Howerter, D.W., Joynt, B.L. and Sankowski, T.P. (1997). Effects of transmitters on reproduction and survival of wild mallards. *Journal of Wildlife Management* **61**, 953–961.

Parish, T. and Kruuk, H. (1982). The uses of radio tracking combined with other techniques in studies of badger ecology in Scotland. In *Telemetric Studies of Vertebrates* (C.L. Cheeseman and R.B. Mitson, eds), 291–299. Academic Press, London.

Parker, N.C., Giorgi, A.E., Heidinger, R.C., Jester, D.B., Prince, E.D. and Winans, G.A. (1990). *Fish-marking Techniques*. American Fisheries Society, Bethesda, Maryland.

Patric, E.F., Husband, T.P., McKiel, C.G. and Sutherland, W.M. (1988). Potential of LORAN-C for wildlife research along coastal landscapes. *Journal of Wildlife Management* **52**, 162–164.

Patric, E.F. and Serenbetz, R.W. (1971). A new approach to wildlife position finding telemetry. *New York Fish and Game Journal* **18**, 1–14.

Paton, W.C., Zabel, C.J., Neal, D.L., Steger, G.N., Tilghman, N.G.and Noon, B.R. (1991). Effects of radio tags on spotted owls. *Journal of Wildlife Management* **55**, 617–622.

Patton, D.R., Beaty, D.W. and Smith, R.H. (1973). Solar panels: an energy source for radio transmitters on wildlife. *Journal of Wildlife Management* **37**, 236–238.

Pearson, N.D. (1986). Automated telemetry systems. In *Animal Telemetry in the Next Decade*, 49–50. Summaries of papers from a meeting organized by the Fisheries Laboratory, Lowestoft. Ministry of Agriculture, Fisheries and Food.

Pennycuick, C.J. (1975). Mechanics in flight. In *Avian Biology,* Vol. 5 (D.S. Farner and J.R. King, eds), 1–75. Academic Press, New York.

Pennycuick, C.J. (1978). Identification using natural markings. In *Animal Marking: Recognition Marking of Animals in Research* (B. Stonehouse, ed.), 147–159. University Park Press, Baltimore, Maryland.

Pennycuick, C.J. (1989). *Bird Flight Performance*. Oxford University Press, Oxford, UK.

Pennycuick, C.J. and Fuller, M.R. (1987). Considerations of effects of radio transmitters on bird flight. In *Biotelemetry IX* (H.P. Kimmich and M.R. Neuman, eds), 327–330.

Pennycuick, C.J., Fuller, M.R. and L. McAllister. (1989). Climbing performance of Harris' hawks (*Parabuteo unicinctus*) with added load: implications for muscle mechanics and for radiotracking. *Journal of Experimental Biology* **142**, 17–29.

Pennycuick, C.J., Schaffner, F.C., Fuller, M.R., Obrecht, H.H. and Sternberg, L. (1990). Foraging flights of the white-tailed tropicbird (*Phaethon lepturus*): radiotracking and doubly-labelled water. *Colonial Waterbirds* **13**, 96–102.

Penzel, T., Salmons, S. and Neuman, M. (eds) (1998). *Biotelemetry XIV*. Tectum Verlag, Marburg.

Perry, M.C. (1981). Abnormal behaviour of canvasbacks equipped with radio transmitters. *Journal of Wildlife Management* **45**, 786–789.

Perry, M.C., Haas, G.H. and Carpenter, J.W. (1981). Radio transmitters for mourning doves: a comparison of attachment techniques. *Journal of Wildlife Management* **45**, 524–527.

Peters, R.H. (1991). *A Critique for Ecology*. Cambridge University Press, UK.

Philo, L.M., Follman, E.H. and Reynolds, H.V. (1981). Field surgical techniques for implanting temperature-sensitive transmitters in grizzly bears. *Journal of Wildlife Management* **45**, 772–775.

Pietz, P.J., Krapu, G.L., Greenwood, R.J. and Lokemoen, J.T. (1993). Effects of harness transmitters on behaviour and reproduction of wild mallards. *Journal of Wildlife Management* **57**, 696–703.

Pollock, K.H., Winterstein, S.R., Bunck, C.M. and Curtiss, P.D. (1989). Survival analysis in telemetry studies: the staggered entry design. *Journal of Wildlife Management* **53**, 7–14.

Poulle, M.L., Artois, M. and Roeder, J.J. (1994). Dynamics of spatial relationships among members of a fox group (*Vulpes vulpes: Carnivora*). *Journal of Zoology, London* **23**, 93–106.

Priede, I.G. (1980). An analaysis of objectives in telemetry studies of fish in the natural environment. In *A Handbook on Biotelemetry and Radio Tracking* (C.J. Amlaner and D.W. Macdonald, eds), 105–118. Pergamon Press, Oxford.

Priede, I.G. (1986). Satellite systems and tracking of animals. In *Animal Telemetry in the Next Decade*, 51–56. Summaries of papers from a meeting organized by the Fisheries Laboratory, Lowestoft. Ministry of Agriculture, Fisheries and Food.

Priede, I.G. (1992). Wildlife telemetry: an introduction. In *Wildlife Telemetry – Remote Monitoring and Tracking of Animals* (I.G. Priede and S.M. Swift, eds), 3–25. Ellis Horwood, Chichester, UK.

Priede, I.G. and French, J. (1991). Tracking of marine animals by satellite. *International Journal of Remote Sensing* **12**, 667–680.

Priede, I.G. and Swift, S.M. (1992). *Wildlife Telemetry – Remote Monitoring and Tracking of Animals*. Ellis Horwood, Chichester, England.

Prince, P.A., Wood, A.G., Barton, T. and Croxall, J.P. (1992). Satellite tracking of wandering albatrosses (*Diomedea exulans*) in the South Atlantic. *Antarctic Science* **4**, 31–36.

Proud, J.C. (1969). Wild turkey studies in New York radio-telemetry. *New York Fish and Game Journal* **16**, 46–83.

Putaala, A., Oksa, J., Rintamäki, H. and Hissa, R. (1997). Effects of hand-rearing and radiotransmitters on flight of gray partridge. *Journal of Wildlife Management* **61**, 1345–1351.

Quade, D. (1969). Using weighted rankings in the analysis of complete blocks with additive block effects. *Journal of the American Statistical Association* **74**, 680–683.

Rado, R. and Terkel, J. (1989). A radio-tracking system for subterranean rodents. *Journal of Wildlife Management* **53**, 946–949.

Raim, A. (1978). A radio transmitter attachment for small passerine birds. *Bird Banding* **49**, 326–332.

Ralls, K., Siniff, D.B., Williams, T.D. and Kuechle, V.B. (1989). An intraperitoneal radio transmitter for sea otters. *Marine Mammal Sciences* **5**, 376–381.

Ramakka, J.M. (1972). Effects of radio-tagging on breeding behaviour of male woodcock. *Journal of Wildlife Management* **36**, 1309–1312.

Randolf, S.E. (1977). Changing spatial relationships in a population of *Apodemus sylvaticus* with onset of breeding. *Journal of Animal Ecology* **46**, 653–676.

Rappole, J.H. and Tipton, A.R. (1991). New harness design for attachment of radio-transmitters to small passerines. *Journal of Field Ornithology* **62**, 335–337.

Rasmussen, D.R. and Rasmussen, K.L. (1979). Social ecology of adult males in a confined troop of Japanese macacques (*Macaca fuscata*). *Animal Behaviour* **27**, 434–445.

Rawson, K.S. and Hartline, P.H. (1964). Telemetry of homing behaviour by the deermouse, Peromyscus. *Science* **146**, 1596–1598.

Reading, C.J. and Davies, J.L. (1996). Predation by grass snakes (*Natrix natrix* L.) at a site in southern England. *Journal of Zoology* **239**, 73–82.

Recht, M.A. (1992). The role of the photosensitive transmitter in wildlife behavioural studies. In *Wildlife Telemetry – Remote Monitoring and Tracking of Animals* (I.G. Priede and S.M. Swift, eds), 106–110. Ellis Horwood, Chichester, UK.

Redpath, S.M. (1995). Habitat fragmentation and the individual: tawny owls *Strix aluco* in woodland patches. *Journal of Animal Ecology* **64**, 652–661.

Reid, D.G., Melquist, W.E., Woolington, J.D. and Noll, J.M. (1986). Reproductive effects of intraperitoneal transmitter implants in river otters. *Journal of Wildlife Management* **50**, 92–94.

Rempel, R.S. and Rodgers, A.R. (1997). Effects of differential correction on accuracy of a GPS location system. *Journal of Wildlife Management* **61**, 525–530.

Rempel, R.S., Rodgers, A.R. and Abraham, K.F. (1995). Performance of a GPS animal location system under boreal forest canopy. *Journal of Wildlife Management* **59**, 543–551.

Reynolds, T.D. and Laundré, J.W. (1990). Time intervals for estimating pronghorn and coyote home ranges and daily movements. *Journal of Wildlife Management* **54**, 316–322.

Ricci, J.-C. and Vogel, P. (1984). Nouvelle methode d'etude en nature des relations spatiales et sociales chez *Crocidura russula* (*Mammalia, Soricidae*). *Mammalia* **48**, 281–286.

Riley, J.R., Smith, A.D., Reynolds, D.R., Edwards, A.S., Osborne, J.L. Williams, I.H., Carreck, N.L. and Poppy, G.M. (1996). Tracking bees with harmonic radar. *Nature* **379**, 29–30.

Riley, T.Z., Clark, W.R., Ewing, D.E. and Vohs, P.A. (1998). Survival of ring-necked pheasant chicks during rearing. *Journal of Wildlife Management* **62**, 36–44.

Robertson, P.A., Aebischer, N.J., Kenward, R.E., Hanski, I.K. and Williams, N.P. (1998). Simulation and jack-knifing assessment of home-range indices based on underlying trajectories. *Journal of Applied Ecology* **35**, 928–940.

Robinson, B.J. (1986). Data storage tags. In *Animal Telemetry in the Next Decade*, 57–59. Summaries of papers from a meeting organized by the Fisheries Laboratory, Lowestoft. Ministry of Agriculture, Fisheries and Food.

Rooney, S.M., Wolfe, A. and Hayden, T.J. (1998). Autocorrelated data in telemetry studies: time to independence and the problem of behavioural effects. *Mammal Review* **28**, 89–98.

Rotella, J.J., Howerter, D.W., Sankowski, T.P. and Devries, J.H. (1993). Nesting effort of wild mallard with 3 types of radio transmitters. *Journal of Wildlife Management* **57**, 690–695.

Saltz, D. (1994). Reporting error measures in radio location by triangulation: a review. *Journal of Wildlife Management* **58**, 181–184.

Saltz, D. and Alkon, P.U. (1985). A simple computer-aided method or estimating radio-location error. *Journal of Wildlife Management* **49**, 664–668.

Saltz, D. and White, G.C. (1990). Comparison of different measures of the error in simulated radio-telemetry locations. *Journal of Wildlife Management* **54**, 169–174.

Samuel, M.D. and Fuller, M.R. (1994). Wildlife radiotelemetry. In *Research and Management Techniques for Wildlife Habitats,* 5th edn (T.A. Bookhout, ed.), 370–418. The Wildlife Society, Bethesda, Maryland.

Samuel, M.D. and Garton, E.O. (1985). Identifying areas of concentrated use within the home range. *Journal of Wildlife Management* **49**, 513–519.

Samuel, M.D. and Green, R.E. (1988). A revised test procedure for identifying core areas within the home range. *Journal of Animal Ecology* **57**, 1067–1068.

Samuel, M.D., Pierce, D.J. and Garton, E.O. (1985). Identifying areas of concentrated use within the home range. *Journal of Animal Ecology* **54**, 711–719.

Samuel, M.D. and Kenow, K.P. (1992). Evaluating habitat selection with radio-telemetry triangulation error. *Journal of Wildlife Management* **57**, 725–734.

Samuel, M.D., Rusch, D.H. and Craven, S.R. (1990). Influence of neck bands on recovery and survival rates of Canada geese. *Journal of Wildlife Management* **54**, 45–54.

Sanderson, G.C. (1966). The study of mammal movements – a review. *Journal of Wildlife Management* **30**, 215–235.

Sanderson, G.C. and Sanderson, B.C. (1964). Radio-tracking rats in Malaya – a preliminary study. *Journal of Wildlife Management* **28**, 752–768.

Sawby, S.W. and Gessamen, J.A. (1974). Telemetry of electrocardiograms from free-living birds: a method of electrode placement. *Condor* **76**, 479–481.

Sayre, M.W., Baskett, T.S. and Blenden, P.B. (1981). Effects of radio tagging on breeding behaviour of mourning doves. *Journal of Wildlife Management* **45**, 428–434.

Schmidt, D.F., Shaffery, J.P., Ball, N.J., Loenneke, D. and Amlaner, C.J. (1989). Electrophysiological sleep characteristics in bobwhite quail. In *Biotelemetry X* (C.J. Amlaner, ed.), 339–344. University of Arkansas Press, Fayetteville.

Schober, F., Bogel, R., Bugnar, R.M., Burchard, D., Fluch, G. and Rohde, N. (1992). Automatic direction finding and location system based on Doppler effect. *Biotelemetry XII*, 145–155.

Schober, F., Bugnar, W.M. and Wagner, J. (1989). A software package for acquisition and evaluation of biotelemetry data from domestic and wild animals. In *Biotelemetry X* (C.J. Amlaner, ed.), 700–708. University of Arkansas Press, Fayetteville.

Schober, F. and Oery, B. (1987). Automatic RF receiving system for carrier frequency pulses. In *Biotelemetry IX*. (H.P. Kimmich and M.R. Neuman, eds), 351–354.

Schoener, T.W. (1968). Sizes of feeding territories among birds. *Ecology* **49**, 123–141.

Schoener, T.W. (1981). An empirically based estimate of home range. *Theoretical Population Biology* **20**, 281–325

Schooley, R.L. (1994). Annual variation in habitat selection: patterns concealed by pooled data. *Journal of Wildlife Management* **58**, 367–374.

Schubauer, J.P. (1981). A reliable radio-telemetry tracking system suitable for studies of chelonians. *Journal of Herpetology* **15**, 117–120.

Schulz, J.H., Bermudez, A.J., Tomlison, J.L., Firman, J.D. and Zhuoqiong, H. (1998). Effects of implanted radio transmitters on captive mourning doves. *Journal of Wildlife Management* **62**, 1451–1460.

Schwartz, A., Weaver, J.D., Scott, N.R. and Cade, T.J. (1977). Measuring the temperature of eggs during incubation under captive falcons. *Journal of Wildlife Management* **41**, 12–17.

Schweinsberg, R.E. and Lee, L.J. (1982). Movement of four satellite-monitored polar bears in Lancaster Sound, Northwest Territories. *Arctic* **35**, 504–511.

Scott, D.K. (1978). Identification of individual Bewick's swans by bill patterns. In *Animal Marking: Recognition Marking of Animals in Research* (B. Stonehouse, ed.), 160–168. University Park Press, Baltimore, Maryland.

Seaman, D.E. and Powell, R.A. (1996). An evaluation of the accuracy of kernel density estimators for home range analysis. *Ecology* **77**, 2075–2085.

Seaman, D.E., Griffith, B. and Powell, R.A. (1998). KERNELHR: a program for estimating animal home ranges. *Wildlife Society Bulletin* **26**, 95–100.

Seaman, D.E., Millspaugh, J.J., Kernohan, B.J., Brundige, G.C., Raedeke, K.J. and Gitzen, R.A. (1999). Effects of sample size on kernel home range estimates. *Journal of Wildlife Management* **63**, 739–747.

Seber, G.A.F. (1982). *Estimation of Animal Abundance and Related Parameters*, 2nd edn Griffin, London.

Seegar, W.S., Cutchis, P.N., Fuller, M.R, Suter, J.J., Bhatnagar, V. and Wall, J.S. (1996). Fifteen years of satellite tracking development and application to wildlife research and conservation. *John Hopkins Advanced Physics Laboratory Technical Digest* **17**, 305–315.

Seidensticker, J.C., Hornocker, M.G., Knight, R.R. and Judd, S.L. (1970). Equipment and Techniques for Radiotracking Mountain Lions and Elk. University of Idaho Forest and Range Experimental Station Bulletin 6.

Seitz, A., Faller-Doepner, U. and Reh. W. (1992). Radio-tracking of the common frog (*Rana temporaria*). In *Wildlife Telemetry – Remote Monitoring and Tracking of Animals* (I.G. Priede and S.M. Swift, eds), 484–489. Ellis Horwood, Chichester, UK.

Ser? heen, C., Thier, T.T., Jonkel, C.J. and Beaty, D. (1981). An ear-mounted transmitter for bears. *Wildlife Society Bulletin* **9**, 56–57.

Seton, E.T. (1909). *Life-histories of Northern Animals. An Account of the Mammals of Manitoba*. Charles Schribner's Sons, New York.

Sibly, R.M. and McCleery, R.H. (1980). Continuous observation of individual herring gulls during the incubation season using radio tags: an evaluation of the technique and a cost-benefit analysis of transmitter power. In *A Handbook on Biotelemetry and Radio Tracking* (C.J. Amlaner and D.W. Macdonald, eds), 345–352. Pergamon Press, Oxford.

Siegel, S. (1957). *Non-Parametric Statistics*. McGraw-Hill, New York.

Siegfried, W.R., Frost, P.G.H., Ball, I.J. and McKinney, D.F. (1977). Effects of radio packages on African black ducks. *South African Journal of Wildlife Research* **7**, 37–40.

Sievert, P.R. and Keith, L.B. (1985). Survival of snowshoe hares at a geographic range boundary. *Journal of Wildlife Management* **49**, 854–866.

Silverman, B.W. (1986). *Density Estimation for Statistics and Data Analysis.* Chapman and Hall, London, UK.

Simpson, V.R., Walls, S.S., Cooper, J.E. and Kenward, R.E. (1997). Causes of mortality in radio-tracked Eurasian buzzards (*Buteo buteo*) in Dorset. In *Proceedings of the 4th Conference of the European Chapter of the Association of Avian Veterinarians* (B. Stockdale, ed.), 188–193. Association of Avian Veterinarians, Loughborough, UK.

Siniff, D.B. and Tester, J.R. (1965). Computer analysis of animal movement data obtained by telemetry. *BioScience* **15**, 104–108.

Siniff, D.B., Tester, J.R. and Keuchle, V.B. (1969). Population studies of Weddell seals at McMurdo Station. *Antarctic Journal of the United States* **4**, 120–121.

Skirnisson, K. and Feddersen, D. (1985). Erfahrungen mit der Implantation von Sendern bei freilebenden Steinmardern. *Zeitschrift für Jagdwissenschaft* **30**, 228–235.

Small, R.J. and Rusch, D.H. (1985). Backpacks versus ponchos: survival and movements of radio-marked ruffed grouse. *Wildlife Society Bulletin* **13**, 163–165.

Small, R.J. and Rusch, D.R. (1989). The natal dispersal of ruffed grouse. *Auk* **106**, 72–79.

Smith, A. and Willebrand, T. (1999). Mortality causes and survival rates of hunted and unhunted willow grouse. *Journal of Wildlife Management* **63**, 722–730.

Smith, E.N. (1974). Multichannel temperature and heart rate radio-telemetry transmitter. *Journal of Applied Physiology* **36**, 252–255.

Smith, E.N. (1980). Physiological radio telemetry of vertebrates. In *A Handbook on Biotelemetry and Radio Tracking* (C.J. Amlaner and D.W. Macdonald, eds), 45–55. Pergamon Press, Oxford.

Smith, E.N. and Moore, S.E. (1989). Inexpensive magnetically switched temperature and egg biotelemetry system. In *Biotelemetry X* (C.J. Amlaner, ed.) 125–130. University of Arkansas Press, Fayetteville, Arkansas, USA.

Smith, E.N. and Worth, D.J. (1980). Atropine effect on fear bradycardia of the eastern cottontail rabbit. In *A Handbook on Biotelemetry and Radio Tracking* (C.J. Amlaner and D.W. Macdonald, eds), 549–555. Pergamon Press, Oxford.

Smith, H.R. (1980). Growth, reproduction and survival in *Peromyscus leucopus* carrying intraperitoneally implanted transmitters. In *A Handbook on Biotelemetry and Radio Tracking* (C.J. Amlaner and D.W. Macdonald, eds), 367–374. Pergamon Press, Oxford.

Smith, R.M. and Trevor-Deutsch, B. (1980). A practical, remotely-controlled, portable radio telemetry receiving apparatus. In *A Handbook on Biotelemetry and Radio Tracking* (C.J. Amlaner and D.W. Macdonald, eds), 269–273. Pergamon Press, Oxford.

Smith, S.G., Skalski, J.R., Schlechte, J.W., Hoffmann, A. and Cassen, V. (1994). *SURPH.1. Manual. Statistical Survival Analysis for Fish and Wildlife Tagging Studies.* Bonneville Cower Administration Division of Fish and Wildlife, Portland, Oregon.

Smits, A.W. (1984). Activity patterns and thermal biology of the toad *Bufoboreas halophilus. Copeia* (1984), 689–696.

Snyder, N.F.R., Beissinger, S.R. and Fuller, M.R. (1989). Solar radio-transmitters on snail-kites in Florida. *Journal of Field Ornithology* **60**, 171–177.

Snyder, W.D. (1985). Survival of radio-marked hen ring-necked pheasants in Colorado. *Journal of Wildlife Management* **49**, 1044–1050.

Sodhi, N.S., Warkentin, I.G., James, P.C. and Oliphant, L.W. (1991). Effects of radiotagging on breeding merlins. *Journal of Wildlife Management* **55**, 613–616.

Solomon, D.J. and Storeton-West, T.J. (1983). *Radio Tracking of Migratory Salmonids in Rivers: Development of an Effective System.* Ministry of Agriculture, Fisheries and Food, Fisheries Research Technical Report 75.

Sorenson, M.D. (1989). Effects of neck collar radios on female redheads. *Journal of Field Ornithology* **60**, 523–528.

Southwood, T.R.E. (1966). *Ecological Methods*. Methuen, London.

Spencer, W.D. and Barrett, R.H. (1984). An evaluation of the harmonic mean measure for defining carnivore activity areas. *Acta Zoologica Fennica* **171**, 255–259.

Spencer, H.J., Lucas, G. and O'Conner, P.J. (1987). A remotely switched passive null-peak network for animal tracking and radio direction finding. *Australian Wildlife Research* **14**, 311–317.

Spencer, H.J. and Savaglio, F.P. (1996). An automatic small animal radio-tracking system employing spread-spectrum concepts. In *Biotelemetry XIII* (C. Cristalli, C.J. Amlaner and M.R. Neuman, eds), 195–191. Williamsburg.

Springer, J.T. (1979). Some sources of bias and sampling error in radio triangulation. *Journal of Wildlife Management* **43**, 926–935.

Standora, E.A. (1977). An eight-channel radio telemetry system to monitor alligator body temperature in a heated reservoir. In *Proceedings of the First International Conference on Wildlife Biotelemetry* (F.M. Long, ed.), 70–78. University of Wyoming.

Standora, E.A., Spotila, J.R., Keinath, J.A. and Shoop, C.R. (1984). Body temperatures, diving cycles, and movement of a subadult leatherback turtle, *Dermochelys coriacea*. *Herpetologica* **40**, 169–176.

Stasko, A.B. and Pincock, D.G. (1977). Review of underwater biotelemetry, with emphasis on ultrasonic techniques. *Journal of the Fisheries Research Board of Canada* **34**, 1262–1285.

Stebbings, R.E. (1982). Radio tracking greater horseshoe bats with preliminary observations on flight patterns. In *Telemetric Studies of Vertebrates* (C.L. Cheeseman and R.B. Mitson, eds), 161–173. Academic Press, London.

Stebbings, R.E. (1986). Bats. In *Animal Telemetry in the Next Decade*, 20–22. Summaries of papers from a meeting organized by the Fisheries Laboratory, Lowestoft. Ministry of Agriculture, Fisheries and Food.

Stickel, L.F. (1954). A comparison of certain methods of measuring ranges of small mammals. *Journal of Mammalogy* **35**, 1–15.

Stoddart, L.C. (1970). A telemetric method of detecting jackrabbit mortality. *Journal of Wildlife Management* **34**, 501–507.

Stohr, W. (1989). Long term heart rate telemetry in small mammals. In *Biotelemetry X* (C.J. Amlaner, ed.), 352–375. *University of Arkansas Press*, Fayetteville.

Stonehouse, B. (ed.) (1978). *Animal Marking: Recognition Marking of Animals in Research*. Macmillan, London.

Storm, G.L., Andrews, R.D., Phillips, R.L., Bishop, R.A., Siniff, D.B. and Tester, J.R. (1976). Morphology, reproduction, dispersal and mortality of mid-western red fox populations. *Wildlife Monographs* **49**, 1–82.

Stouffer, R.H., Gates, J.E., Holutt, L.H. and Stauffer, J.R. (1983). Surgical implantation of a transmitter package for radio-tracking endangered hellbenders. *Wildlife Society Bulletin* **11**, 384–386.

Strathearn, S.M., Lotimer, J.S., Kolenosky, G.B. and Lintack, W.M. (1984). An expanding break-away collar for black bear. *Journal of Wildlife Management* **48**, 939–942.

Strikwerda, T.E., Fuller, M.R., Seegar, W.S., Howey, P.W. and Black, H.D. (1986). *John Hopkins Advanced Physics Laboratory Technical Digest* **7**, 203–208.

Stüwe, M. and Blohowiak, C.E. (1987). *MCPAAL, Microcomputer Programs for the Analysis of Animal Locations*. Conservation and Research Center, National Zoological Park, Smithsonian Institution, Front Royal, Vancouver.

Sutherland, W.J. (1996). *From Individual Behaviour to Population Ecology*. Oxford University Press, UK.

Swanson, G.A. and Keuchle, V.B. (1976). A telemetry technique for monitoring waterfowl activity. *Journal of Wildlife Management* **40**, 187–189.

Swenson, J.E., Wallin, K., Ericsson, G., Cederlund, G. and Sandegren, F. (1999). Effects of ear-tagging with radiotransmitters on survival of moose calves. *Journal of Wildlife Management* **63**, 354–358.

Swihart, R.K. and Slade, N.A. (1985a). Testing for independence of observations in animal movements. *Ecology* **66**, 1176–1184.

Swihart, R.K. and Slade, N.A. (1985b). Influence of sampling interval on estimates of home-range size. *Journal of Wildlife Management* **49**, 1019–1025.

Swihart, R.K. and Slade, N.A. (1986). The importance of statistical power when testing for independence of animal movements. *Ecology* **67**, 255–258.

Swihart, R.K. and Slade, N.A. (1988). Relating body size to the rate of home range use in mammals. *Ecology* **69**, 393–399.

Swihart, R.K. and Slade, N.A. (1997). On testing for independence of animal movements. *Journal of Agricultural, Biological and Environmental Statistics* **2**, 1–16.

Swingland, I.R. and Frazier, J.G. (1980). The conflict between feeding and overheating in the Aldabran giant tortoise. In *A Handbook on Biotelemetry and Radio Tracking* (C.J. Amlaner and D.W. Macdonald, eds), 269–273. Pergamon Press, Oxford.

Sykes, P.W., Carpenter, J.W., Holzman, S. and Geissler, P.H. (1990). Evaluation of three miniature radio transmitter attachment methods for small passerines. *Wildlife Society Bulletin* **18**, 41–48.

Taberlet, P. and Bouvet, J. (1992). Bear conservation genetics. *Nature* **358**, 197.

Taillade, M. (1992). Animal tracking by satellite. In *Wildlife Telemetry – Remote Monitoring and Tracking of Animals* (I.G. Priede and S.M. Swift, eds), 149–160. Ellis Horwood, Chichester, UK.

Taulman, J.F., Smith, K.G. and Thill, R.E. (1998). Demographic and behavioral responses of southern flying squirrels to experimental logging in Arkansas. *Ecological Applications* **8**, 1144–1155.

Taylor, I.R. (1991). Effects of nest inspections and radiotagging on barn owl breeding success. *Journal of Wildlife Management* **55**, 312–315.

Taylor, K.D. and Lloyd, H.G. 1978. The design, construction and use of a radio-tracking system for some British mammals. *Mammal Review* **8**, 117–141.

Tester, J.R. (1971). Interpretation of ecological and behavioural data on wild animals obtained by telemetry with special reference to errors and uncertainties. *Proceedings of a Symposium on Biotelemetry*, 385–408. CSER, Pretoria, South Africa.

Thirgood, S.J. (1995). The effects of sex, season and habitat availability on patterns of habitat use by fallow deer (*Dama dama*). *Journal of Zoology*, **235**, 645–659.

Thirgood, S.J. and Redpath, S.M. (1997). Red grouse and their predators. *Nature* **390**, 547.

Thomas, D.W. (1980). Plans for a lightweight inexpensive radio transmitter. In *A Handbook on Biotelemetry and Radio Tracking* (C.J. Amlaner and D.W. Macdonald, eds), 175–179. Pergamon Press, Oxford.

Thomas, D.L. and Taylor, E.J. (1990). Study design and tests for comparing resource use and availability. *Journal of Wildlife Management* **54**, 322–330.

Timko, R.E. and Kolz, A.L. (1982). Satellite sea turtle tracking. *Marine Fisheries Review* **44**, 19–24.

Tonkin, J.M. (1983). Ecology of the red squirrel (*Sciurus vulgaris*) in mixed woodland. PhD Thesis, University of Bradford.

Trent, T.T. and Rongstad, O.J. (1974). Home range and survival of cottontail rabbits in southwestern Wisconsin. *Journal of Wildlife Management* **38**, 459–472.

Troy, S., Coulson, G. and Middleton, D. (1992). A comparison of radio-tracking and line transect techniques to determine habitat preferences in the swamp wallaby (Wallabia bicolor) in south-eastern Australia. In *Wildlife Telemetry – Remote Monitoring and Tracking of Animals* (I.G. Priede and S.M. Swift, eds), 651–660. Ellis Horwood, Chichester, UK.

Tufto, J., Andersen, R. and Linnell, J. (1996). Habitat use and ecological correlates of home range size in a small cervid: the roe deer. *Journal of Animal Ecology* **65**, 715–724.

Tyack, A.J., Walls, S.S. and Kenward, R.E. (1998). Behaviour in the post-nestling dependence period of radio-tagged common buzzards. *Ibis* **140**, 58–63.

Uchiyama, A. and Amlaner, C.J. (eds) (1991). *Biotelemetry XI*. Macmillan, Tokyo.

Uda, S. and Mushiake, Y. (1954). *Yagi-Uda Antenna*. Sasaki Printing and Publishing, Sendai, Japan (cited in Beaty and Swapp, 1978).

VanVuren, D. (1989). Effects of intraperitoneal transmitter implants on yellow-bellied marmots. *Journal of Wildlife Management* **53**, 320–323.

Van Winkle, W. (1975). Comparison of several probabalistic home-range models. *Journal of Wildlife Management* **39**, 118–123.

Vekasy, M.S., Marzluff, J.M., Kochert, M.N., Lehmann, R.N. and Steenhof, K. (1996). Influence of radio transmitters on prairie falcons. *Journal of Field Ornithology* **67**, 680–690.

Verts, B.J. (1963). Equipment and techniques for radiotracking striped skunks. *Journal of Wildlife Management* **27**, 325–339.

Voight, D.R. and Broadfoot, J. (1983). Locating pup-rearing dens of red foxes with radio-equipped woodchucks. *Journal of Wildlife Management* **47**, 858–859.

Voight, D.R. and Tinline, R.R. (1980). Strategies for analysing radio-tracking data. In *A Handbook on Biotelemetry and Radio Tracking* (C.J. Amlaner and D.W. Macdonald, eds), 387–404. Pergamon Press, Oxford.

Walker, A.F. and Walker, A.M. (1992). The Little Gruinard Atlantic salmon (*Salmo salar* L.) catch and release tracking study. In *Wildlife Telemetry – Remote Monitoring and Tracking of Animals* (I.G. Priede and S.M. Swift, eds), 434–353. Ellis Horwood, Chichester, UK.

Wallace, M.P. and Temple, S.A. (1987). Releasing captive-reared Andean condors to the wild. *Journal of Wildlife Management* **51**, 541–550.

Wallace, M.P., Fuller, M. and Wiley, J. (1994). Patagial transmitters for large vultures and condors. In *Raptor Conservation Today* (B.-U. Meyburg and R.D. Chancellor, eds), 381–387. World Working Group on Birds of Prey, Berlin.

Walls, S.S. and Kenward, R.E. (1994). The systematic study of radio-tagged raptors: II Sociality and dispersal. In *Raptor Conservation Today*. (B.-U. Meyburg and R.D. Chancellor, eds) 317–324. World Working Group on Birds of Prey, Berlin.

Walls, S.S. and Kenward, R.E. (1998). Movements of radio-tagged common buzzards in early life. *Ibis* **140**, 561–568.

Wanless, S., Harris, M.P. and Morris, J.A. (1988). The effect of radio transmitters on the behavior of common mures and razorbills during chick rearing. *Condor* **90**, 816–823.

Wanless, S., Harris, M.P. and Morris, J.A. (1989). Behaviour of alcids with tail-mounted radio transmitters. *Colonial Waterbirds* **12**, 158–163.

Wanless, S., Harris, M.P. and Morris J.A. (1991). Foraging range and feeding locations of shags *Phalacrocorax aristotelis* during chick rearing. *Ibis* **133**, 30–36.

Ward, D.H. and Flint, P.L. (1995). Effects of harness-attached transmitters on premigration and reproduction of brant. *Journal of Wildlife Management* **59**, 39–46.

Warner, R.E. and Etter, S.L. (1983). Reproduction and survival of radio-marked hen ring-necked pheasants. *Journal of Wildlife Management* **47**, 369–375.

Watkins, W.A., Moore, K.E., Wartzok, D. and Johnson, J.H. (1981). Radio-tracking of finback (*Balaenoptera physalus*) and humpback (*Megaptera novaeangliae*) whales in Prince William Sound, Alaska. *Deep-Sea Research* **28A**, 577–588.

Wauters, L. and A.A. Dhondt. (1992). Spacing behaviour of red squirrels, *Sciurus vulgaris*: variation between habitats and the sexes. *Animal Behaviour* **43**, 297–311.

Weatherhead, P.J. and Anderka, F.W. (1984). An improved radio transmitter and implantation technique for snakes. *Journal of Herpetology* **18**, 264–269.

Weber, J.-M. and Meia, J.S. (1992). The use of expandable radio collars for radio-tracking fox cubs. In *Wildlife Telemetry – Remote Monitoring and Tracking of Animals* (I.G. Priede and S.M. Swift, eds), 698–700. Ellis Horwood, Chichester, UK.

Weeks, R.W., Long, F.M., Lindsay, J.E., Bailey, R., Patula, D and Green, M. (1977). Fish tracking from the air. *Proceedings of the First International Conference on Wildlife Telemetry*, 63–69.

Wheeler, W.E. (1991). Suture and glue attachment of radio transmitters on ducks. *Journal of Field Ornithology* **62**, 271–278.

White, G.C. (1983). Numerical estimation of survival rates from band-recovery and biotelemetry data. *Journal of Wildlife Management* **47**, 716–728.

White, G.C. and Garrott, R.A. (1990). *Analysis of Wildlife Radio-tracking Data*. Academic Press, New York, USA.

Whitehouse, S. (1980). Radio tracking in Australia. In *A Handbook on Biotelemetry and Radio Tracking* (C.J. Amlaner and D.W. Macdonald, eds), 733–739. Pergamon Press, Oxford.

Whitehouse, S.J.O. and Steven, D. (1977). A technique for aerial radio tracking. *Journal of Wildlife Management* **41**, 771–775.

Widèn, P. (1982). Radio monitoring the activity of goshawks. In *Telemetric Studies of Vertebrates* (C.L. Cheeseman and R.B. Mitson, eds), 153–160. Academic Press, London.

Willebrand, T. and Marcström, M. (1988). On the danger of using dummy nests to study predation. *Auk* **105**, 378–379.

Williams, T.C. and Williams, J.M. (1967). Radio tracking of homing bats. *Science* **155**, 1435–1436.

Williams, T.C. and Williams, J.M. (1970). Radiotracking of homing and feeding flights of a neotropical bat, *Phyllostomus hastatus*. *Animal Behaviour* **18**, 302–309.

Wilson, R.P. and Culik, D.M. (1992). Packages on penguins and device induced data. In *Wildlife Telemetry – Remote Monitoring and Tracking of Animals* (I.G. Priede and S.M. Swift, eds), 573–580. Ellis Horwood, Chichester, UK.

Wilson, R.P., Ducamp, J.-J., Rees, W.G., Culik, B.M. and Niekamp, K. (1992). Estimation of location: global coverage using light intensity. In *Wildlife Telemetry – Remote Monitoring and Tracking of Animals* (I.G. Priede and S.M. Swift, eds), 131–143. Ellis Horwood, Chichester, UK.

Wilson, R.P., Spairani, H.J., Coria, N.R., Culik, B.M. and Adelung, D. (1990). Packages for attachment to seabirds: what color do Adelie penguins like least? *Journal of Wildlife Management* **54**, 447–451.

Wilson, R.P. and Wilson, M.P. (1989). Dead reckoning – a new technique for determining penguin movements at sea. *Meeresforschung-reports on Marine Research* **32**, 155–158.

Wilson, W.L., Montgomery, W.I. and Elwood, R.W. (1992). Range use in female woodmice (*Apodemus sylvaticus*) in deciduous woodland. In *Wildlife Telemetry – Remote Monitoring and Tracking of Animals* (I.G. Priede and S.M. Swift, eds), 549–560. Ellis Horwood, Chichester, UK.

Winter, J.D., Keuchle, V.B., Siniff, D.B. and Tester, J.R. (1978). *Equipment and Methods for Tracking Freshwater Fish*. Agricultural Experimental Station of the University of Minnesota Miscellaneous Report 152.

Woakes, A.J. (1992). An implantable data logging system for heart rate and body temperature. In *Wildlife Telemetry – Remote Monitoring and Tracking of Animals* (I.G. Priede and S.M. Swift, eds), 120–127. Ellis Horwood, Chichester, UK.

Woakes, A.J. and Butler, P.J. (1975). An implantable transmitter for monitoring heart rate and respiratory frequency in diving ducks. In *Biotelemetry II*, 153–160.

Woakes, A.J. and Butler, P.J. (1989). Wildlife studies in the laboratory. In *Biotelemetry X* (C.J. Amlaner, ed.), 317–324. University of Arkansas Press, Fayetteville.

Wolcott, T.G. (1977). Optical tracking and telemetry for nocturnal field studies. *Journal of Wildlife Management* **41**, 309–312.

Wolcott, T.G. (1980). Heart rate telemetry using micropower integrated circuits. In *A Handbook on Biotelemetry and Radio Tracking* (C.J. Amlaner and D.W. Macdonald, eds), 279–286. Pergamon Press, Oxford.

Wooley, J.B. and Owen, R.B. (1978). Energy costs of activity and daily energy expenditure in the black duck. *Journal of Wildlife Management* **42**, 739–745.

Worton, B.J. (1987). A review of models of home range for animal movement. *Ecological Modelling* **38**, 277–298.

Worton, B.J. (1989). Kernel methods for estimating the utilization distribution in home-range studies. *Ecology* **70**, 164–168.

Worton, B.J. (1995a). A convex hull-based estimator of home-range size. *Biometrics* **51**, 1206–1215.

Worton, B.J. (1995b). Using MonteCarlo simulation to evaluate kernel-based home-range estimators. *Journal of Wildlife Management* **59**, 794–800.

Wray, S., Cresswell, W.J. and Rogers, D. (1992a). Dirichlet tesselations: a new, non-parametric approach to home range analysis. In *Wildlife Telemetry – Remote Monitoring and Tracking of Animals* (I.G. Priede and S.M. Swift, eds), 247–255. Ellis Horwood, Chichester, U.K.

Wray, S., Cresswell, W.J., White, P.C.L. and Harris, S. (1992b). What, if anything, is a core area? An analysis of the problems of describing internal range configurations. In *Wildlife Telemetry – Remote Monitoring and Tracking of Animals* (I.G. Priede and S.M. Swift, eds), 256–271. Ellis Horwood, Chichester, U.K.

Yerbury, M.J. (1980). Long range tracking of *Crocodylus porosus* in Arnhem Land, Northern Australia. In *A Handbook on Biotelemetry and Radio Tracking* (C.J. Amlaner and D.W. Macdonald, eds), 765–776. Pergamon Press, Oxford.

Zar, J.H. (1974). *Biostatistical Analysis*. Prentice-Hall, Englewood Cliffs, USA.

Zicus, M.C., Schultz, D.F. and Cooper, J.A. (1983). Canada goose mortality from neckband icing. *Wildlife Society Bulletin* **11**, 70–71.

Ziesemer, F. (1981). Methods of assessing predation. In *Understanding the Goshawk* (R.E. Kenward and I.M. Lindsay, eds), 144–151. International Association of Falconry and Conservation of Birds of Prey, Oxford.

Zinnel, K.A. and Tester, J.R. (1984). Non-intrusive monitoring of plains pocket gophers. *Bulletin of the Ecological Society of America* **65**, 166.

Glossary

In this glossary of the main technical terms used throughout the book, a reference to another item in the glossary is in **bold** text.

Activity centre: a single pair of coordinates estimated for a set of animal **locations**, typically as an arithmetic mean, geometric mean, **harmonic mean** or median of their x and y values.

Adcock antenna: a pair of **dipole antennas** coupled in phase in an H shape, typically used for **direction finding**.

Adequate accuracy: the estimation of a **location** for a radio-tagged animal by moving so that the estimated precision of the location is less than the desired **tracking resolution**.

ASCII, American Standard Code for Information Interchange: is used to turn bytes in a digital file into text characters.

Ambiguous antenna: an antenna that cannot discriminate between a **true-bearing** and a **back-bearing** at 180° to the **true-bearing**.

Antenna accuracy: the expected spread of estimates when measuring bearings with an antenna; usually the standard deviation between a sample of estimated **bearings** and the **true-bearing**.

Autocorrelation: an analysis that estimates an index of spatiotemporal relatedness as a function of time intervals in a set of **locations**.

Availability (of a resource): a measure of habitat, food or other resource within an area on a map, a **home range** outline or an **opportunity circle**.

Azimuth: **bearing**.

Back-bearing or reverse-bearing: a bearing at 180° to the **true-bearing**.

Balun match: a length of conducting wire used to bring about matching of **phase** in a composite antenna.

Band or ring: a circular strip of metal or plastic attached round the leg of a bird for identification.

Bandwidth: the number of radio frequency units between, for example, the upper and lower limits of an allocated **RF** band.

Bearing: the angle between 0° and the estimated direction from a receiver to a transmitter; the **true-bearing** is the real angle from the receiver coordinates to the transmitter coordinates.

Boundary strip: an area of half the **tracking resolution** that is added to a **minimum convex polygon** to compensate for tracking inaccuracy; on a single **location**, it creates a **grid cell**.

Box-counting: the summation of **locations** that occur in rectangular areas, typically from the size of a **grid cell** up to several multiples of that minimum size.

Brown-out: the malfunction of a circuit as a result of low voltage, typically with unpredictable consequences for operation of a **microcontroller**.

Cell capacity: the energy capacity of a battery, measured either in Ampere-hours at a given voltage or in Watt-hours.

Cell life: the duration of operation of a **radio tag** with a particular type of battery.

Close approach: radio-tracking to within the minimum **detection distance** of the **radio tag** with a given receiver.

Cluster analysis: the creation of groupings among **locations** according to the distances between them in two or more dimensions.

Cluster polygons: **minimum convex polygons** round separate groups of **locations** defined by **cluster analysis**.

Compositional analysis: a method for data in proportions, typically for habitats, that avoids the **unit-sum-constraint** and any assumption that observations are **statistically independent**.

Continuous monitoring: **radio tracking** a single animal to record its **location** at very frequent intervals or its every movement.

Crystal: a thin slice of quartz that is cut to vibrate at a required frequency.

CSV (comma-separated-value): data in **ASCII** text with each value separated by a comma.

Decibel (dB): a logarithmic measure used either as a ratio (e.g. a three-element **Yagi antenna** has 6–7 dB more gain than a dipole) or as a measure of power in dBW or dBm (dB-milliWatts).

Density contouring: plotting contours across a **density matrix** estimated from a set of **locations** by a **harmonic mean** or other **kernel functions**.

Density matrix: a set of indices of **location density** at nodes, usually on a rectangular grid, that extends across and beyond a set of **home range** locations.

Detection distance, detection range: the distance at which a given **radio tag** can be detected by a given **receiving system** under specified conditions.

Differential correction: the application of a compensation to **location** estimates from the Navstar **GPS** system to correct for errors introduced for military security purposes.

Diffraction: when an electromagnetic radiation is deflected round the edge of an object; this can create a pattern of **interference** with radiation on a different path from the same source.

Dipole antenna: a linear antenna, with length a precise fraction of the required wavelength, that is coupled in the middle to a transmitter or receiver.

Direction-finding: the estimation of **bearings** from **receiving systems** to **radio tags**.

Discriminant function analysis: the use of a number of independent variables to separate cases into two or more categories of dependent variable.

Dispersal: movement without immediate return from an area repeatedly traversed by an animal.

Doppler principle: the use of a shift in frequency of radiation, for which the frequency of emission is known, to estimate the relative speed of the emitter and receiver.

DXF: a widely-used **ASCII** Data eXchange Format developed by Autocad Inc. to transfer vector graphics data.

Dynamic interaction: spatiotemporal relationships between animals defined by analysing **locations** recorded at more or less the same time for each individual.

Edge correction: at the edge of a trapping grid, an area that needs to be added when estimating density if some trapped animals live partly off the grid.

Effective radiated power: the output power of a **radio tag** that is measured independent of the tag, typically in microVolts.

End-fire array: a linear array of matched **dipole antennas**, as used on an aircraft wing for **radio tracking**.

Energy density (of a cell): the **cell capacity** as a function of either the volume or weight of the battery.

Error ellipse: an **ellipse** generated by **Lenth estimators** to represent the area within which a **location** estimated from three or more bearings is likely to occur.

Error polygon: an area created by the overlap of arcs defined by the **antenna accuracy** for two or more **bearings**, within which the location of a **radio tag** is likely to occur.

Error triangle: the triangle formed between three bearings that intersect close to the **location** of a **radio tag**.

Event alert: a system for interpreting life events from **radio-tag** signals and alerting remote members of a study team.

Excursion: the movement to, and immediate return from, an isolated **location** beyond an area that is repeatedly traversed by an animal.

Focal site: a site, such as a nest, feeding station or release point, that is used for estimating distances in movement analyses.

Fresnel zone: an area in front of an object creating **diffraction** of radiation in which strong **interference** patterns occur.

Fundamental frequency: the frequency at which a quartz **crystal** is cut to vibrate most strongly.

Gap-gain search: the process of **homing** to a **radio tag** by passing through gaps between buildings to maximise the strength of received signals.

GridAscii: a file format used for transferring **raster map** data.

Grid cell: an area, usually a square with sides the size of the **tracking resolution**, that is assigned to an animal **location**.

Grid search: a search along parallel paths, which are usually separated by the maximum **detection distance** of **radio tags**.

Harmonic mean: a mean estimated by inverting the sum of reciprocal (inverse) values.

Harness-mount: the attachment of a **radio tag** by harness straps around the body of an animal.

Hazard rate: an expression of death or other adverse event as a function of the number of days of exposure.

Home range: an area repeatedly traversed by an animal.

Homing: either the tendency of an animal to return to a **focal site** or the process of deliberately approaching an animal while **radio tracking**.

Impedance matching: the process of tuning a transmitter or receiver to an antenna.

Implant: a **radio tag** that is inserted inside an animal, typically in a **subcutaneous** or **intra-peritoneal** position.

Incremental analysis: a process of plotting the area of a **home range** as consecutive **locations** are added, to determine when sampling further locations will cause little change in area.

Interference: a change in signal amplitude as a result of differences in **phase** of radiation that has reached a point along different paths.

Intra-peritoneal: a position within the body cavity.

Jack-knifing: a procedure for estimating variation in a measure, by omitting one observation at a time from a number of observations of the measure.

Kernel functions: mathematic functions of distances of **locations** from a point that provide an index of **location density**.

Kurtosis: the variation in a measure as a function of its mean, that indicates whether values are mostly close to the mean (leptokurtic) or spread widely about the mean (platykurtic).

Least Squares Cross Validation: a method for estimating the value of a smoothing parameter to be applied when estimating a **density matrix** based on **kernel functions**.

Lenth estimator: one of a family of similar methods for defining the area within which a **location** estimated from three or more **bearings** is likely to occur.

Line-transect: the process of estimating animal density by travelling along a route and recording observations of animals within a distance of maximum observation efficiency on either side.

Location: the estimated x and y coordinates, representing east and north coordinates on a map.

Location density: an index of density, usually estimated by a function of the distance of **locations** from a point.

Logistic regression: a regression of a dependent variable, typically in two categories, on one or more independent variables, with the assumption that the relationship between them is logistic.

LORAN: LOng RAnge Navigation system that uses **time of arrival** of low-frequency signals from coastal transmitters in North America to estimate the **location** of a receiver.

Loop antenna: a circular antenna that either has a circumference with a precise fraction of the desired wavelength, or is tuned with a capacitor to **resonance** at the equivalent frequency.

Marking bias: the attachment of tags to an atypical sample of animals, for example because those individuals were especially easy to trap.

Microcontroller: a 'computer on a chip' that can be used to control functions of a **radio tag**.

MCP, Minimum Convex Polygon: a polygon in which the linkage distances between peripheral **locations** sum to a minimum.

Minimum-linkage: a process for grouping **locations** within an outline, based on minimising the distances between them.

Mononuclear: a **home range** with one area of high **location density**.

Multinuclear: a **home range** with more than one separate area of high **location density**.

Multiple Response Permutation Procedure: a randomisation procedure that produces an index of spatial coincidence for two sets of **locations**.

Multivibrator: an electronic circuit that can generate signal pulses.

Nearest-neighbour: the **location** nearest to another location.

Nearest neighbour distance: the distance between **nearest-neighbours**, which is 0 if they have the same coordinates.

OMEGA: a European equivalent of the **LORAN** system used in North America.

Opportunity circles: circular areas of resource **availability** around animal **locations** or an **activity centre**.

Outlier-detection: a process of estimating the probability that the distance to a **location** is part of a normal distribution based on the other locations, using **nearest-neighbour distances**.

Passivation: a chemical change within a battery that supplies a low current for a long period, that makes it unable to provide high power again unless the change can be reversed.

Passive integrated transponders: **transponders** without a power supply, which use energy from an incoming signal to emit a weak coded signal on a different frequency.

Patchiness index: the summed area of polygons round separate groups of **locations**, defined by **cluster analysis**, as a proportion of the area of a single polygon round the same locations.

Peeled polygon: **MCP**s from which outlying locations are excluded on the basis of their distance from a chosen **activity centre**.

Phase: radio signals are 'in phase' when they arrive at the same point in the cycle of amplitude variation of each wavelength, which strengthens the signal by constructive **interference**.

Pilot study: an initial period of observation in which techniques are tested and calibrated.

Point sampling: the recording of **locations** from **radio tags** at intervals, so that many tagged animals can be sampled in the same period.

Polarisation: a tendency of radiation to propagate most strongly in either a vertical or horizontal plane.

Potting: application of a fluid that sets to enclose and protect components.

Primary software: an instruction set, typically machine-code, that controls hardware used in radio tagging.

Probe antenna: a receiving antenna with very short range, often with a long handle, that is used for very precise location of **radio tags**.

Proportional-hazards model: regression of a **hazard-rate** function on covariates that are assumed to have the same relative effect through time.

Pseudoreplication: the use of observations as if they were **statistically independent** data, when in fact they are related in some way.

PTT: Platform Transmitter Terminal that is tracked by a satellite using the **Doppler principle**.

Pulse modulation: change in the length of brief radio signals, or in the interval between signals.

Qualifying variable: a variable, such as an activity observation, a habitat measure or a time, that is associated with **location** coordinates from **radio tracking**.

Radio-corrected transect: a density estimate from **line transects** which is adjusted by the frequency of missed sightings for radio-tagged animals known or estimated to be present.

Radio location: the process of estimating coordinates for an individual with a **radio tag**.

Radio surveillance: the use of **radio tracking** to approach an animal and make visual observations.

Radio tag: the combination of a radio transmitter, battery, antenna and mounting materials that creates a package for attachment to an animal.

Radio telemetry: the process of recording data at a distance from a **radio tag**.

Radio tracking: the process of recording location data from a **radio tag**.

Range core: an outline encompassing a proportion of the **locations** in a **home range** that are at a higher density than in an outline encompassing all the **locations**.

Raster map: a map composed of a grid of identical shapes, typically squares, with different characteristics.

Receiving system: the combination of radio receiver, receiving antenna and other ancillary equipment, such as headphones or computing equipment.

Reference bandwidth: initial of a **smoothing parameter** in **kernel** contouring.

Reflection: when an electromagnetic wave is turned back at the interface between two media, leaving at the same angle as it arrived.

Refraction: when an electromagnetic wave passes an interface between two media but leaves at a different angle to its angle of arrival.

Resonance: the adoption by a system of a vibration at the same frequency as a source or at a multiple of the source frequency.

Resolution: see **tracking resolution**.

Resource dependence analysis: an investigation of relationships between home-range characteristics and habitat or other resources available to an animal.

Reverse bearing: see **back-bearing**.

RF: radio frequency, expressed in cycles per second (Hertz) or MegaHertz (MHz).

Right-censored: in **staggered-entry** analyses of **survival**, when an animal that is assumed not to have died is omitted from the data set at the time it left the population.

Ring or band: a circular strip of metal or plastic attached round the leg of a bird for identification.

Robust analyses: statistical tests based on as few assumptions as possible about the structure of the data.

Sampling saturation: when the recording of additional **locations** will not increase the size of a **home-range** estimate in the short term.

Secondary software: an instruction set that is used to collect data from **radio tags**.

Selective availability: the process in the Navstar **GPS** system whereby timing signals are modified to reduce the accuracy of estimating **locations** for purposes of military security.

Selectivity: the ability of a receiver to discriminate between signals that are on nearby frequencies.

Sensitivity: the ability of a receiver to detect low-power signals.

Smoothing parameter: a variable that modulates the density estimated by a **kernel function** to vary the tightness with which contours conform to **locations**.

Soft tagging: the fitting of tags to animals at a time when they are protected from factors that could elicit adverse effects from the tags.

Spectrum analyser: equipment for analysing the frequency output of a radio transmitter.

Spread spectrum: a radio transmission with high **bandwidth** that is used for **radio location** by **time difference of arrival**.

Staggered-access: **survival** analysis that accommodates entry of animals, and exit for causes other than death, at different times.

Static interaction: spatial relationships between animals defined by analysing overlap of **home range** outlines.

Statistical independence: lack of any relationship between separate observations.

Subcutaneous: a position under the skin.

Tag-day estimation: the calculation of **survival** rates on the basis of the number of days for which each **radio tag** is monitored.

Team tracking: **radio tracking** in which two or more observers take **bearings** for the same animal from different sites at the same time.

Temperature drift: change in frequency of a receiver or transmitter as a result of variation with temperature.

Territory: an area defended by an animal.

Tertiary software: an instruction set that is used to derive a best estimate from raw data values, typically by estimating mean values of samples but also by using error-rejection procedures.

Tesselation: a polygon formed from perpendicular lines that bisect links between adjacent locations.

Thermistor: a semiconductor with resistance that varies with temperature.

Time difference of arrival, TDOA: a process used to estimate the **location** of a transmitter from the time of receipt of its single signal at several different receiving stations.

Time of arrival, TOA: strictly, a process used to estimate the **location** of a receiver from several signals that indicate their time and place of origin, as in the Navstar **GPS** system.

Time To Independence: the minimum time interval between **location** records within a **home range** beyond which the distance between them cannot be predicted

Timetabling: the tendency of animals to return to a similar location at the same time each day.

Tracking resolution: the smallest distance that can be recorded between adjacent locations.

Transcutaneous: passage through the skin.

Transponders: electronic systems that transmit on receipt of another signal.

Triangulation: the estimation of **locations** from the intersection of a pair of **bearings**.

True-bearing: at the coordinates of a receiver, the angle between 0° and the direction to the transmitter coordinates.

Truncated-mark-resighting: estimating population density by using records of animals with and without **radio tags** within a maximum-efficiency distance from the centre of a **line transect**.

TSV (tab separated variable): data in **ASCII** text with each value separated by a tab character (ASCII code 9).

Type Approval: a measure of quality which is legally required for sale or use of equipment in the UK.

UHF: ultra-high frequencies, above 300 million cycles per second (300 MHz).

Unit-sum constraint: lack of **statistical independence** between proportions, for example of habitat in an area, because the sum of proportions is 1.

Universal Transverse Mercator: a global rectangular coordinate system.

Utilisation distribution: plot of range area against frequency of occurrence, as a proportion of the **location density** or of the **locations** (assuming these reflect occurrence) in the area.

Vector map: a map composed of lines and closed shapes defined by a sequence of x and y coordinates.

VHF: very high frequencies, between 30 and 300 million cycles per second (30–300 MHz).

Whip antenna: a linear antenna coupled at one end to a transmitter or receiver.

Window width: **smoothing parameter**.

Yagi antenna: an antenna with a reflector element and one or more directors fixed parallel on either side of a **dipole antenna**; it can discriminate **true-bearings** from **back-bearings**.

Appendix I

This appendix lists some suppliers of hardware that has been used in radio tagging-projects, without implying any endorsement of their products.

RECEIVING, TAGGING AND LOGGING EQUIPMENT

ATS (Advanced Telemetry Systems, Inc.)
470 First Ave NO, Box 398
Isanti MN 55040, USA
tel: +1 612 444 9267 fax: +1 612 444 9384
70743.512@compuserve.com
http://www.biotelem.org/ats/index.html
VHF receivers, loggers, tags

AVM Instrument Company Ltd.
2356 Research Drive Livermore, California 94550, USA
tel: +1 925 449 2286 fax: +1 925 449 3980
bckermeen@avminstrument.com
http://www.biotelem.org/manufact.htm
VHF tags, receivers

Bio Telemetry Tracking (Australia)
18 Magill Road
Norwood, South Australia
tel: +61 8 8362 6666 fax: +61 8 8362 7955
biotelemetry@senet.com.au
VHF tags

Biotrack Ltd
52 Furzebrook Road
Wareham
Dorset BH20 5AX, UK
tel: +44 1929 552992 fax: +44 1929 554948
brian@biotrack.co.uk; sean@biotrack.co.uk
http://www.biotrack.co.uk
VHF tags, receivers

Clark Masts Teksam Ltd
Binstead
Isle of Wight PO33 3PA, UK
tel: +44 1983 563691 fax: +44 1983 566643
pneumatic and fixed masts

Custom Electronics Of Urbana, Inc.
2009 Silver Ct. W.
Urbana, IL 61801, USA
tel: +1 217 344 3460 fax: +1 217 344 3460
customel@aol.com
http://members.aol.com/~customel/
VHF receivers, tags

Custom Telemetry
1050 Industrial Drive
Watkinsville GA 30677, USA
tel: +1 706 769 4024 fax: +1 706 769 4026
VHF tags

Entwicklungsbüro Rohde
Johann-Schill-Str. 22
D 79232 March, Germany
tel: +49 7665 3885
VHF receivers, tags,
Doppler receiving system

Holohill Systems Ltd
3387 Stonecrest Road
Woodlawn, Ontario, Canada K0A 3M0
tel: +1 613 832 3649 fax: +1 613 832 2728
info@holohil.com
VHF tags

Konisberg Instruments Inc
2000 Foothill Boulevard Pasadena
California 91107-3294, USA
physiological telemetry

Lotek Engineering Inc.
115 Pony Drive
Newmarket, Ontario, Canada L3Y 7B5
tel: +1 416 836 6680 fax: +1 416 836 6455
telemetry@lotek.com
http://www.lotek.com
VHF receivers, loggers, tags
GPS tags, storage tags

Mariner Radar Ltd
Bridleway, Camps Heath
Lowestoft, Suffolk NR32 5DN, UK
tel: +44 1502 567195 fax: +44 1502 508762
VHF receivers, tags, PTTs

Microwave Telemetry Inc.
10280 Old Columbia Road, Suite 260
Columbia, Maryland 21046, USA
tel: +1 410 290 8672 fax: +1 410 290 8847
Microwt@aol.com
PTTs

Mini-Mitter Co., Inc
P.O. Box 3386
Sunriver, OR 97707, USA
tel: +1 503 593 8639 fax: +1 503 593 5604
http://fairway.ecn.purdue.edu/~ieeeembs/companies/minimitter.html
physiological telemetry equipment

Sirtrack Ltd
Private Bag 1402, Goddard Lane
Havelock North, New Zealand
tel: +64 6 877 7736 fax: +64 6 877 5422
http://goddess.hb.landcare.cri.nz/sirtrack/sirtrack.html
VHF tags
PTTs

Televilt International AB
Box 53
S-711 22 Lindesberg, Sweden
tel: +46 581 17195 fax: +46 581 17196
per-arne.lemnell@televilt.se
http://www.televilt.se/
VHF receivers (logging), tags
GPS tags

Telonics
932 E. Impala Avenue
Mesa, AZ 85204, USA
tel: + 1 602 892 4444 fax: +1 602 892 6699
http://www.telonics.com/index.html
VHF receivers, loggers, tags
GPS tags, PTTs

Titley Electronics Pty Ltd
PO Box 19
Ballina NSW 2478, Australia
tel: +2 6681 1017 fax: +2 6686 6617
titley@nor.com.au
http://www.titley.com.au
VHF receivers, tags

Vemco
3895 Shad Bay, RR #4
Armdale, Nova Scotia, Canada B3L 4J4
tel: +1 902 852 3047 fax: +1 902 852 4000
sonar tagging equipment

Wildlife Computers
20630 N.E. 150th Street
Woodinville, WA 98072, USA
tel: +1 206 881 3048 fax: +1 206 881 3405
aquatic data logging tags

WMI (Wildlife Materials Inc.)
1031 Autumn Ridge Road
Carbondale, IL 62901, USA
tel: +1 618 549 6330 fax: +1 618 457 3340
VHF receivers, loggers, tags, capture aids

TAG AND MOUNTING COMPONENTS

Bally Ribbon Mills
23 N. 7th Street
Bally, PA 19503, USA
tel: +1 215 845 2211 fax: +1 610 845 8013
Teflon ribbon

BS Lewis Ltd
6 Newcroft Rd., Woolton, Liverpool L25 6EP, UK
tel: +44 151 448 1230 fax: +44 151 486 2423
lithium/thionyl chloride cells

Ethicon Ltd
PO Box 408
Bankhead Avenue
Edinburgh EH11 4HE, UK
absorbable sutures

Farnell (Premier Farnell plc)
Canal Road
Leeds, Yorkshire LS12 2TU, UK
tel: +44 113 2636311 fax: +44 113 263411
http://www.farnell.com
general electronic components and tools
magnetic reed switches

Heyco Ltd
Uddens Trading Estate
Wimborne, Dorset, BH21 7NL, UK
tel: +44 1202 861000 fax: +44 1202 861086
cable ties

Itchen Adhesives
4 Dean Close
Winchester, Hampshire, SO22 5LP
tel: 01962 853679 fax: 01962 866489
hot-melt glue

Minerva Dental
1 The Beaver Industrial Estate
Airfield Road, Christchurch, Dorset, BH23 3TG, UK
dental acrylic

Plasti Dip
Unit 1, Harvesting Lane
East Meon, Petersfield, Hampshire, GU32 1QR, UK
tel: 01730 823823 fax: 01730 823321
Plastidip

Maplin Electronics
Freepost smu 94, PO Box 777, Rayleigh, Essex, SS6 8LU, UK
tel: +44 1702 556 001
http://www.maplin.co.uk
general electronic components and tools

Rayfast
2 Westmead
Swindon, Wiltshire, SN5 7SY, UK
tel: +44 1793 616700 fax: +44 1793 644304
http://www.devicelink.com/company/euro/r/r0011.html
specialist heat-shrink tubing

RS Components Ltd
PO Box 99
Corby, Northants NN17 9RS, UK
tel: +44 1536 201201 fax: +44 1536 201501
http:/rswww.com
general electronic components and tools

Saft (UK) Ltd
Castle Works, Station Road
Hampton, Middlesex TW12 2BY, UK
tel: +44 181 979 7755 fax: +44 181 783 0494
lithium/thionyl chloride cells

Sarstedt (UK) Ltd
68 Boston Rd
Beaumont Leys, Leicestershire LE4 7AW UK
tel: 01533 359023 fax: 01533 366099
polystyrene tubes

Techmate Ltd
10 Bridgeturn Ave., Old Wolverton, Milton Keynes, MK12 5QL, UK
tel: 01908 322222 fax: 01908 319941
polycarbonate tubes

Titan Corporation
5629 208th St SW, USA
tel: +1 425 775 2582 fax: +1 425 778 7680
physiocompatible epoxy

Appendix II

This appendix lists some sources of software for collecting and analysing data from radio tags.

Note that some suppliers listed in Appendix I also supply software to use with their equipment.

DATA COLLECTION

Service ARGOS
18, Avenue Edouard Belin
31055 Toulouse Cedex, France
tel: +33 5 61 39 47 00 fax: +33 5 61 75 10 14
http://www.argosinc.com/
or
1801 McCormick Dr., Suite 10
Landover, Maryland, USA 20785
tel: 301-925-4411 fax: 301-925-8995
jw@argosinc.com
location data from PTTs

LOAS (Location Of A Signal)
Ecological Software Solutions
http://www.ecostats.com/software/software.htm
data entry software

LOCATE II
Pacer, P.O. Box 1767
Truro, Nova Scotia, Canada B2N 5Z5
tel: +1 902 893 6607
data entry software

ZYLOG4
(see Dodge and Steiner 1986)
US Fish and Wildlife Service
Fisheries and Wildlife Technical Report 4
data entry software

ANALYSIS SOFTWARE

CALHOME (CALifornia HOME-range)
John G. Kie
Forestry Sciences Lab
2081 East Sierra Avenue, Fresno CA 93710, USA
tel: +1 209 487-5589
http://www.biotelem.org/software.htm
home range analyses

HOMERANGE
Robert Huber, Jack Bradbury (see Ackerman *et al.,* 1990)
http://wwwzoo.kfunigraz.ac.at/~lobsterman/analysis.html
home range analyses

KERNELHR
D. Erran Seaman
National Biological Service
Forest and Rangeland Ecosystem Science Center
Olympic Field Station, 600 E. Park Ave., Port Angeles WA 98362-6798, USA
tel: +1 360 452-0303 fax: +1 360 452-0348
Erran_Seaman@NPS.GOV
http://nhsbig.inhs.uiuc.edu/wes/khr427.html
home range analyses using kernels

MARK
Gary White
Department of Fishery and Wildlife Biology
Colorado State University, Fort Collins, CO 80523, USA
gwhite@cnr.colostate.edu
survival analysis software

McPAAL (Micro-Computer Programs for Analysis of Animal Locations)
Michael Stuwe (see Stuwe and Blohowiak, 1985)
http://www.biotelem.org/software.htm
home range analyses

MOVEMENT
P.N. Hooge and B. Eichenlaub
Alaska Biological Science Center
US Geological Survey, Anchorage, AK, USA
http://www.absc.usgs.gov/glba/gistools/animal_mvmt.htm
animal movement extension to Arcview

RANGES V
Robert Kenward (see Kenward and Hodder, 1996)
ITE Furzebrook Research Station
Furzebrook Road, Wareham, Dorset, BH20 SAS, United Kingdom
tel: +44 1929 551518 fax: +44 1929 551087
email: r.kenward@wpo.nerc.co.uk
http://www.anatrack.com
analysis system for biological location data

SURPH (SURvival by Proportional Hazards)
(see Smith *et al.*, 1994)
survival analyses

TRACKER
(*see Camponotus AB and Radio Location Systems AB 1994*)
contact Televilt International AB
Box 53, S-711 22 Lindesberg, SWEDEN
tel: +46 581 17195 fax: +46 581 17196
http://www.televilt.se/
home range analyses

WILDTRACK
Ian Todd
University of Hertfordshire, Hatfield, UK
http://www.geocities.com/RainForest/3722/
home range analyses for MACs

GENERAL PURPOSE GIS

ESRI (Environmental Systems Research Institute)
380 New York St.
Redlands, CA 92373-8100, USA
tel: +1 909 793 2853
http://www.esri.com/software/index.html
Arc/Info, ArcView GIS software

GRASS (Geographic Resources Analysis and Support System)
http://www.hpcc.nectec.or.th/GIS/grass.html
US Army developed Geographic Information System

IDRISI
Mag. Eric J. LORUP
IDRISI Resource Center Salzburg
Institut für Geographie an der Universität Salzburg, Hellbrunner Straße 34, A-5020
Salzburg, AUSTRIA
tel: +43 662 8044 5235 fax: +43 662 8044 525
http://www.sbg.ac.at/geo/idrisi/irchome.htm
raster-based Geographic Information System

MAPMAKER PRO
Map Maker Limited
The Pier, Carradale, Argyll, Scotland, PA28 6SQ, UK
tel: +44 1583 431 358 fax: +44 1583 431 728
email: pro@mapmaker.com
http://www.mapmaker.com
map-making suite

SOURCES OF MAP DATA

Land Cover Map of Great Britain
Robin Fuller
Monks Wood, Abbots Ripton, Huntingdon, Cambridgeshire PE17 2LS, UK
tel: +44 1487 773381-8 fax: +44 1487 773467
RF@wpo.nerc.ac.uk
landcover map of UK at 25 m resolution

Eros Data Center
Customer Services
U.S. Geological Survey, EROS Data Center, 47914 252nd Street
Sioux Falls, SD 57198-0001, USA
tel: +1 800 252 4547 fax: +1 605 594 6589
custserv@edcmail.cr.usgs.gov
http://edcwww.cr.usgs.gov/content_products.html
satellite images from around the world

OTHER USEFUL WEB SITES

HOME RANGE SOFTWARE
http://nhsbig.inhs.uiuc.edu/www/home_range.html
http://www.biotelem.org/software.htm

RADIO TELEMETRY EQUIPMENT
http://detritus.inhs.uiuc.edu/wes/equipment_suppliers.html
http://www.biotelem.org/manufact.htm

BIOTELEMETRY CLEARINGHOUSE
http://www.biotelem.org/index.html

INDEX